Science and the Human Prospect

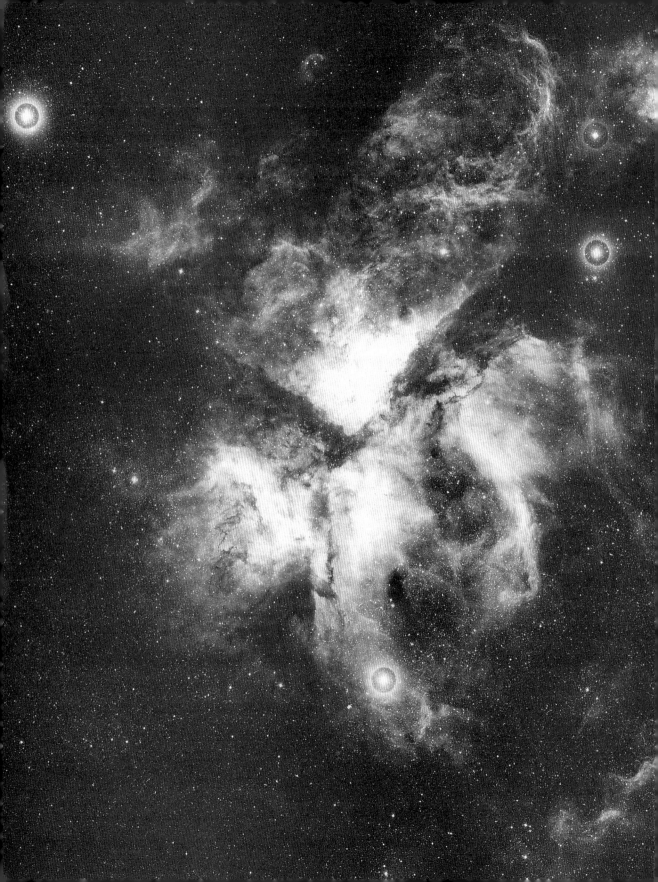

SCIENCE
and the
HUMAN
PROSPECT

Ronald C. Pine
Honolulu Community College

Wadsworth Publishing Company
Belmont, California
A Division of Wadsworth, Inc.

To the students of Honolulu Community College,
who taught me how to explain big thoughts clearly,
and to enjoy it.

Philosophy Editor: Ken King
Editorial Assistant: Cheri Peterson
Production Editor: Gary Mcdonald
Managing Designer: James Chadwick
Print Buyer: Karen Hunt
Designer: Peter Martin/Design Office
Copy Editor: Pat Tompkins
Compositor: Graphic Composition, Inc.
Cover: Kandinsky, "Composition," 1930. Stadtische Kansthalle
Mannheim. Copyright ARS N.Y./ADAGP, 1988. Used by permission.
Signing Representative: Karen Buttles

Printed in the United States of America   xx

1 2 3 4 5 6 7 8 9 10—92 91 90 89

Library of Congress Cataloging-in-Publication Data

Pine, Ronald C.
   Science and the human prospect.

   Bibliography: p.
   Includes index.
   1. Science—History. 2. Science—Philosophy.
3. Science—Methodology. I. Title.
Q125.P6 1989   509   88-20697
   ISBN 0-534-09756-1

A human being is part of the whole, called by us "Universe"; a part limited in time and space. He experiences himself, his thoughts and feelings as something separated from the rest—a kind of optical delusion of his consciousness. This delusion is a kind of prison for us, restricting us to our personal desires and to affection for a few persons nearest us. Our task must be to free ourselves from this prison by widening our circle of compassion to embrace all living creatures and the whole of nature in its beauty. Nobody is able to achieve this completely but the striving for such achievement is, in itself, a part of the liberation and a foundation for inner security.

ALBERT EINSTEIN

The creative scientist studies nature with the rapt gaze of the lover, and is guided as often by aesthetics as by rational considerations in guessing how nature works.

TIMOTHY FERRIS

Contents

7 Our Time:
1. Understanding the Theory of Relativity

8 Our Time:
2. Quantum Physics and Reality

Preface

Albert Einstein supposedly once remarked that everything should be as simple as possible, but no simpler. His remark is well suited to the task of writing a book on some of the major scientific and philosophical issues of our time, of simplifying complex and specialized subjects for the nonspecialist and the inquisitive reader, without oversimplifying the careful conclusions of contemporary science and insulting the many undaunted and persevering men and women who have attempted to understand this marvelously strange universe.

Several years ago a report from the Association of American Colleges, *Integrity in the College Curriculum,* lamented the fragmented state of a college education. Among other things, the report noted that most college programs have failed to keep pace in providing a broad understanding of science, that in the light of the overwhelming recent scientific and technological developments, a college education should produce graduates who are "intellectually at ease with science." This cannot be accomplished, the report noted, with a quick exposure to skills and content or with more emphasis on specialized research, but must involve a wider invitation to philosophical and intellectual growth, to critical inquiry, and to understand the crucial connections of science to other fields, our historical roots, the human prospect, and life itself. It is one thing to have the important and irreplaceable experience of the laboratory, learning how to observe, experiment, and infer and another to understand the imposing human questions generated by laboratory results. As Charles Stores, a master science teacher has stated, "Few curricula are not compartmentalized . . . to the point of suffocation and few teachers dare to venture beyond the safe harbors of their own disciplines to bring together . . . the sciences into a coherent and historical whole and fewer still relate the sciences to the humanities."

This book addresses this problem by providing a supplement to an introductory science, history of science, philosophy of science, or interdisciplinary course in which the world view of modern science plays a major role. Its aim is to honor science by helping the reader understand not only the unexpected beauty of the modern scientific picture of the universe but also the remarkable process by which this picture has been, and is being, achieved. In addition to reinforcing the goal of scientific literacy in terms of content and method, *Science and the Human Prospect* will concentrate on the philosophical implications and questions generated by the method and its stunning results. It will show that philosophical analysis, reflection, and contemplation are an inseparable part of the continual process to understand what science has discovered.

This book is also dedicated to the proposition that thinking about and exchanging opinions regarding the big questions of life are very much related to dealing with our more immediate practical problems. Why is there something rather than nothing? What is the nature of reality? What is the nature of a reasonable belief? Are we alone in the universe? Will the human species survive this present state of technological adolescence? Can such questions be answered? Can something important be learned, even if such questions cannot be answered? Although the connection with such lofty questions may seem tenuous, recent studies of the imaginative and innovative management of successful businesses have revealed that a significant percentage of its leaders have a broad liberal arts background rather than degrees in business administration. This indicates that people who have addressed the larger questions of life have not wasted their time with ivory tower pursuits, but have learned to think critically and to be adaptable to new situations and, most importantly, have learned how to learn. The connection between the philosophical contemplation of science and the human prospect is similar. The abstract quest for the truth of nature and the nature of truth is pursued by scientists and philosophers not only for its own sake but also because such a venture has a crucial bearing on understanding who we are and what choices we should make in determining our future.

We have discovered only recently that our Earth is but a fragile speck of dust floating in an unimaginably vast sea of space and time and that, from a biological perspective, human life is precious and unique but not necessarily special. In a world beset by so many pressing global problems, requiring so many daily decisions, we should all make it our business to discuss what this means. In a world of lasers, space shuttles, and computers, in a world culture linked by instantaneous telecommunications, we should have made some progress since a more primitive time when people burned witches and books and indulged in human sacrifice. Yet today we witness the plague of terrorism spreading throughout the world, where the psychology of righteousness has degenerated so that one person's "freedom fighter" is another's terrorist. Also, the mass hysteria over AIDS is producing the same psychology of fear, prejudice, and bigotry that produced witch and book burning. Daily news clips from around the world demonstrate repeatedly the tendency to commit the logical fallacy of ignorance and the average person's lack of understanding of scientific method. People who would not think twice of smoking cigarettes or of not wearing seat belts in driving their children to school demand that science prove beyond any doubt that AIDS cannot be acquired by breathing the same air as an AIDS victim. And because science cannot prove this absolutely, these same people too often conclude that it is true, and another round of persecution begins.

The truth is, in spite of the tremendous success of applying the scientific method to the problems of starvation and disease, in spite of the astounding vision produced by science of our home in the universe, most human beings on this Earth still require a popular, quasi-religious security-psychology for belief and action. Even informing the public about the dangers of smoking requires a media blitz replete with Madison Avenue advertising techniques; we assume that no one would pay attention if the scientific evidence and the controlled experiments were presented instead. Religion has contributed much to life, and in this book we will see that it has played a very positive role in the development of scientific ideas, but scientists should not need to become priests or authoritative TV personalities before their message is understood.

Because this book attempts to provide perspective to a general science education, let me make clear from the beginning what my perspective is. Although there is an inherent risk in life and this is reflected in the uncertainty in the application of the scientific method, and although there has been much bad science[1] in the pursuit of ideology and the same need for a security-psychology, science, as an inherently self-corrective process, remains one of the most effective methods for dealing with risk and uncertainty. Problems never end, and a problem such as AIDS is not a new test for our humanity; it is the same test. To solve such difficult problems we need to bring out the best of our "humanness," not the worst. And science, in spite of its human elements, in the long run is a powerful method for bringing out the best in us; in spite of the puzzling nature and arguability of its self-correctiveness, science brings out our ability to be humbled by the world, to figure out our mistakes, and in the words of Melvin Konner, to feel "a sort of wonderment at the spectacle of the world, and its apprehensibility by the mind: a focusing, for the purpose of elevation."

This book is for any curious person, but it is best suited for an "honors" high school or freshman or sophomore college course. Its hope is to generate in the reader an understanding of the philosophical drama that lies behind humanity's attempt to embrace nature with a human conceptual framework. In the process it will survey the major developments of this attempt, both modern and historical, to allow the reader to be intellectually and emotionally at ease with science.

It is also a textbook with a tangential theme, no doubt a controversial one. In addition to the usual presentation of the necessary rigor of scientific method, it presents science as part of the process, noted by Einstein in the opening epigraph of this book, of "widening our circle of compassion." It will show science to be far from a cold, heartless, sterile probing of nature but a passionate enterprise, involving all that we are as human beings. This book will openly flaunt what Robert Oppenheimer used to call the "sentimental" aspect of science; it will present *science as a romance with the universe,* a fervent striving to understand, where ideas are cast out like romantic gestures for acceptance from this great presence that is our home. It will show that

---

[1] For instance, within just the past 100 years in the United States, science has been used to justify "niggerology" and "polygenism," the supposed separate evolution of race, the exploitation of the weak by the strong (social Darwinism), and even the idea that some people are "born" criminals, justifying compulsory sterilization and legal castration.

science stirs the heart as well as the mind. Most importantly, it will claim that without the nonrational elements expressed in this human romance with the universe, scientific progress as we know it would not have been possible.

Although this book adopts controversial perspectives and presents controversial arguments, readers should profit even if they disagree with the validity of the concepts. Because the book cuts across many disciplinary boundaries, normally jealously guarded by the scholar specialist, its content is meant to be argued about, in particular to provoke discussion at the introductory level. Each chapter is deliberately open-ended and should be thought of as a guide to discussion or a point of departure. And each chapter includes a Concept Summary and Suggested Readings, a reviewlike annotated bibliography presented in an order from introductory to more technical. Some of these books will have been cited in the chapter preceding the bibliography, and some will be recommended reading on the topics discussed.

One final prefatory comment: The reader should not be put off by such terms as *epistemology, ontology,* and *relativity theory.* Each philosophical and scientific term is defined and developed in a step-by-step manner, assuming only curiosity and an average reading level as a prerequisite. However, in following the development of the factual and philosophical details of each chapter, keep in mind that there are different levels of understanding and confusion. In the many years of teaching a philosophy course covering the topics of this book, I have often had students complain to me, "It's all so amazing, but I do not understand." Students will complain that they do not understand how a subatomic particle that is supposed to be an object in one place at one time can also be a wave that is in many places at the same time. Or how can the entire universe, with its incomprehensible amount of matter, have started from a single dimensionless point, a point that is not even surrounded by space or in time, because space and time are also "in" the point? Or how is it possible that the universe can be expanding but not have a center? Or how could natural selection produce the anglerfish? On such occasions, after a brief discussion, I will often find that the student has understood the lecture or reading material. Then we both realize that the problem does not involve reading or listening comprehension but that the student wants to have a complete "feel" for the material; he or she wants to feel as comfortable with these new ideas as with the idea that a circle is not a square, that two plus two is four, not five, or that Americans have parties on Super Bowl Sunday. What the reader must understand about science is that there are many examples of scientific knowledge that no one understands in this sense. As the English writer G. K. Chesterton once stated, "We have seen the truth, and the truth makes no sense."

Many great scientists have declared that most of the scientific knowledge of the twentieth century can be understood only mathematically, and like mystics professing that they cannot tell the rest of us the truth because such truths expressed in ordinary language will sound foolish, they often choose public silence as the most prudent course. Strip the great mathematical physicists of their mathematics, ask them to confront the universe as regular human beings who must explain to their children what it all means, and they will be equal to the student in humble perplexity. As physicist Nick Herbert in his book *Quantum Reality* has confessed:

When my son asks me what the world is made of, I confidently answer that deep down, matter is made of atoms. However, when he asks me what atoms are like, I cannot answer though I have spent half my life exploring this question. How dishonest I feel—as "expert" in atomic reality—whenever I draw for schoolchildren the popular planetary picture of the atom; it was known to be a lie even in their grandparents' day.

If science has taught us anything, it is that nature is not in the business of making us feel comfortable in this sense, of presenting a static and predictable personality, one consistent with our provincial expectations. Our romance with this great Being is not like those affairs of an initial great passion that inevitably lapse into a boring, stable routine. Rather, it is like those rare affairs that remain vibrant and exciting for a lifetime, alternating often between confidence and misgiving, ecstasy and despair, with each emergence from despair and uncertainty making the relationship stronger, producing a greater understanding, but never one that is complete. In our relationship with the universe there will always be the next misunderstanding and the next reconciliation. Thus, there is an inevitable element of struggle in our attempt to embrace the universe with human concepts, and no matter how much effort has been made to simplify the topics discussed in this book, some amount of struggle will be required of the reader. Nevertheless, there is something in our nature that makes this alluring task most enjoyable.

Acknowledgments

No man is an island, and no book is the result of a single mind. The influences that contributed to the thoughts contained in this book have been many. Any mistakes are, of course, mine alone, but many people need to be thanked for their helpful suggestions and insights. First and foremost are the students of Honolulu Community College whose cosmopolitan spirit and multicultural representation never cease to amaze me, and whose struggle to know and understand in many ways parallels the intellectual struggle to embrace the universe with a human conceptual scheme. When I reflect on a teaching career that is approaching 20 years, I realize more each day how much a teacher learns from his students. To name but a few who I am very grateful for having the privilege of being a fellow traveler and guide in our cosmic journey, I wish to thank Sharyn Abe, Kirk Arndt, Robert Bell, Michael Benson, Chris Breininger, Patricia Calabrese, Alfredo Carma, Jerry Ching, Dan Colon, Nancy Cooper, Barbara Corbin, Ildiko Csontos, Craig Cundiff, Carol Dipiazza, Sharon Fischer, Donnie Hakuole, Alvin Hansen, Laurie Heath, Jennifer Higgins-Ross, Joseph Huster, Chris Jones, Masae Kiyota, Stephanie Kokernak, Grover Lamb, Ted Lieberman, Steve Love, Joanne Muraoka, Wendy Paakuala, Philip Pebenito, Dan Petersen, Lewis Poe, Frank Salidivar, Diane Sasaki, Suzette Smetka, Guy Sueoka, Fred Sutter, George Szamosi, Mie Takeda, Laine Uyeno, Nancy Vallely, Jennifer Venezia, Tad Wilkinson, Harry Wong, and Brian Young.

Next are my colleagues. To name but a few who contributed either directly or indirectly to the manuscript, a special thanks to physicist Mark Schindler, who read the entire manuscript several times; his expertise and devil's advocate role were indispensable as a backdrop for Chapters 7 and 8; computer science instructor Sam Rhoads for his helpful editing suggestions and thoughts on artificial intelligence and the Copernican revolution; mathematics and astronomy instructor Jim Reeder for his help in corroborating the quantitative astronomical data and with the math analogy in Chapter 5; psychology instructor Caroline Blanchard for her expertise on aggression and evolution theory; philosophy instructor Terry Haney and English instructor Gloria Hooper for attending my cosmology course; political science instructor Rick Ziegler for his encouragement of the thoughts in Chapter 10; and Helen Rapozo for her help in telecommunication data transfers.

I am also grateful to Gary Mcdonald and the production team at Wadsworth for their tireless handling of the myriad publication tasks, and to Joyce S. Tsunoda, chancellor for the Community Colleges of the State of Hawaii, and Peter R. Kessinger, provost of Honolulu Community College, for their leadership and institutional support.

A very special thanks also to Charles Stores of Highline Community College, who seems to have most understood the lonely path traveled by the thoughts contained in this book.

Finally, I owe much indebtedness to my wife, Carmen, and children, David, Kym, and Reyna, who weathered my obsessiveness, stimulated me intellectually with their perspectives on life, and taught me that understanding a black hole and electron tunneling bestow no guarantee of understanding a single human being.

Science and the Human Prospect

CHAPTER 1

Introduction:
Our Cosmological Roots

In all of us there is a hunger, marrow deep, to know our heritage, to know who we are, and where we have come from. Without this enriching knowledge, there is a hollow yearning. No matter what our attainments in life, there is a vacuum, an emptiness, and a most disquieting loneliness.

ALEX HALEY, *ROOTS*

The effort to understand the universe is one of the very few things that lifts human life a little above the level of farce, and gives it some of the grace of tragedy.

STEVEN WEINBERG, *THE FIRST THREE MINUTES*

The realities of nature surpass our most ambitious dreams.

RODIN

Our Home in the Universe

By the time you finish reading this sentence you will have moved over 150 miles. The seemingly stable ground beneath your feet is actually hurtling through space at a colossal speed in many different relative directions. We live, in the words of the poet Walt Whitman, on "a great round wonder rolling through space," a spinning sphere whirling around at approximately 1,000 miles per hour at its midsection, the equator. This gyrating home completes one rotation in 24 hours and continues this process 365 times before sweeping out an enormous elliptical path around our Sun. To complete this path the Earth moves at 18½ miles per second, or 66,000 miles per hour.[1] And our Sun is also moving at the incredible speed of over 150 miles per second, or 540,000 miles per hour, taking the Earth and other planets of our solar system on a mysterious journey through space and time, as it sweeps a 250-million-year path around our galaxy.

On a small piece of this spinning ball, on the island of Hawaii, the largest in the chain of islands that make up the State of Hawaii, are two of the largest volcanoes on Earth, Mauna Kea and Mauna Loa. On the top of Mauna Kea are human structures symbolizing much that is right about human nature (Figure 1-1). Because this location offers the best astronomical viewing conditions in the world, the United Kingdom, Canada, France, the Netherlands, and the United States have assembled here multimillion-dollar monuments to that human characteristic responsible for much of our present success on this planet, our curiosity. Here astronomers peer out into the blackness of the night, attempting to fathom the secrets of places trillions of miles away, attempting to understand the details of our cosmological roots, of who we are and where we have come from. At 14,000 feet above sea level (above nearly half of the Earth's obscuring atmosphere), the summit of Mauna Kea is 32,000 feet from the ocean floor. These massive mountains were created by lava pouring out of a hot spot in the ocean floor, a hole in the Earth's crust estimated to be 186 miles long and 124 miles wide. Few who visit these volcanoes fail to be awed by this silent testament to the potential power of nature, and astronomers are grateful for nature's construction of these formidable platforms.

In relative terms, however, these volcanoes are very small. On Mars there is a single volcano larger than the entire Hawaiian island chain. Figure 1-2 shows a superposition of this monstrous extraterrestrial volcano, called Olympus Mons, the Hawaiian island chain, and Hurricane Iwa. Hurricane Iwa

Men and women are not content to comfort themselves with tales of gods and giants, or to confine their thoughts to the daily affairs of life; they also build telescopes and satellites and accelerators, and sit at desks for endless hours working out the meaning of the data they gather.

STEVEN WEINBERG, *THE FIRST THREE MINUTES*

---

[1] We will use the common *miles* rather than kilometers. To convert the number of miles to kilometers, multiply the number of miles by 1.62. Thus 66,000 miles per hours $\times$ 1.62 = 106,920 kilometers per hour.

FIGURE 1-1

A State of Hawaii artist's rendering of the observatories on top of the volcano Mauna Kea on Hawaii, the "Big Island" of the Hawaiian islands. At 13,600 feet this is one of the best astronomical platforms in the world from which humankind attempts to figure out the details of our cosmological roots. (Reprinted by permission of the California Association for Research in Astronomy)

The unexpected and the incredible belong in this world. Only then is life whole.

CARL GUSTAV JUNG

hit the islands of Kauai and Oahu in 1982. Terrorizing these islands for only a single day, its size and power were an unexpected shock to people who took for granted gentle breezes and balmy weather. Some pampered residents were indignant that they had no electricity for a few days afterward. Perhaps they should have kept things in better perspective. The next picture (Figure 1-3), taken by the *Voyager* spacecraft, shows the famous Red Spot on the planet Jupiter. This gigantic hurricane, the size of several Earths, has been observed from Earth for over 300 years, and some estimate that it may be a million years old! The top speeds of the winds of Hurricane Iwa reached 110 miles per hour. Those on Jupiter blow at over 300 miles per hour.

We are all astronauts. As we sail on through an immense sea of seemingly indifferent space, as we work and play, experience success and defeat, as the trials and tribulations of humankind are carried out on this little bluish sphere, we fail to recognize for the most part our cosmic plight. Consider the unheralded fact that it is impossible for any of us to go home again to an exact mathematical spatial point of our childhoods. Recently, I returned to where I lived as a child, to Southern California, a town called Garden Grove about five miles from the original Disneyland. I sought and found my old street and house.

It was a haunting experience walking down that street. Everything seemed so much smaller, except for the plum trees in each of the front yards. When my family first moved in, the contractor planted a small plum tree in the front of every home—a token gesture to the thousands of orange trees ripped from the ground to plant miles of tract houses. Now the plum trees were large with maturing fruit, except in front of the homes where the maturing fruit became a nuisance to its owner. These front yards now had either no

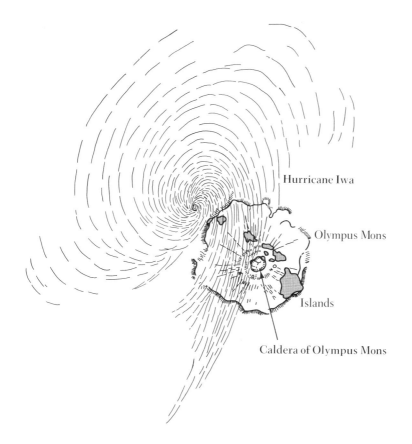

FIGURE 1-2

A superposition of the islands of Hawaii, the Martian volcano Olympus Mons, and Hurricane Iwa as it passed the islands at about 3 p.m., November 23, 1982. At 80,000 feet, Olympus Mons is almost 6 times taller than Mauna Kea, and although ferocious in appearance as it engulfs the entire state of Hawaii, compare this Pacific hurricane to that on the planet Jupiter in Figure 1-3.

Hurricane Iwa

Olympus Mons

Islands

Caldera of Olympus Mons

trees at all or some kind of tree that neither bore fruit nor dropped its leaves. I could remember playing with the other children in the huge piles of dead orange trees that the workmen piled up. In every backyard they left a few standing so that the new residents, arriving from all over the country, could feel like they were really living in sunny California. None were left now. Shortly after the tract homes were completed, most of the remaining orange trees died of a disease. The old-timers in the area claimed it was due to shock from the cumulative death of all the other trees.

I found my old house and stopped in front of the short cement driveway. Here was where I had found my dog dead in the street, apparently run over by a car. I was sure he was only sleeping and carried him to the backyard and put a blanket over him. I had no concept of death yet and could not understand why he did not get up the next day and why his body was so stiff. Here was where I parked my first car and experienced the uncertainty of adolescence. It was a strange experience being at a place that I had not been to for over 30 years.

Stranger still is that, in a sense, I had actually not been "here" before at all. Since I had lived on this street, the Earth had spun on its axis like a

The deflation of some of our more common conceits is one of the practical applications of astronomy.

CARL SAGAN

FIGURE 1-3

The Great Red Spot, an immense hurricane, on the planet Jupiter. In contemplating its size (at least the size of several Earths) and age (many centuries old), we begin to understand our fragile place in a universe of colossal physical processes. (top, NASA; bottom, Hale Observatories)

top 11,000 times, completing over 30 sweeping journeys around the Sun, all the while being carried by the Sun over 140,000,000,000 miles from the original relative mathematical point[1] of my childhood experiences! This was not home; it was only a reminder of home.

Such an astonishing distance, but actually no more than a trivial speck of what scientists call space-time—a mere 0.00000012 of the distance our Sun will travel in its course of one revolution around our galaxy. To have some idea of the vastness of our cosmic home, let's leave our Earth and take a brief imaginary trip into deep space. Not far from our Earth, a mere 250,000 miles, is the Moon. Although that distance is equivalent to driving a car around the entire Earth 10 times, the Moon is part of our cosmic backyard. We can travel to this place now in a few days and witness Earthrises (Figure 1-4).

Approximately 93 million miles away is an object of great importance to our continued existence, our Sun. If we could drive in a magic car at a steady speed of 60 miles per hour, it would take us 175 years to reach the Sun from the Earth. The Sun is over 300,000 times more massive than the Earth, and a million Earths could easily fit within its volume. We are very fortunate that it is just this size and just this distance away, or our Earth could have suffered a climatic catastrophe similar to those of the planets Venus and Mars. A little closer and our planet would not be the biological haven that it is. Perhaps it would be more like Venus (about 70 million miles from the Sun), where the temperature is over 800° Fahrenheit. A little farther and our planet might be like Mars (140 million miles from the Sun), a cold, apparently lifeless desert.

Life is possible on Earth because of the particles of light from the Sun that scientists call photons. We should all treat these little particles with deep respect. They have had a very long, difficult journey to make everything possible for us. Created through cataclysmic nuclear processes in the deep interior of the Sun, they require tens of thousands of years to finally escape this massive turmoil. Once they escape, they display their exhilaration by speeding away from the Sun at 186,000 miles per second. At this speed they reach the Earth in a little over eight minutes, allowing tourists on Waikiki Beach to get tans, and plants and microorganisms to harvest this energy through photosynthesis, starting the food chain that is essential to life.

Particles of light are very small.[2] Millions are required to make up a faint stream of light. Neutrinos, another kind of particle generated by the Sun's

FIGURE 1-4

As the *Apollo* astronauts orbited the Moon in 1968, our beautiful but lonesome Earth appeared to rise over the lunar landscape. (NASA)

---

[1] As we will see in Chapter 7, "Understanding the Theory of Relativity," there are no stable spatial mathematical points. Thus, there is no such thing as the same place.

[2] As we will see in Chapter 8, "Quantum Physics and Reality," it is a mistake to think of particles of light as ordinary objects that have a small, but localized, size. For now consider that billions of atoms make up the head of a pin, and that an atom is over 1,000 times more massive than a photon.

nuclear furnace, are even smaller. Billions are at this moment passing through your body and surrounding Earth, either through the top of your head, if it is daytime, or through the bottoms of your feet, if nighttime. They pass through our bodies and the solid Earth as if our bodies and the Earth were a universe of empty space. Nuclear physicists estimate that approximately 4 million tons of the Sun's nuclear fuel are consumed every second to make all this possible and that this process has been ongoing for about 5 billion years and will continue for at least another 4 billion years.

As important as this star is to us, it is simply a grain of sand in an enormous sea of stars and galaxies. Consider that the next nearest star to us is about 300,000 times more distant than our Sun, which is part of a single galaxy of stars. No one, of course, knows the precise number, but just within our galaxy there may be up to 400 billion stars. Our star is an average star, one that astronomers call a yellow dwarf star. Figure 1-5 shows what astronomers call the Eta Carinae nebula. This is a view from within our galaxy; stars that appear close together are actually trillions of miles apart. On this scale our entire solar system, the area within the orbit of the planet Pluto, an area approximately 7½ billion miles across, would be but a tiny dot on this photograph. We have begun to send our spacecraft out into this colossal cosmic neighborhood, but it will be a long time before they travel very far. *Pioneer 10* is such a spacecraft. Launched in March 1972 and traveling now at a little over seven miles per second (25,000 miles per hour), it did not cross the orbit of the outermost planet of our solar system until June 1983, and it will not reach the vicinity of a relatively close star for 33,000 years.[1] Because this star is moving toward us at a very rapid closing speed, *Pioneer's* journey is shortened by over 200,000 years.

Talking about large astronomical distances in terms of miles or kilometers is very inconvenient. The distances between cosmological structures such as stars and galaxies are more easily judged by using the standard of a *light-year,* the distance light travels in one year. At about 186,000 miles per second, 670 million miles per hour, or 16 billion miles in one day, a single light-year is equal to 6 trillion miles. The distance then from the Earth to the nearest star is about four light-years.[2]

With the flights of *Pioneer 10* and *11,* and *Voyagers 1* and *2,* we have begun to explore some of the objects within our solar system backyard. Launched in the fall of 1977, *Voyager 1* arrived at Jupiter in March 1979, about 500

The sun rises. In that short phrase, in a single fact, is enough information to keep biology, physics, and philosophy busy for all the rest of time.

LYALL WATSON, *LIFETIDE*

[1] According to a report from the Jet Propulsion Laboratory, "Prospects for the *Voyager* Extra-Planetary and Interstellar Mission," *Journal of the British Interplanetary Society,* March 1984, *Pioneer* will then make a "remote" flyby (it will still be 19 trillion miles away) of the star Ross 248.

[2] Astronomers also use the measurement known as a *parsec* (3.26 light years). Thus, the distance to the Eta Carina nebula is 2,800 parsecs, or over 9,000 light-years (54,000,000,000,000,000 miles).

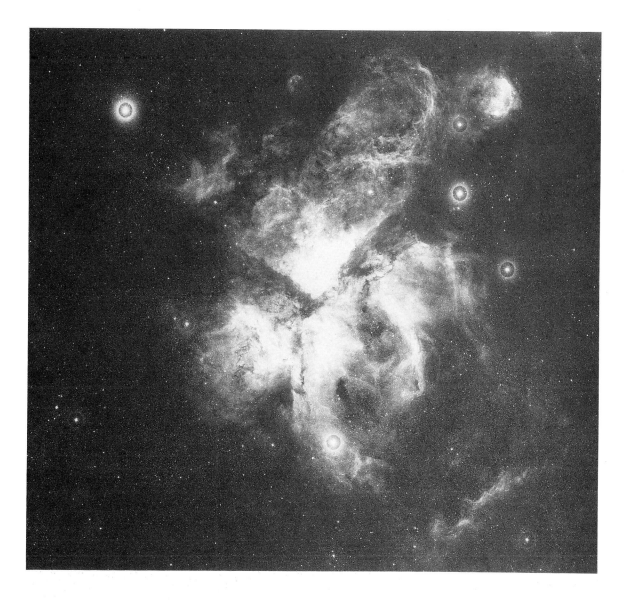

million miles from the Sun and 400 million miles from the Earth. This year-and-a-half journey was a marvelous technological achievement, but what human technology accomplished in a year and a half, nature accomplishes in about 36 minutes at the speed of light. *Voyager* then traveled to Saturn, a distance of almost a billion miles from the Sun. It arrived in November 1980 and is now in the process of leaving our solar system. *Voyager 2* followed a different path, arriving at Jupiter in July 1979, Saturn in August 1981, and Uranus in January 1986, a distance of about 1.8 billion miles; if all continues to go well, it will reach Neptune in 1989, 2.8 billion miles from the Sun.

FIGURE 1-5

The Eta Carinae nebula. At many thousands of billions of miles from Earth, the stars in this picture that appear close together are actually trillions of miles apart. A single dot of space would represent our entire solar system. (National Optical Astronomy Observatories)

Jupiter is the largest planet. Over 1,000 Earths could fit within its volume. Some scientists speculate that this gigantic world has no solid surface, that deep beneath its frigid ammonia clouds is an enormous ocean of hydrogen and, very deep and under enormous pressure, a core of liquid metallic hydrogen. The four, historically important, Galilean moons of Jupiter—Io, Europa, Callisto, and Ganymede—so called because Galileo discovered them in 1609, were surveyed close up for the first time by the *Voyager* flights. Io is the first body in the solar system other than our Earth where active volcanoes have been observed. Pictures from *Voyager* revealed at least eight active volcanoes and enormous volcanic explosions, spewing solid material up over 60 miles, a distance much greater than that of the eruption of Mt. St. Helens. Some of the eruptions were observed to last for four months. Io is essentially a large ball of sulfur, with brief sulfur dioxide clouds caused by the volcanic eruptions, sulfur snowstorms, and perhaps even lakes of liquid sulfur.

Europa, the brightest of the Galilean moons, is about the size of the Earth's Moon, but unlike our Moon its surface shows few impact craters. Instead it appears as a smooth billiard ball full of fractures, again implying an active surface environment. Although water volcanoes have not been observed, planetary scientists believe that such volcanoes probably exist on Europa, spewing into space pressurized water that freezes and then falls back to the surface, in the process erasing any craters.

Ganymede, perhaps the largest moon in the solar system, is larger than the planet Mercury and 40 percent larger than our Moon. Density measurements indicate a large amount of water and ice as a basic constituent. Planetary scientists have discussed seriously the following physical possibility: With a surface temperature of $-338°$ F., a mostly ice crust would form extending downward from the surface 30 to 60 miles; then a large underground ocean, 310 miles deep and several billions of years old—25 times more liquid water than on the entire Earth—would extend down to a hot rocky core with temperatures ranging from 80° to 1,800° F. This intriguing model has fired the imagination of many scientists. Could there be some form of life in such an ocean? What would it be like?

Finally, there's poor, heavily cratered Callisto. Unlike Io and Europa, which seem to have recovered from the meteoric bombardment that scientists believe struck our entire solar system billions of years ago, Callisto remains scarred and wounded, like the Earth's Moon. Its surface has probably changed little in the past 4 billion years.

Just over two years later, and another 500 million miles, *Voyager* reached majestically ringed Saturn. Although Saturn is 15 percent smaller than Jupiter, *Voyager*'s encounter with this planet was not without many surprises. Many more rings were discovered than Earth-based observations had indicated, and a spokelike, braiding effect between some of the rings was observed that initially had no known gravitational explanation. Saturn's beautiful rings stretch 171,000 miles tip to tip, but are so thin that for a few days every 15 years the rings become difficult to see from the Earth when their edges align.

Somewhere, there is something incredible waiting to be known.

CARL SAGAN

Saturn is essentially a huge ball of weather with winds up to 1,000 miles an hour. It requires almost 30 Earth years to revolve around the Sun, but its day is only 10½ hours long.

Saturn's major moon is Titan, a very large moon, perhaps larger than Jupiter's Ganymede. It is difficult to tell, because the atmosphere is so dense (much denser than that of Earth) that planetary scientists, much to their disappointment, were unable to make out any surface features of Titan as *Voyager* flew by. With the exception of the basic chemical facts—this moon has large amounts of nitrogen and methane and small amounts of ethane, acetylene, ethylene, and hydrogen cyanide—scientists were left with only their imaginations and various paths of speculation. Might the surface contain the same biochemical soup that made the origin of life possible on Earth? How about rivers, waterfalls, and oceans of methane? Perhaps on Titan there would never be an energy problem, because the rain consists of oil rather than water.

When *Voyager 2* encountered Uranus, it discovered on the moons some of the strangest geological features known in the solar system, plus a gigantic 5,000-mile-deep ocean on the planet itself. On the moon Oberon is a mountain at least 12½ miles high, over twice the height of Mt. Everest. Miranda's geology is so strange that planetary scientists speculate that—like a Humpty Dumpty put together again—it was blasted apart in the past by a cataclysmic impact with an object about half its size and then was reformed (reaccreted) by gravity. After Nepture, *Voyager 2*'s next milestone will be the star Sirius. It will pass this, the brightest star in the Northern Hemisphere, some 350,000 years from now, missing it by only about 5 trillion miles. It is time too for us to leave our backyard. The 3.7 billion miles to Pluto, the farthest planet[1] from the Sun, is just a speck of space within our galaxy, the distance light travels between breakfast and lunch.

Great nebulae such as Eta Carina abound throughout our galaxy. Astronomers use the word *nebula*, from the Sanskrit "nabhas" meaning cloud, to refer to both places where stars are born and the remnants of the death throes of a star. Some stars like our Sun die by gas expansion, like sloughing off skin, and become dense black dwarf stars. The death of a larger star is marked by a cataclysmic explosion and is called a *supernova*.[2]

The word *nova* (from the Latin "novus" meaning new) indicates that early observers of the universe thought these objects were new stars. Some stars

Everything that you could possibly imagine, you will find that nature has been there before you.

JOHN BERRILL

---

[1] Because of its eccentric orbit, Pluto will be closer to the Sun than Neptune is until 1999.

[2] Astrophysicists believe that for a star to be a supernova it must be at least a dozen or more times as massive as our Sun. In 1987 a supernova was observed in the Large Magellanic Cloud, a relatively small satellite galaxy of stars to our own galaxy. The original star, previously observed and catalogued, is estimated to have been 15 times more massive than our Sun.

are so massive—some are as large as half our entire solar system—and erupt upon their deaths so violently that they appear from Earth as if a bright star had emerged from nowhere. The past death of a star in our cosmic neighborhood was absolutely essential to life as we know it, since it is in the interior of large stars that the basic atomic elements necessary for life are created. Without stars there would be no gold to kindle humankind's greed, iron to build our steel civilization, no carbon to form the basic building blocks of life and to generate our fossil fuels, and no oxygen to breathe. Just as the universe has experienced a structural evolution, and at least one form of biological and cultural evolution (on Earth), the universe has also undergone a chemical evolution. During the formative years of the universe, only the elements hydrogen and helium existed.[1] Over a period of billions of years the nuclear processes that take place in stars have converted hydrogen to heavier elements and these heavier elements to still heavier elements. When these stars explode, they seed the universe with the chemical elements. Our Sun is at least a second- and perhaps a third-generation star. In the vicinity of our solar system, another star previously existed. Its death made our lives possible.

Upon their deaths, the most massive of these stars turn into neutron stars or the notorious black holes. Neutron stars are supercondensed worlds where even the atoms have been crushed. So compact are these remnants of an intense gravitational collapse that a spoonful of matter from one of these stars would weigh more than all the Earth's continents. A black hole is an astronomical object that stretches our imagination to even greater limits. It is essentially a hole in our three-dimensional reality, where the future and the past coexist and the concepts of direction and distance become meaningless. The mass of a star destined to be a black hole is so large, and its eventual gravitational collapse so extreme, that the star's entire material existence disappears into what mathematicians call a "singularity," a one-dimensional point.

The Crab nebula (Figure 1-6) is a famous example of a supernova remnant. In A.D. 1054 the original explosion of the supernova star was recorded by the Chinese, Arabs, and American Indians. In spite of being over 40,000 trillion miles from Earth, for the first few months it was so bright that a person could read by it at night. Since the nebula is about 7,000 light-years from Earth, what we are looking at now occurred in 5,000 B.C. The explosion is still expanding at a rate of several hundred miles per second. Any planet that might have existed within a solar system vicinity (billions of miles) at the time of the initial explosion would have been entirely vaporized. Any

Everything around us is filled with mystery and magic. I find this no cause for despair, no reason to turn for solace to esoteric formulae or chariots of gods. On the contrary, our inability to find easy answers fills me with a fierce pride in our ambivalent biology . . . with a constant sense of wonder and delight that we should be part of anything so profound.

LYALL WATSON, *LIFETIDE*

---

[1] Next time you have a glass of water or take a swim or a bath, consider that the hydrogen atoms in the water were formed within the first million or so years of the existence of the universe, some 15 billion years ago.

FIGURE 1-6

The Crab nebula, a super-
nova remnant 7,000 light-
years from Earth, is still expanding at over 800 miles
per second and now is over 8
light-years in diameter. (Pal-
omar Observatory)

planet within a few light-years would have been completed sterilized.[1] At the
center of this explosion is a neutron star spinning around 30 times a second.
Astronomers refer to such a spinning neutron star as a *pulsar*, because the
star emits very precise pulses of radiation as it spins dizzily around its axis.
So precise and regular is this radiation that when first detected by radio telescopes on Earth, it was thought that we were receiving a message from an
intelligent cosmic neighbor.

---

[1] The supergiant star Betelgeuse is about 540 light-years from Earth,
and it is so large that if it could be magically transported to the location
of our Sun, it would extend to the orbit of Jupiter, approximately a billion miles in diameter. If such a star exploded today, it could cause
harmful radiation to strike the Earth. At the earliest, however, this radiation would not reach us for over 500 years.

FIGURE 1-7

Like the galactic merry-go-round of our home galaxy, billions of stars swirl around the great galaxy of Andromeda. At over 2 million light-years from Earth, massive stars, many trillions of miles apart, appear as packed-together white dust. (Lick Observatory)

All that we have surveyed thus far occurs within our galaxy, one galaxy out of billions, another grain of sand. If we were able to leave our galaxy, and travel a few million light-years away and look back, astronomers believe we would see something similar to Figure 1-7. This is one of our closest galactic neighbors, the Andromeda galaxy. It is 2.2 million light-years away, a mere 13,200,000,000,000,000,000 miles. What we are seeing now occurred when our evolutionary ancestor, *Homo erectus,* walked through the grasslands of Africa. Spiked points of light visible on the picture frame are stars from our galaxy as we look through it. If this were a picture of our galaxy, our Sun and Earth would be located on the inner edge of one of the outermost visible spiral arms. Our galaxy is approximately 100,000 light-years in diameter, and even though we are spinning around this galactic merry-go-round at over 500,000 miles per hour, it will take 250 million years to complete one revolution. Some galaxies spin their stars at an even faster rate, over a million miles per hour.

Just as people and stars are different, so are galaxies. There are dwarf elliptical, giant elliptical, globular, spiral, and irregular galaxies. Figure 1-8 shows the charming Sombrero galaxy.

Further out into the universe, clusters of galaxies form. The Virgo cluster, a group of over 2,500 galaxies, is 70 million light-years from Earth. It stretches 20 million light-years in diameter and is the closest to what astronomers refer to as the Local Group, the cluster of galaxies in which our galaxy and the Andromeda galaxy are located.

Still farther away we see a picture of the Coma cluster (Figure 1-9). Astronomers have counted 800 major galaxies in this cluster, which is 500 million light-years from our galaxy. The light that made this photograph left these galaxies about the time land animals first evolved on Earth. At 700 million light-years is the Hercules cluster, 50 million light-years across, encompassing perhaps as many as 5 million galaxies. These clusters are receding from us at great speeds—the further away, the greater the speed. The Virgo cluster is receding at some 700 miles per second, the Coma cluster at 4,000 miles per second, and the Hercules cluster at 6,000 miles per second. Quasars, which most astronomers now think are extremely energetic galactic nuclei

(some are 100,000 times brighter than the Andromeda galaxy), are billions of light-years away and recede from us at speeds of 100,000 miles per second. Finally, we see a map (Figure 1-10) of over a million galaxies. Astronomers estimate that there are at least 10 billion galaxies within the range of our earthly telescopes (about 8 billion light-years) and that this immense outpouring of activity has been expanding and evolving for about 15 billion years.[1]

FIGURE 1-8

Twenty million times farther away than the Andromeda galaxy, the Sombrero galaxy reminds us that universal processes produce variety as well as commonality. It also reminds us of the balloonlike nature of the universe; this galaxy is receding from us at over 2 million miles per hour. (National Optical Astronomy Observatories)

---

[1] The methods astronomers use to measure these distances are, of course, subject to error. The farther away an object is, the more the probable error. At the edge of the visible universe, the error is estimated to be 50 percent. Thus a quasar that is 8 billion light-years away could be as close as 4 billion or as far away as 12 billion light-years. Some astronomers believe that a recently discovered galaxy is 12 billion light-years away. Although crucial for evaluating theories on the origin of the universe, in terms of the stirring philosophical considerations these distances produce, the probable error hardly seems to matter.

Romancing the Universe

No matter how preoccupied we are with more practical matters on our little planet, every human being must at some time or another confront profound philosophical questions, questions impossible to ask without deeply felt emotion. Why does all this exist? Why are we a part of it? If we are honest with ourselves, is there any philosophy or religion completely consistent with such a perspective? Such an insignificant speck that we occupy, such an enormous

FIGURE 1-9

At a distance between 350 and 500 million light-years from Earth, the Coma cluster of galaxies, containing over 800 galaxies, teaches us how quantitatively insignificant is a single galaxy. The recessional velocity of this cluster is over 15 million miles per hour. (National Optical Astronomy Observatories)

extension of time compared to our fleeting existence—yet we have been able to figure it out. We occupy such a small place, but we are capable of being aware that it is a small place. Imagine a small, intelligent insect in a small, isolated forest, that lives for only a few days and whose species is only a few hundred years old, but that nevertheless has been able to indirectly figure out the size and age of the Earth, that it consists of mountains, rivers, continents, oceans, cities, and people. Imagine such an insect that has not, and perhaps cannot, travel to these other places, but has expanded its little mind to encompass these realities nevertheless. In many ways we are this creature.

According to Albert Einstein, this is the greatest cosmological mystery: Not our knowledge of the universe, but why and how do we know it? Today some scientists believe that we are very close to knowing the basic details of the origin of the entire universe. What does this mean in terms of who we are? Are there other forms of life capable of this awareness? Are there other kinds

A scientist lives with all of reality. There is nothing better. To know reality is to accept it and eventually to love it.

GEORGE WALD

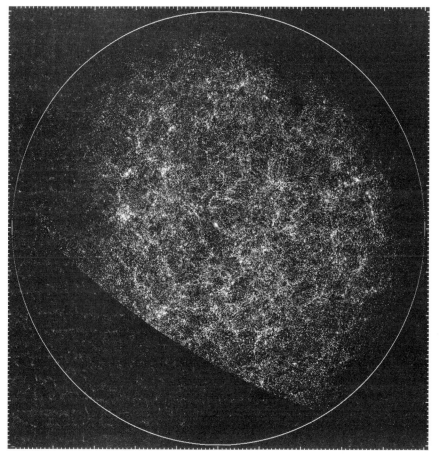

FIGURE 1-10

At billions of light-years from Earth, the galaxies and galaxy clusters would begin to take on the appearance of a flimsy film of cobwebs. This is a computerized plotting of the location of over a million galaxies. (Courtesy Michael Seldner)

of awareness? Does the universe have a purpose? Are we an important part of this purpose?

As we address these questions, we will see that the intellectual evolution of our present understanding is as fascinating as its results. In many ways the meandering path of our understanding is like an enduring but somewhat turbulent romance in which each idea is a gesture to the universe for acceptance. Before we can discover anything, we must have an idea of what we are looking for. How we pick the right idea at the right time puzzled Einstein all his life, and much of this book will address this question. But why we seek to understand is no mystery. It is as natural for our species as love.

Infinite Textures and the Microcosmos

Before we address such questions, we must note that our introductory story is only half told. We have only gone in one direction. The size of a printed letter on this page is about 10 billion times smaller than the Earth. Yet a printed letter is 10 billion times larger than an atom and 100 billion times larger than a neutrino. There is a universe in any direction that we look, an infinite texture to every object.

There are insects small enough to fly through window screens of the best-kept houses and breed in drains of bathtubs and sinks. To something small enough, the cracks in an egg are like a grand canyon.

Usually when we think of insects, we think of them as a rather insignificant part of the life of this planet, at most a nuisance we could do without. Yet, three-fourths of all the species of life in the world are insects. For every human being there are 1 million insects. There are 500 species of fleas alone and 100 different species of mosquitoes; species of flies and cockroaches have existed for 300 million years. The human species and our closest ancestors have existed for at most about 3 million. Modern humankind has existed for only about 30,000 years. Remarkably, compared to our fragile existence on this planet, all our sprays, chemicals, and poisons have failed to eliminate a single species of nature's insect inventions. For several decades pineapple growers on the island of Oahu used a chemical called Heptachlor to control ants from ruining the pineapple plants and fruit. The ants are still doing fine as a species, but recently residents were shocked to find out that almost all of the milk produced on Oahu contained unacceptable levels of this pesticide. After the fruit was harvested, the pineapple plants were chopped up and fed to milk-producing cows, and the pesticide ended up in the milk.

The average house accumulates about 40 pounds of dust a year, and dust mites (Figure 1-11) inhabit every house. Over 40,000 of these creatures could survive in an ounce of mattress dust. Next, we see a picture of a marijuana flower (Figure 1-12) and a spider mite stuck in a broken resin nodule (Figure 1-13). Capturing insects in its sticky resin is the evolutionary strategy the marijuana plant has adopted to defend itself. In the State of Hawii law enforcement officials are having as much luck eradicating this number one agricultural cash corp as farmers did eradicating ants. Perhaps they would have better success if they showed marijuana smokers electron microscope pictures of what they are smoking.

Another insect that we have unsuccessfully tried to eliminate with powerful poisons is the termite. It is somewhat fitting, however, that insects have their own insect problems. Figure 1-14 shows a mite on the neck of a termite. If you look carefully, you can just make out some bacteria in the upper righthand corner. A bacterium is so small (0.000039 of an inch long, less than half this distance wide) that a swim across a drop of sweat for one would be roughly the same as having a human being swim across the English Channel. Billions of bacteria make each of our bodies their home, their planet. There are more bacteria on each of our bodies than there are people in the world. Every time

Nothing is rich but the inexhaustible wealth of nature. She shows us only surfaces, but she is a million fathoms deep.

RALPH WALDO EMERSON

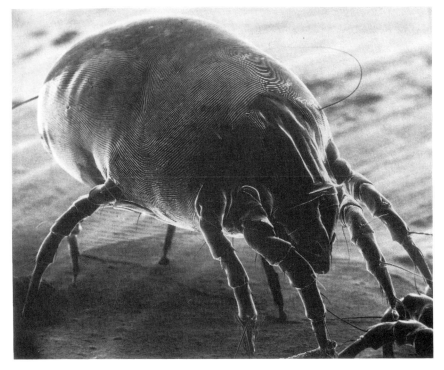

FIGURE 1-11

Dermatophagoides farinae, the house dust mite. Thousands inhabit the secret, microscopic nooks of even the best-kept homes. (Acarology Laboratory, The Ohio State University)

FIGURE 1-12

Resin nodules on a female *Cannabis sativa* flower ("bud"). These sticky, globular secretions have the greatest concentrations of the marijuana drug. (Photo from *Magnifications.* Copyright © 1977 by David Scharf. Reprinted by permission of Schocken Books, published by Pantheon Books, a Division of Random House, Inc. and Century Hutchinson, London)

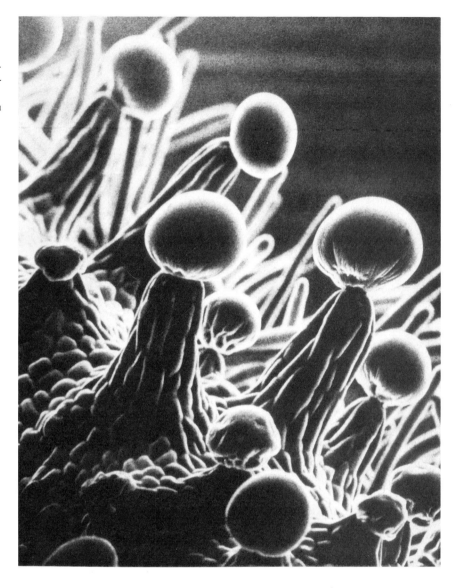

we take a hot shower we attempt a little genocide, but within a few hours those that survive will have reproduced to their original numbers. Bacteria have been on this planet for at least 3½ billion years, and they are not about to leave.

From an evolutionary point of view we are not only surrounded by bacteria but in a sense we also are composed of them! Nine out of 10 of the trillions of cells that make up the human body are bacterialike cells. All the cells of our bodies are descendants from the first bacteria. From this viewpoint, the

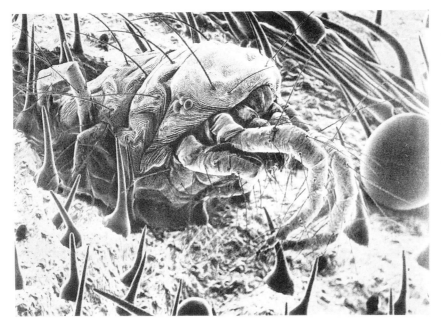

FIGURE 1-13

At this microscopic level, a world of treacherous, exploding stickum greets uninvited prowlers. A spider mite is entrapped in the resin of a broken nodule. (Photo from *Magnifications*. Copyright © 1977 by David Scharf. Reprinted by permission of Schocken Books, published by Pantheon Books, a Division of Random House, Inc. and Century Hutchinson, London)

human body is thus just a colony of cooperating animals, and life in general is not a struggle of the survival of the fittest but one of networking and expanding cooperation.[1]

Insects and bacteria are huge compared to the microcosmic spaces physicists routinely discuss. With only the language of mathematics for pictures, physicists are able to investigate spaces called quantum vacuums filled with "virtual" particles, objects that do not deserve the name "object" because they both exist and do not exist. In every material object there is another universe. All material objects and all life forms consist of mostly empty space. Most nonliving material objects consist of crystals and atoms. All living things consist of cells, molecules, and atoms. Although it is impossible to picture what an atom looks like,[2] to at least have a vague idea of the relative sizes of its parts, imagine if we could blow up the size of an ordinary hydrogen atom to the size of the Earth. The nucleus of the atom would be about the size of

Mitochondria seem to be able to exist, in the form of free-living bacteria, without our help. But without them, we die in a matter of seconds.

LYALL WATSON, *LIFETIDE*

[1] For an in-depth discussion of this theme, see Lynn Margulis and Dorion Sagan, *Microcosmos: Four Billion Years of Evolution from Our Microbial Ancestors* (New York: Summit Books, 1986).

[2] Recent photographs of atoms are actually pictures of the effects of our attempts to see atoms. As we will see in Chapter 8, the atom and its parts are not ordinary things that can be pictured in a normal camera image.

FIGURE 1-14

An insect with a bug problem: a mite on the neck of a
termite and some bacteria in
the dark hollow in the upper
right section of the picture.
(Photo from *Magnifications*.
Copyright © 1977 by David
Scharf. Reprinted by per-
mission of Schocken Books,
published by Pantheon
Books, a Division of Random
House, Inc. and Century
Hutchinson, London)

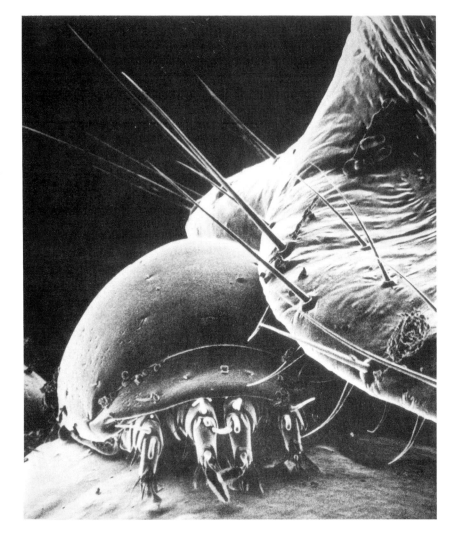

a basketball located in the center of the Earth, and in the outermost portion of the Earth's atmosphere we would find an electron, smaller than the smallest cherry. If the nucleus were the size of our Sun, then the size of the atom would be 20 times larger than our solar system, about 75 billion miles in diameter. The nucleus of an atom represents but a mere one-trillionth of the atom's volume. An electron is about 2,000 times smaller in mass than a pro- ton, a particle that makes up the nucleus of an atom. Thus, atoms are almost entirely empty space, and remember, billions of atoms can fit within the space of a single printed letter.

The Cosmic Calendar

If the universe has evolved over a period of approximately 15 billion years, then the Earth and our Sun are relatively new arrivals on the cosmic scene, being only about 5 billion years old. Life began on Earth shortly after its formation (4 billion years ago), but complex, multicellular forms of life did not begin until about 600 million years ago, a period of time geologists call the Cambrian explosion. Some of the first land animals were insect creatures of which the millipedes and centipedes are representative (450 million years ago). We know from the fossil record that some of these first millipedes were as long as cows! Some other noteworthy developments are flies and cockroaches (first appearing 300 million years ago), the dinosaurs (200 to 60 million years), dophins (30 million years), the immediate ancestors to the human species (3 million), modern man (*Homo sapiens sapiens*, 30, perhaps 90 thousand years), the first technological achievements by humankind (10,000 years), and our modern technological life from the industrial revolution to the present (200 years).

The time scales with which we are familiar in our daily existence are so different from what these numbers represent that it is very difficult to comprehend the significance of such enormous times. To better understand this, scientists make use of an analogy, a cosmic calendar in which every second in the calendar corresponds to 475 normal years and every 24 calendar days is equal to a billion regular years. In this way the entire 15 billion years of cosmic development can be shortened into a single calendar year. Accordingly, the following list represents some important dates relevant to the human species:

Origin of the universe—January 1
Origin of our galaxy—May 1
Origin of our solar system—September 9
Formation of the Earth—September 14
Life on Earth—September 25
Sexual reproduction by microorganisms—November 1
Oxygen atmosphere—December 1
Cambrian explosion (600 million years ago when most complex organisms appear, fish, trilobites)—December 17
Land plants and first insects (millipedes)—December 19, 20
First amphibians—December 22
First reptiles and trees—December 23
First dinosaurs—December 25
Dinosaur extinction, rise of mammals, first birds, flowers—December 28

If we are still here to witness the destruction of our planet some five billion years or more hence . . . , then we will have achieved something so unprecedented in the history of life that we should be willing to sing our swan song with joy.

STEPHEN JAY GOULD

First primates (monkeylike creatures)—December 30
Australopithicenes (Lucy, etc.)—10:00 p.m., December 31
Homo habilis—11:25 p.m., December 31
Homo erectus—11:40 p.m., December 31
Early *Homo sapiens*—11:50 p.m., December 31
Neanderthal man—11:57 p.m., December 31
Cro-Magnon man—11:58:38 p.m., December 31
Homo sapiens sapiens (modern man)—11:58:57 p.m.
Ancient Greeks to present—last five seconds
Average human life span—a little over one-tenth of a second

One-tenth of a second! As a child I had the good fortune of being part of a family that enjoyed exploring nature. I can vividly remember one day when an aunt took me on our annual hike in a mountain range of Southern California to what we called the golden forest. Not far from a small lake was an oasis of large old oak trees in the midst of mile after mile of pine trees. Every fall the leaves would turn from green to gold, and the startling effect of emerging finally from the green of the pine forest to this magically golden place was well worth what seemed to a little boy a long hike. On the way back this day we came upon a most beautiful flower. My aunt casually remarked that it was an unusual flower to bloom at this time of year and that it bloomed for only a day and then the whole plant died. I remember feeling very sad. I felt so sorry for this poor little flower. Life was so full of freshness and fun for me. It seemed like it would go on forever, yet this poor flower had only a day to romance the universe.

Children can be fascinated by almost any object. Perhaps they feel the universe that each one contains. Most adults seem to lose the enthusiastic intensity of exploring ordinary objects. Ironically, an important condition for being a scientist or a philosopher is that one not lose a childlike mind, or at least maintain that curiosity for much of each day. Today, scientists believe that locked within the physical matter of every earthly object is a fossil record, a self-contained, endless detective story that can be traced back to the origin of our universe. Fueled by a childlike fascination for our cosmic roots, harmonious mathematical equations, and huge, expensive machinery, astrophysicists believe we now have a reasonable basis for knowing not only how our universe began but also the details of what happened within the first fraction of a second, 15 billion years ago.

The most commonly accepted view among scientists today of the origin of the universe is known as the Big Bang theory. At time zero nothing existed but a point, a singularity. Not a point "in" space and time, but only a one-dimensional point. Space and time were inside the point! Do not try to imagine this; your mind will not let you create an internal picture of it—it can only be described mathematically. From this point space, time, and energy emerged.

A scientist is . . . a learned child. There is something of the scientist in every child. Others must outgrow it. Scientists can stay that way all their lives.

GEORGE WALD

At 10^{-43} seconds,[1] what physicists call the Planck time, the temperature of this energy was 100 thousand billion billion billion degrees. At this time the separate forces of nature did not exist, such as the gravitational force that holds us to our spinning Earth, or the electromagnetic force responsible for magnetic repulsion and attraction, or what is called the strong force, responsible for keeping the nucleus of an atom together, or the weak force responsible for atomic radiation. At this time there was only a single, unified superforce. At 10^{-35} seconds, the temperature cooled enough to freeze out the identity of the strong and weak forces. At 10^{-14}, 100 trillionths of a second, the identity of the weak and the electromagnetic forces emerged. As the completion of the first second of the universe approached, the temperature was still too hot for the existence of individual subatomic particles such as protons, electrons, and neutrinos. There were only the puzzling, forever invisible, and inseparable quarks. At three minutes the temperature finally cooled enough for subatomic particles to form, but even after 500,000 years there were no atoms, only loose nuclei and individual electrons. Once atoms finally formed, only hydrogen and a little helium existed. The rest is a history of structural and chemical evolution. As the universe expanded, galaxies and the stars within them formed.

On the Fourth of July when I was a teenager, my cousins and I used to do what spirited, mettlesome teenagers are apt to do: We harassed my neighborhood with fire crackers. On one Fourth of July night in particular, I remember that a heavy Southern California fog unexpectedly aided our mischievous running from house to house, throwing fire crackers onto people's porches. At one point a cousin ignited a Mexican cherry bomb in the middle of the street, a sort of grand finale to our night of terrorism. The effect startled all of us. The particles of fog allowed us to see the extent of the explosion as a yellowish-orange semicircle with what seemed to be a radius of at least 15 feet. It probably took only a few seconds for the visual effects to dissipate, but the unexpected scene caused a slow-motion picture in my mind. At first the colors were very intense and expanded quickly, then gradually they faded and vanished over what seemed like a long time. For what was probably only a fraction of a second, individual speckles of fog throughout the semicircle were heated and seemed to flash with light and life.

We are living in the midst of an explosion, luckily in a relatively mild, somewhat frozen, period. We can measure the temperature of this explosion

Sit down before fact as a little child, be prepared to give up every preconceived notion, follow humbly wherever and to whatever abysses nature leads, or you shall be nothing.

T. H. HUXLEY

---

[1] Scientific notation allows scientists to make better use of their time than writing a lot of zeros. The exponent 43 represents the number of zeros that would be added if this number were written out normally. Since a minus sign is used in this number, the zeros would be added to the front of the 10 with a decimal point at the beginning. Thus, the Planck time is a tenth of a thousandth of a millionth of a billionth of a trillionth of a trillionth of a second after the Big Bang.

by detecting a faint radio whisper with our radio telescopes, view it backward 8 billion years or so with our optical telescopes, and even calculate its rate of expansion. Its temperature is 3° Kelvins (3° above absolute zero, or −454° F.), and its expansion is estimated to be 10^{18} cubic light-years per day, such that the distance between any pair of galaxies will double every 10 billion years or so.[1]

It all sounds like a grand myth. We have discovered, however, that the truth is much more fantastic than our most imaginative myths. It is also serious business with billions of dollars spent each year by competing countries and teams of scientists to unravel the smallest details of nature's secrets, especially as to what occurred in the first fraction of a second 15 billion years ago. The Conseil European des Recherches Nucleaires (CERN), the research center of the European Council for Nuclear Research near Geneva, Switzerland, is a place where childlike fascinations with ordinary objects are expressed in grown-up form. This physics laboratory is the size of a small city, cost billions of dollars, and has a staff of over 3,000. Here 2,000 physicists play with the microcosmos.

About 100 of these scientists are a special breed. Their colleagues call them theorists. These men and women may be some of the happiest people in the world, although no psychological study has ever investigated this notion. They get fairly good salaries to spend their waking moments doing nothing but mathematical speculation and deduction. For many it is a passionate obsession that never ends. Formally, their work is done with complex supercomputers, but it is also often done on the back of cocktail napkins and in their dreams. For average students struggling to get through their first algebra class, it is probably difficult to imagine that such activity can be meaningful and fun. But for these men and women, following an ancient tradition that began with the Greeks, mathematics is the key to unveiling the secrets of the universe, and they dream of Nobel Prizes and some, following another tradition popular in the sixteenth century, of being the first to read the mathematical mind of God.

The Nobel Prizes, however, usually go to the physicist-administrator.[2] So large and complex is the experimental machinery used here to probe nature's secrets that it also takes a special breed of human being to manage it all. At CERN there is a Super Proton Synchrotron, 4.2 miles in circumference, where matter and antimatter are made to smash into each other; a Large

> *Unless his mind soars above his daily pursuits, it is fairly natural that the working scientist should characterize his business as a welter of different techniques. In the same spirit, the woodsman might claim that there are only trees but no forests.*
>
> HENRY MARGENAU

---

[1] It is a strange explosion, however. Like raisins in a baking muffin, the space between galaxies expands (like heated dough) without each galaxy moving much with respect to the space right around it.

[2] Who may not be as cheerful and lighthearted as the detached theorists. See the book by Gary Taubes, *Nobel Dreams*, in the suggested readings at the end of this chapter.

Electron-Positron collider that is 17 miles in circumference; and numerous support facilities—all for unlocking the fossil relics of ordinary physical matter. Throughout the world, in many other citylike laboratories, other physicists are competing earnestly with those at CERN—at TRISTAN in Japan and HERA in Germany, for example.

Although the United States has the Tevatron collider at Fermilab, in Illinois, and the Stanford Linear Collider in California, American physicists are somewhat unhappy with the facilities currently available for elementary particle research in their own country. Accordingly, to end an embarrassing Nobel Prize drain by the Europeans, they have proposed a Superconducting Super Collider. Estimated to cost $3 to $6 billion, it has been nicknamed the Desertron, because it will be so big it will need to be built in a desert. At a circumference of 90 miles, it will encircle an area the size of the State of Rhode Island. This is admittedly a large price to pay, but it is advertised to be able to mimic experimentally the conditions of the early universe to within a thousand of a millionth of a millionth of a second from time zero.

The Relevance of the Cosmic Perspective

Would an independent nonhuman observer, unmoved by the fact that this is a human accomplishment, view our spending billions of dollars to know what happened 15 billion years ago as evidence of our unique intelligence or our madness? Do we really need to know what happened when the universe was only one-trillionth of a second old? Why spend billions of dollars on space exploration and massive accelerators to find out what matter does under conditions similar to the first stages of the universe when we have so many urgent earthly problems? At least one-fourth of the U.S. population and three-fourths of the world population live without their basic needs being met. Our environment is dangerously polluted and we have enough nuclear weapons to destroy the world many times over. Should not our more immediate concerns be primarily social, economic, and political?

The answer to this question is very long. In fact, this entire book is an answer to this question. In the end we will argue that the ancient Greek philosophers were right: A connection exists, though neither rational nor logical, between knowledge and the good life, both on the individual level and the social and political as well. We will try to show that an understanding of our cosmological, biological, cultural, and philosophical roots is a necessary condition for understanding ourselves and making crucial choices affecting the future of the human race. The human species is at a crucial crossroad, and a cosmic perspective is a critical variable in our struggle for survival.

Our intelligence has brought us far, but it has also brought us to the brink of total destruction. It cuts both ways, its application sometimes terrifies us, but it also reveals a humbling, sobering perspective of our cosmological home.

Although we are most interested in philosophical considerations, we can here briefly discuss a few preliminary points. First of all, there is a well-known *technological spin-off* from adopting the cosmic perspective. Studies have shown that for every dollar we spend on exploring space, fifteen dollars are returned to the economy in terms of revolutionary products and economic development.[1] To name only a few, the space program has given us Teflon, integrated circuits, computer chips the size of a thumbnail, advanced microprocessors, direct broadcasting satellites, solar energy, revolutionary medical technology, and soon, manufacturing and industrial parks in orbit around the Earth. Among other things that could be made more efficiently because of microgravity and the superior vacuums attainable in space, these industrial parks will manufacture medicines that would cost millions of dollars per pound if manufactured on Earth; amounts that would take 30 years to produce on Earth would take only 30 days in space. Remote satellite sensing of the Earth's surface to map mineral deposits, vegetation patterns, and land use is predicted to generate billions of dollars in annual revenues by the year 2000. In 1986 a study by a committee of the U.S. House of Representatives showed that by 2010 a trillion-dollar global economy could be created in space, the equivalent of 35 million jobs. For many economists the question is not whether this will actually happen but how. If the human species does not destroy itself, it is only a matter of who will make this economic success story a self-fulfilling prophecy. As William Claybaugh, space investment analyst, has stated, "Some of the next-generation fortunes will be made in space."

The ancient Greeks believed that all knowledge was interrelated and valuable. Some people are often cynically amused upon reading that someone has just received a PhD degree for studying what happens to drops of water when they strike the surface of a pool. Those with the appropriate scientific background know, however, that such research has applications for the design and analysis of turbines, cooling towers, steam generators in nuclear reactors, and many chemical processes.

The exploration of space has also given us what might be called an *Earth environment spin-off*. The more we have studied our astronomical neighbors, the more we have learned about our Earth. By studying Venus we have been alerted to the climatic danger of a runaway greenhouse effect. Because of the dense atmosphere of that planet, radio mapping was developed to "see" the surface. This technology is now applied to finding mineral resources on Earth and ocean mapping for a better understanding of continental drift. Studying Mars has led to a better understanding of volcanism, weathering, and chemistry. The *Voyager* flights past Saturn, Jupiter, and Uranus have taught us

---

[1] Every year the U.S. government's Technology Utilization Office publishes a spinoff summary. The latest is called *Spinoff '87*. And although they are free to the public, the number acquired by businesses soon makes it difficult to get a copy.

about electromagnetic fields and different types of weather, and by studying the moons of these planets, we have learned more about the varieties of planetary geology.

Most important for the focus of this book, however, is what might be called a *philosophical and psychological spin-off*. That human beings have accomplished much materially is not doubted by even the most cynical critics of science and technology. What is doubted is whether our understanding has made us better as individuals and as a society. Do new technological feats benefit the poor or only the rich? Have human beings made significant moral progress, or have the tools of manipulation, exploitation, and control just become more sophisticated? Is our potential for awareness, for "figuring things out," an asset, or is it an evolutionary mistake, a dead end? Does knowledge make life simpler or more complicated?

Few optimists claim that the present scientific revolution will change human nature. Animal species are biologically programmed to be self-centered under certain conditions, and the human species is no exception. However, when confronted by a common problem, an outside force, or a threat, human beings are capable of great cooperation, of unique altruistic acts, of seeing apparent major differences as a confining illusion.

An old Humphrey Bogart movie illustrates this point. In *The Treasure of Sierra Madre,* Bogart plays a man down on his luck who eventually links up with a couple of similarly inclined cohorts who just happen to finally strike it rich prospecting for gold in some isolated mountains of Mexico. They have a big problem, however: how to keep their claim protected and how to move the gold and turn it into cash without revealing their secret.

One day one of the men must make the long and dangerous journey to the nearest town for supplies. Being a stranger in a Mexican town, he will have difficulty not calling attention to himself. There are bandits everywhere who could easily follow him back to the gold. Ironically, his biggest problem when he gets to town is that another U.S. citizen is already there. This man naturally expects a reaction of companionship and is surprised by the partner's aloofness. The partner hurriedly gets his supplies and leaves town, but despite his attempts to avoid being followed, the stranger finds their camp.

The stranger was interested only in the companionship of a fellow American (and perhaps a little gold), but when he sees what they are doing, he realizes immediately that his life is at stake. He pleads with the three men to take him in as a partner. There is plenty of gold for all, and an extra gun could be handy. The original three withdraw to discuss the situation. They decide that they have three alternatives: They can persuade him to leave and not tell anyone of their find in return for not killing him, they can take him in as a fourth partner, or they can kill him. For a brief moment they look at each other and then without any further discussion one partner asks simply, "Who's going to do it?" None of them has ever killed a man before. The matter is settled by concluding that they should all shoot him at the same time.

Man differs from the animal only a little; most men throw that little away.

CONFUCIUS

They return to the edge of a cliff where the stranger awaits their decision. All that needs to be done is to shoot him and push him over the edge. "Well? . . ." he asks. The men pull their guns. The stranger shows little surprise, not an unexpected decision, another item in a disastrous string of bad luck. At the last minute, however, the stranger turns to look down the cliff and calmly remarks, "Before you pull the trigger, you better look down there." Far down the mountain is a large cloud of dust. Shortly, the cause of the dust is clear. Coming up the mountain on horseback at a furious pace is a large contingent of Mexican bandits. Fear instantly registers on the faces of the men, and a gun that a moment ago was to shoot the stranger is uncocked and tossed to him; the four men get ready to defend themselves.

The four men ambush the bandits and force them to retreat by inflicting heavy casualties. On their side, only the stranger is killed. Before they bury him they find a letter from his wife. They discover that he was just like them: a man down on his luck, looking for some way to strike it rich so he could return to his wife and children.

Our situation as a species today is not much different. In many ways we are surrounded by cosmological bandits, and our differences, trumpeted daily on the front pages of our newspapers, vanish in the face of this outside threat. A cosmic perspective reveals a cosmic predicament, an awesome predicament. We occupy a very small space and have lived a very short time. Our cosmic home is a place where monstrous violent physical processes are the norm. The cosmos could burp in our little corner and we, and all our accomplishments, would be gone. A star close to us could become a supernova and fry the Earth with its radiation. As our little star circles our galaxy, another star could pass close to our solar system, disrupting what astronomers call the Oort cloud, a cloud of billions of comets that orbit our Sun. Comet and meteorite debris would then strike the Earth, causing death and mass animal extinctions—a process that some scientists believe is a relatively normal process, perhaps happening every 25 million years or so and perhaps what caused the extinction of the dinosaurs.

Philosophers use the word *contingent* to refer to the fact that our continued existence depends on events that we have no control over. We are not necessary beings; there is no guarantee that any of us will survive the next few seconds. There is no special guarantee for our individual lives or the life of our species.

An understanding of our biological roots demonstrates that we are indeed a unique and precious species, but we are not special. Life may exist elsewhere, but it will not be like us. An artist once complained to me that in a single human face there is an infinite texture and beauty and that there is no need to study the universe of the astronomer. The appropriate answer to him is that by exploring the cosmic perspective and the awesomeness of our physical situation, we can more readily appreciate the uniqueness and beauty of a single human face. Nowhere else in the billions of galaxies will human beings evolve again to share our loneliness.

The dangers that face the world can, every one of them, be traced back to science. The salvations that may save the world will, every one of them, be traced back to science.

ISAAC ASIMOV

The cosmological calendar earlier in this chapter is very misleading in one sense. It implies that the 15-billion-year history of the universe has had but one purpose: the evolution of the human species. An understanding of natural selection, the mechanism of biological evolution discussed in Chapter 3, and evolutionary biology, the science that fills in the details of natural selection on Earth, demonstrates that even on Earth we are not special. No species is special, the majority of life forms that have existed on this planet are extinct; the average lifetime of a species is 10 million years, and a vertebrate species like ourselves lives a mere 2 to 3 million years.

A study of biological evolution is a sobering and humbling experience, revealing no direction or purpose. Life evolves not with a design in mind or lofty plans for the development of a ruling creature but by offering variety, spinning the wheel of chance with every birth, gesturing with each unique creature for acceptance from the environment. In the process it produces a messy equality, a tree of successful branches with no branch any more fit than another. Human beings do some things well, but insects also do some things well, and they have been here much longer and are much more likely to survive a nuclear holocaust. As humans, we boast and celebrate that we alone possess "intelligence," but even if this is true, there is no guarantee from nature that this characteristic is any better than the body structure of a mosquito; it is just another experiment, another gesture for acceptance, perhaps only an evolutionary afterthought. All we can honestly say is that both the characteristics possessed by human beings and mosquitoes have served each well, so far.

In the nature of life and in the principles of evolution we have had our answer. Of men elsewhere, and beyond, there will be none, forever.

LOREN EISELEY, *THE IMMENSE JOURNEY*

Cosmic Brethren

We are all related. We are all cousins, floating together on this fragile blue biological refuge. No supremacy exists amongst biological brethren, only creatures adapted to their environments for the moment. What evolutionary biologists call the fossil record is sobering in this respect. The trilobite, an animal that flourished for over 150 million years in the ancient seas, some 400 million years ago, is now exinct; it demonstrates that although life has thrived on Earth for a long time, durability and adaptability do not ensure future success. The trilobite is an example of a radical invention that took place in these ancient seas, the shell. It is also the first known example of sophisticated eyesight; the eyes of one species consisted of 15,000 separate lenses. Its closest living relative is the horseshoe crab.

The fossil record also shows transitions between life and death. Archaeopteryx, or "ancient wing," shows transitional features between those of dinosaurs and birds. A bird's feather is made of the same material as that of a reptile's scales (so are our fingernails!), but more striking is that in archaeopteryx we still see examples of a reptile claw and tail. Most evolutionary biologists today have concluded that this creature was not a true bird. Because of its body structure, its wings would have been very weak, and it was probably just a glider.[1]

When most people in the developed world think of a bird, they probably think of a little sparrowlike creature. The cassowary of New Guinea, which can be as tall as a small person and be very dangerous, still exists today. It is an example of a similar evolutionary strategy as that of diatryma, a huge flightless bird (also revealed by the fossil record) that was over six feet tall and lived 65 million years ago. Because flight requires the expenditure of a great deal of energy and consequent food gathering, birds will sometimes evolve in the direction of losing the characteristic of flight and increase in size, if there are no predators. Because nature has no preference for survival strategies, sometimes losing what seems like a valuable survival trait actually has survival value. The modern ostrich is another example of this.

Moles are another example. Many species have very reduced eyesight; some are almost blind. Instead they have sense organs at each end of their bodies, large hands, and powerful forelegs for a life spent predominantly underground, digging tunnels and harvesting worms that fall into the tunnels. Moles are also an example of another evolutionary revolution, the mammal. When the dinosaurs became extinct about 65 million years ago, mammals developed in many directions. They now come in many shapes and sizes, possessing various survival strategies.

The variety of our brethren is endless. Variety is nature's way of making sure something works. Many species of anglerfish flourish in the seas, some at great depths. All possess little fishing poles, fleshy appendages that grow from the tops of their heads, with which to entice a dinner close to their mouths. The pink flower mantis looks exactly like a most beautiful flower, which is very convenient when you eat insects that are attracted to such colorful displays.

The echidna is a mammal that looks like a ball of spikes. With powerful claws, this spiny anteater can dig into the hardest ground in seconds for protection. The tamandua is a tree anteater; it looks like a big dog with a very long nose. The long-eared bat uses sonar to experience the world. It sends out clicks 20 to 30 times a second and can navigate quickly around objects at night, a navigational feat that no human machine could dare mimic. The tarsier, which still lives in the forests of Southeast Asia, is an example

> *It is important to realize that life on this planet has spent about three-quarters of its existence in single-celled form, and even today the majority of organisms still exist as single cells. The evolutionary pressure to become complex is evidently not very great.*
>
> GERALD FEINBERG and ROBERT SHAPIRO

---

[1] Bishop Usher has calculated from the Bible that all life began in 4004 B.C., and many who call themselves scientific creationists agree, although some, risking heresy, have pushed the date back to 10,000 years. Archaeopteryx lived over 100 million years ago. We will discuss scientific creationism in Chapter 3.

of early mammal development. Like most early mammals, it is nocturnal and has developed large eyes to gather light at night. It looks like a strange little bear with funny suctionlike feet for grasping the tree branches in which it lives, and its eyes are so large that they are fixed in their sockets. To overcome this disadvantage, it can swivel its head 180 degrees and look directly backward without turning its body.

All mammals, including the human species, evolved from a common ancestor, a ratlike nocturnal creature, similar to the modern tree shrew. Like the modern tree shrew, this mammalian ancestor was probably a tough little guy that had to be very careful. The daytime world was dominated by dinosaurs that reached heights of a six-story building, including the recently discovered superdinosaur, seismosaurus (120 feet long and weighing 100 tons). Although hyperactive and jittery and smaller than a mouse, a modern shrew can kill a large rat.

In spite of this ancestry, not all mammals are fierce and jittery. The three-toed sloth, which resembles a huge, elongated dog with long, spiked claws, reveals an evolutionary development that parallels the loss of flight in some birds. Because of a nonthreatening environment, this creature has not only lost its ancestral nervousness but also lives as if in a slow-motion time warp. It has a top speed of half a mile an hour, sleeps 18 hours per day, and has a disgusting hygiene—algae, moths, and even caterpillars make its fur their home. It even defecates and urinates only once a week. As obnoxious as this creature may be to us, however, it displays the same revolutionary reproductive strategy common to all mammals: love and care for their young. Attraction to our babies is part of our mammalian heritage. This ability to relate positively to a young appearance may well account for the popularity of pygmy marmosets at zoos (only five inches tall, these monkeys have a squirrellike behavior), puppies, and even, according to Harvard naturalist Stephen Jay Gould, the progressively more childish appearance of the cartoon character Mickey Mouse over the years.

Evolution not only *diverges* into greater and greater diversity but also *converges*. A characteristic with superior survival value, relative to the general environment found on Earth, can evolve in creatures that are not closely related. Birds and insects are not very closely related, but they both fly. In Hawaii there is a hummingbird moth that resembles a hummingbird and even gets its food in a similar manner, feeding on the nectar of flowers while in flight. In South America there are frogs that use the enlarged webbing in their feet to glide from tree to tree, recalling the flying squirrels of North America and Australia.

An awareness of the endless variety and the fated fragility of life on Earth is crucial for our times. Much of human history has been dominated by a philosophical and religious outlook that has repeatedly granted a central status to human nature and questionably professed a special dominion for humankind over the Earth and its resources. As we will see in the pages that follow, our existence on this planet is the result of a meandering path of serendipity.

There is no such thing as a destiny of the human race. There is a choice of destinies.

J.B.S. HALDANE

We need another, wiser, and perhaps a more mystical concept of animals. For the animal shall not be measured by man. . . . They are not underlings. They are other nations, caught with ourselves in the net of life and time, fellow prisoners of the splendor and travail of Earth.

HENRY BESTON

Investing some time in the cosmic perspective is not only practical in a pro-found philosophical sense but it is fun as well. Through chance, evolution has made curiosity an essential part of our nature, and, good or bad, it is part of our happiness. We have evolved to be curious, and science has revealed that if we just look, there is magic everywhere.

We indeed live in the best of times and the worst of times. At a time when we are discovering the most profound of nature's secrets, the human race is poised on the brink of committing suicide as a species. At a time when we are beginning to fully understand how lucky we are to exist at all, we must face the fact that we also live when it is considered "intelligent" to spend $3 billion on a single aircraft carrier ($9 billion fully outfitted with planes and warriors). We live when for the first time the fate of the Earth is in our own hands. In spite of recent better relations, the United States and the Soviet Union, still with many more weapons than needed to destroy the other completely, are like two men facing each other from a few feet away, both with arms extended with loaded .44 magnum pistols pointed at the other's head. Both know that if one fires, he kills himself as well. Both know that this game is pointless and dangerous. Both think disarmament is stupid if it is unilateral. And neither has time to be distracted by the magic of the stars and to romance the universe.

Concept Summary

From a cosmological perspective our fragile Earth is a mere grain of sand whirling through space with another grain of sand, our Sun, at a great speed unnoticed by most of its inhabitants. Although some can be as large as half our solar system, billions of other grains of sand (stars) are just within our galaxy—a single galaxy of an estimated 10 billion within the range of our Earthly telescopes.

Our universe is one of endless variety, infinite microscopic textures, and vast cosmic distances where colossal processes are the norm within, scientists believe, a strange explosion (the Big Bang theory) of matter, space, and time that began from a lonely spaceless point located "nowhere" approximately 15 billion years ago. Strange as these thoughts seem, scientists can describe "how" the universe is here rather than not here. They cannot, however, describe "why" it is here or why there is something rather than nothing.

Cosmologically speaking, our existence and that of our entire species seem insignificant, but cosmologically speaking we are the cosmologists. Although a species for only a relatively short period of time, we have been able to figure out some of the details of our cosmic home. We participate now in an endless detective story, and every success at understanding produces new mysteries. This too is a mystery. Why should we know all this? Why should we be capable of such strange thoughts that work?

The Earthly platform from which we seek understanding is shared with an endless variety of biological brethren. Like us, they are the precious new leaves at the end of a long evolutionary tree. They do not seem to be as smart as we are; they do not seem to be aware of atoms, photons, galaxies, supernovas, neutron stars, black holes, and quasars. But for all we know, it may be smart not to be too smart. Our intelligence has brought us far, but it has also brought us to the brink of total destruction. It cuts both ways, its application sometimes terrifies us, but it also reveals a humbling, sobering perspective of our cosmological home. Unlike most of our biological brethren, our survival is threatened more by our own actions. An awareness of a cosmological perspective may be a critical step in obtaining a successful future for our species.

Suggested Readings

Galaxies, by Timothy Ferris (New York: Stewart, Tabori & Chang, 1982).

Spectacular color photographs of our galactic and intergalactic home with accompanying narration and explanation. An entertaining introduction to the cosmic perspective.

Powers of Ten: A Book About the Relative Size of Things in the Universe and the Effect of Adding Another Zero, by Philip Morrison and Phylis Morrison (San Francisco: Scientific American Library, 1982).

An excellent book version of the dramatic short film *Powers of Ten.* Starting a billion light-years from Earth, through successive powers of 10 we journey to Earth to the hand of a man asleep in Lake Shore Drive Park in Chicago. From there we tour the microcosm, through the mountainous terrain of the skin, individual cells, and finally the interior of an atom.

Magnifications: Photography with the Scanning Electron Microscope, by David Scharf (New York: Schocken Books, 1977), and *The Scanning Electron Microscope: World of the Infinitely Small,* by Clarence Percy Gilmore (Greenwich, Conn.: New York Graphic Society, 1972).

Picture books of the microcosm; astonishing electron microscope photographs of the little hidden dramas and landscapes all around us. These books would go well with *The Secret House,* by David Bodanis (New York: Simon and Schuster, 1986), an amazing tour of the furtive universe in our homes.

Flyby: The Interplanetary Odyssey of Voyager 2, by Joel Davis (New York: Atheneum, 1987).

Intended for a generalist audience, a summary of the *Voyager* mission and scientific discoveries. In addition to the basic survey, two important points are discussed: the number of discoveries made but not yet known (unanalyzed data) and the overshadowing of the greatest event of this decade—the *Voyager 2* flyby of Uranus—by the *Challenger* disaster.

Nobel Dreams: Power, Deceit, and the Ultimate Experiments, by Gary Taubes (New York: Random House, 1986).

Although not the best book for explaining to the nonscientist the intricacies of particle physics—see *The Moment of Creation: Big Bang Physics from before the First Millisecond to the Present Universe,* by James Trefil (New York: Scribner's, 1983)—this book is a fascinating, though disturbing, look at the all-too-human obsessions of some scientists in search of glory and the Nobel Prize.

Microcosmos: Four Billion Years of Evolution from Our Microbial Ancestors, by Lynn Margulis and Dorion Sagan (New York: Summit Books, 1986).

Many books attempt to reveal humankind's anthropocentrism through a comparison with a large-scale cosmic perspective—stars, galaxies, galaxy clusters, cosmic strings, and so on. This book reminds us that a cosmic perspective must involve a microcosmic perspective as well.

Atoms of Silence: An Exploration of Cosmic Evolution, by Hubert Reeves (Cambridge, Mass.: MIT Press, 1984).

Among the many good books by distinguished scientists and science writers that introduce the stirring worldview of modern science to the lay reader—Carl Sagan's *Cosmos,* Harald Fritzsch's *The Creation of Matter: The Universe from Beginning to End,* and Louise B. Young's *The Unfinished Universe*—this book by a distinguished French astrophysicist is clearly one of the best. Translated from *Patience dans l'azur,* it is full of staccatolike French koans that jolt us from our everyday shallow complacency. Consider (p. 31) "Walking home at night, I shine my flashlight up at the sky. I send billions of . . . photons toward space. What is their destination? A tiny fraction will be absorbed by the air. An even smaller fraction will be intercepted by the surface of planets and stars. The vast majority . . . will plod on forever. After some thousands of years they will leave our galaxy; after some millions of years they will leave our supercluster. They will wander through an even emptier, even colder realm. *The universe is transparent in the direction of the future.*"

Cosmology, the Science of the Universe, by Edward Robert Harrison (Cambridge, England: Cambridge University Press, 1981), *Cosmology, Physics, and Philosophy,* by Benjamin Gal-Or (New York: Springer-Verlag, 1983), and *The Return to Cosmology: Postmodern Science and the Theology of Nature,* by Stephen Edelston Toulmin (Berkeley: University of California Press, 1982).

These three books on philosophical cosmology are most appropriate for an upper-division college audience. Although the reading level is more difficult, these books are well worth the effort for anyone seeking a more synoptic vision of the landscapes pictured in the preceding texts.

CHAPTER 2

Philosophical Interlude:
Philosophy and the Scientific Method

The gods did not reveal, from the beginning, all things to us; but in the course of time, through seeking, men find that which is better. But as for certain truth, no man has known it, nor will he know it; neither of the gods, nor yet of all things of which I speak. And even if by chance he were to utter the final Truth, he would himself not know it; for all is but a woven web of guesses.

XENOPHANES, SIXTH CENTURY B.C.

The essence of science is that it is self-correcting.

CARL SAGAN, *COSMOS*

An analysis of natural science is the only path to the central problems of epistemology.

HANS REICHENBACH, *THE RISE OF A SCIENTIFIC PHILOSOPHY*

Chapter 1 presented some very bold claims. We claimed to know some amazing things and that this knowledge has a bearing on our values, our problems, and the actions we choose to solve these problems. How do we know the universe is so big? How do we know our physical Earth and the life on it are the result of 15 billion years of cosmic evolution? How do we know these ideas are true? How do we know that the ideas accepted by scientists have any more validity than those in the following examples?

Science and Certainty

In 1981 the news media descended upon Tucson, Arizona, to interview a small religious sect that was waiting for Jesus Christ to take them to heaven. Bill Maupin, the leader of the Lighthouse Gospel Tract Foundation, had predicted that on June 28th the world would end, and all those who were to be saved by God would be "spirited aloft like helium balloons." About 50 people had gathered to experience the realization of their leader's vision—a vision, according to Maupin, resulting from 16 years of careful Biblical study and meditative prayer. Present were Maupin, who owned an ornamental ironworks business, a doctor, a surgical technician, a painting contractor, and other members of suburbia who had quit their jobs, sold their homes and Porsches, all waiting for the fulfillment of what they called "rapture day."

On the 28th, except for an electrical storm, nothing happened. The news media left, but a few months later a follow-up story appeared. Maupin admitted that obviously he had got the date wrong. However, he wanted to make it clear that God was not to blame, their faith in Jesus was still strong, and "someday" they were "going up." All of the members agreed: It was their mistake, not God's.

A few years later a news item gave another account of a vision, this time involving a bold religious fund-raising appeal by evangelist Oral Roberts. Roberts said Jesus had appeared to him and told him that he had been chosen to find a cure for cancer. This "supernatural breakthrough" would involve asking thousands of "prayer partners" to send $240 each, so that Roberts' Tower of Faith Research Center could be completed. Cures could then be found for cancer, and "other dread diseases," which are the "work of the devil."

One of the major points of this chapter is to show that a major difference

We are trying to prove ourselves wrong as quickly as possible, because only in that way can we find progress.

RICHARD FEYNMAN, *THE CHARACTER OF PHYSICAL LAW*

exists between the astonishing, sometimes uncomfortable, beliefs we surveyed in the first chapter and these "visions," but the difference is not necessarily one of truth. That is, we cannot claim categorically to know that the predictions of Maupin and Roberts are false. Since no one can see the future, it is possible that tomorrow Maupin's group will experience their rapture day and Roberts will find a cure for cancer.

The difference is also not necessarily in the level of strangeness or the degree to which such ideas violate common sense. In science we find ideas much more incredible than most religious ideas. A survey of modern science reveals ideas such as black holes, where astrophysicists tell us time and space disappear and the end of time occurs in an instant; the concept of a singularity where, in the case of the origin of the universe, all matter, space, and time explode from a single point with no physical dimensions; quantum spaces and vacuums, where nothingness consists of virtual, potential existence; the implications of Einstein's relativity theory, in which a mother can take a long space voyage and become younger than her son; and perhaps the most bizarre of all, quantum objects, where the mathematical description of the energy of a single electron reveals that it can be in more than one place at a time and even "tunnel" through reality, popping in and out of existence.

The distinguishing characteristic is what is called "epistemological status," a difference in the way science and religion make their respective claims of knowledge. Ironically, what separates scientific theories from nonscientific ones is how vulnerable the ideas are allowed to be. Here's another example.

Suppose that as a student of philosophy and science I become disillusioned by the fact that a leader of a religious cult can make hundreds of thousands of dollars a year and an educator is lucky to make $30 thousand. So I decide to use my talents to create my own religion. A little market research reveals a sizable untapped portion of the population is disillusioned with traditional religion. These people consider themselves very modern, have a passing interest in science and science fiction, but do not have a belief system that addresses ultimate concerns, one that gives meaning to life. What they need is a philosophical perspective or religion that places the past, present, and future in a meaningful whole: something that combines the security and comfort of traditional religion with modern technology and scientific understanding, making clear the purpose behind historical developments and our potential.

I begin like Bill Maupin by announcing that I have had a vision, a vision based upon years of reading the texts of all the world's major religions. From the Bible, the Koran, and major Buddhist and Hindu texts I find passages that make plausible, with a little creative interpreting, that the human species is now ready, both spiritually and technologically, to fulfill life's ultimate mission. I show that all the world's religions have been hinting at the same conclusion. What was needed was the right time and a spiritual vehicle to read the texts correctly. I announce that God exists in the form of a universal consciousness or energy field; that many higher forms of life exist in the universe, and a basic plan has been established for each one of them. Each must fulfill an innate potential before being reassimilated into this blissful consciousness. Our beacon becomes the quote from the Bible, "In my Father's House there are many mansions." We have been given a spiritual and technological potential to realize. Actualizing this potential will enable us to eventually come face to face (or mind to mind) with God.

In my new religion, we must literally find God, as a demonstration of our development. For us, God will appear in a form we can understand on a planet somewhere in the universe in a resplendent pearl that is exactly 10 times larger than any other pearl in the entire universe. Our goal is to find this planet, build the spacecraft based on present technology, and boldly go where no human being has gone. All we need, of course, is about $240 from each of our "cosmo-partners" to start our final adventure.

Now suppose that my new religion is very successful. I raise large sums of money and build the spaceships; thousands of different groups of cosmo-partners begin to crisscross the galaxy. Finally, one day it happens: A pearl exactly 10 times larger than any other pearl is found. But, except for an electrical storm in Tucson, Arizona, nothing happens. What would I say? Nothing to worry about, a minor setback, I tell my followers. We must keep the faith; someday we will find the right pearl. It is our mistake, not God's.

The problem with all of these ideas is that they are logically *irrefutable;* they have been made invulnerable to disproof. They could be true, but no matter what happens, *they cannot be proved false.* A way out always exists—an explanation to show that an inconsistency is only an apparent inconsistency. If a fact appears to contradict an irrefutable belief, we can always show that it is only an apparent contradiction and even turn this apparent inconsistency into a *confirmation,* a positive instance in favor of the belief. We just did not find the right pearl. But we are on the right track. After all, we did find a pearl 10 times larger than any other pearl. This proves there are pearls 10 times larger than other pearls. So we need to keep looking for another pearl 10 times larger than the one we found. And so on. If a predicted date for a trip to heaven does not materialize, obviously there was human error in calculation. After all, there was an electrical storm. Is this not a sign that we are on the right track? If we build a Tower of Faith Research Center to cure cancer and people continue to get cancer, then obviously the collective faith of the human race is still weak and cannot completely fend off the work of the devil. We need more contributions from prayer partners to take our message to the world. And so on.

Psychologically, irrefutable beliefs are very compelling. Everything is explained: There is a finality to our views, a secure feeling that allows us to face life's contingencies and uncertainties. Whatever happens can always be viewed as part of the plan. But when one irrefutable belief contradicts another, there is then no decisive way for a choice between them. With an irrefutable belief, all that we are left with is a "maybe-belief" that may make us feel good.

But a scientific theory must be refutable in principle; a circumstance or a

Reasoning is the best guide we have to the truth. . . . Those who offer alternatives to reason are either mere hucksters, mere claimants to the throne, or there's a case to be made for them; and of course, that is an appeal to reason.

MICHAEL SCRIVEN,
REASONING

set of circumstances must potentially exist such that if observed it would logically prove the theory wrong. Although there is much more to scientific discovery and justification than this, logicians and philosophers of science refer to a logic in discussing scientific reasoning, although it is seldom followed as rigorously in practice as it is in theory. This logic is often given an impressive name, known as the *hypothetical-deductive method*.

Scientific Method

The aim of science is not to open the door to everlasting wisdom, but to set a limit on everlasting error.

BERTOLT BRECHT, *LIFE OF GALILEO*

A simplified version of the logic of scientific method can be summarized as follows. Scientists begin the encounter with nature by making observations. Somehow through a kind of creativity mill, a *hypothesis* is generated about how some process of nature works. On the basis of this hypothesis, a test or experiment is logically *deduced* that will result in a set of particular observations that should occur, under particular conditions, if the hypothesis is true. For instance, if the hypothesis states that all physically abused children grow up to be abusive parents, then it follows that a particular abused girl should be observed in the future to be an abusive mother. If the predicted observation occurs, then the hypothesis is *confirmed*. If the predicted observation does not take place, then the hypothesis is *disconfirmed*.[1]

This skeletal essence of science may involve a logical process, but the actual process often involves a great deal of insight and creativity. How would child abuse be observed in a family without affecting the result? How would abuse be measured? If we observe an abused child to grow up and not become an abusive parent, do we know our hypothesis is wrong, or were we wrong about the child being truly abused? The need for creative insight and interpretation is an essential part of actual scientific practice, whether it involves a complex social situation or a relatively simple calculation of the Earth's circumference. At some point we will need to address the question of whether scientific theories can also be made invulnerable to disproof through this interpretive process.

For now, simply note that without a disconfirmation being possible in principle, a belief is not acceptable as even a potential scientific hypothesis. It may be true that somewhere in the universe there is a pearl in which a cosmic consciousness will become manifest. But this theory is irrefutable in principle; no concrete test is possible at all. In the example of the child, we can

---

[1] Some philosophers of science prefer to distinguish between a *hypothesis* and a *theory*. Some prefer hypothesis in the sense of a "mere hypothesis," an explanation that is put forth tentatively. As the hypothesis gains testable support, it becomes a confirmed theory. A general, inclusive theory will consist of less general hypotheses. The important factor is that both are contingent, hypothetical propositions and not categorical dogmas.

at least argue critically about whether a child was abused; in the example of the pearl, no argument is possible. (Someone can always say that we have found the wrong pearl.) The vulnerability of scientific ideas forces us to be critical, to observe and test our ideas against the world. Let's explore the critical attitude behind this method further with a few more examples.

Suppose a social scientist reads the following letter in a newspaper:

> Just why is everyone pushing this sex education in schools? Why is it necessary? The worn-out reason is that a lot of parents do not talk about it at home, therefore it must be taught in school. Yet, since this trend started, VD and pregnancy among teenagers and even preteens has sharply gone up. Why then? I thought sex education was supposed to reduce it, not increase it. The answer is that it is a "how-to-do-it" course, nothing else. Sex education is bunk.

What for this person is "proof" that there is a causal connection between sex education in schools and a recent rise in venereal disease and teenage pregnancy is only an observation of a possible *correlation*. Suppose our scientist is intrigued enough to do a little research and finds that the observation is accurate. There has been a significant rise in venereal disease and pregnancies among teenagers simultaneous with the introduction of sex education courses in public schools. Also, he is surprised to find out that the possible causal connection between these correlated events has never been tested scientifically.

What for most people is the conclusion of an investigation is only the beginning of an investigation for a scientist.

It is important to note that what for most people is the conclusion of an investigation is only the beginning of an investigation for a scientist. Many other factors could cause the rise of teenage VD and pregnancies. A rise in the population of teenagers is possible, causing every activity related to teenagers to go up: automobile accidents or purchasing particular types of clothing and albums, for example. Few would claim that teaching sex education in schools has been the cause of increased purchases of acne lotion. There could be an increase in the population of particular types of teenagers, those in an area of the country where sex education is not taught or where early sexual experimentation is encouraged by various social or family pressures. Correlation does not prove causation. A correlation between sex education and teen sex problems does not prove a causal connection, and, by itself, it does not give us a clear indication in which direction there may be a connection. For all we know at this point, an increase in teen sex problems has led to an increase in sex education classes!

However, even though this belief is far from proved, it at least can be given the status of a scientific conjecture. Unlike the alleged visions discussed earlier, this claim is refutable. It is testable. We can imagine a set of possible observable circumstances that would prove this belief false.

Suppose our scientist applies for a government grant to study this. Suppose the time is right. The country is in a conservative mood; a very popular conservative president is in the White House, and he has been championing the notion that government intrusion into what is purely a personal or family matter can bring nothing but harm.

The political and cultural environment are important as a backdrop to scientific "objectivity." Perhaps the social science community has been too liberal to recognize the possible connection between sex education and teenage sex problems. Similarly, it is probably our letter writer's bias against sex education that allowed him to make this possible connection in the first place. Ideas do not emerge from a vacuum or a purely unbiased state of mind or purely from objective observations as portrayed in our brief description of the hypothetical-deductive method. Ideas often must be popular or controversial before they are studied.

In the history of science, ideas frequently have been accepted as true before crucial evidence for them has been found. This has caused many a cynic to wonder how many "truths" are out there that are not popular yet, or never will be, and to claim that scientific objectivity is a myth. In Chapter 5 we will see that insofar as science is an activity carried out by human beings, such human factors as cultural and political influence can never be eliminated. We will also see, however, that such human factors are often unexpectedly helpful to scientific discovery, that they give the logic of science life and direction. Initial biases are not necessarily bad as is often supposed. Biases can help us see things that we otherwise may have missed. For the moment though, let's leave this issue and return to our study.

The scientist receives his grant and begins his study. What he needs is a *controlled study*. As noted, there could be many other causes of the increase in teen pregnancies and VD. The possible causes are called *variables*. The goal of a controlled study is to control as many of these variables as possible, so that given two populations of teenagers all the possible variables are the same overall except one—only one population will have had sex education in school. In this way, if there is a significant difference in the percentages of pregnancies and VD in the group that had sex education, then we will have a reasonable basis for claiming a causal connection between sex education classes in public schools and subsequent teen sex problems. On the other hand, if we find no significant difference between the two groups, it would be reasonable to conclude that a population increase or some other causal factor is involved.

Even though our scientist may receive a large grant from the government, he cannot possibly survey every teenager in the country. He must be content with a sample, one that matches in characteristics the total population of teenagers. Such a sample is called a *representative sample*. Techniques exist for creating such a sample. Political pollsters are able to interview between 1,000 and 1,500 carefully chosen people and, on this basis, predict the overall voting habits of over 50 million people. Suppose then that after a great deal

The ultimate goal of the educational system is to shift to the individual the burden of pursuing his own education.

JOHN GARDNER

of research two representative groups of 1,000 teenagers each are identified. Each group has characteristics evenly distributed. There are as many boys as girls; poor, middle-class, and wealthy; rural, urban, and suburban. The only significant difference known between the two groups is that the teenagers in one group have had a sex education class in either the fifth or sixth grade, and the teenagers of the other group have never had a sex education class. If there is a much larger percentage of problems in the sex education group, the original hypothesis will be confirmed. If there is no significant difference in the percentages, the hypothesis will be disconfirmed. What if there is a larger percentage of problems in the group without sex education? The government sponsors would probably be very unhappy, but many social scientists would likely view this result as evidence for the hypothesis that sex education classes are helping prevent teenage sex problems and that the increase must be due to other factors.

In this study the scientist probably would already have data on which teenagers had become pregnant, and which ones had not, or had an episode of VD, and which ones did not. In this case the study would be a retrospective study. It is usually much more practical to first identify the results or effects—in this instance, teenagers who have become pregnant or have VD—and then trace the histories behind these effects. This method weakens the conclusion somewhat. The past must be reconstructed from data that are available in the present. The scientist must use questionnaires and records that may be biased or inaccurate.

In the case of a questionnaire, the results can easily be biased by the questions asked. For instance, in 1984 Ann Landers conducted a survey in her syndicated column. She asked only one question: "Would you be content to be held close and treated tenderly, and forget about the act?" Not surprisingly, out of 90,000 women who responded, over 60,000 chose "tenderness" and "closeness." Her question implies that sexual intercourse cannot be a vehicle for communication, love, care, and emotional expression between couples and is only an animalistic act needed and desired by the male of our species. Similarly, in our example a slanted question could cause a teenager to distort whether he or she had a sex education class. If the question is asked in such a way as to indicate approval for having had a sex education class, then a student who may already be feeling guilty about his or her situation is less likely to be truthful, not wanting to admit an ignorance of the facts of life. On the other hand, if the question reflects disapproval, then a student is more likely to indicate that he or she has not had a sex education class. What would the results be like if the question asked was, "When you were younger, were you forced by a teacher to learn about the act?"

A better method, but one that is usually more expensive and time-consuming, is known as a prospective study. Rather than recording present incidents of teenage sex problems and inferring from a backward-in-time analysis to the possible cause, a prospective study attempts to gain more control over the present and then allows nature to take its course, the results being

Our whole problem is to make mistakes as fast as possible.

JOHN ACHIBALD WHEELER

observed in the future. In our example, this would mean finding 1,000 students of the fifth and sixth grades who will soon have a sex education class and 1,000 who will not. Again, ideally, there would be no other significant difference between the two groups. We would then wait ten years or so and analyze the percentage of sex problems of the two groups. Such a study would not only be logically tighter, because of greater control and observational closeness to the original groups, but the results would also then corroborate the retrospective study if they were the same. There would then exist an independent study pointing to the same conclusion.

Scientists seldom consider one study conclusive. The results of a single study are only suggestive. A single study is analyzed carefully by other members of the scientific community; they critique it, and on the basis of this critique, conduct other independent studies. The new studies will often have tighter controls, addressing weaknesses overlooked in the initial study. In other words, scientific studies must be *replicated*. They must be repeated many times by different investigators using different approaches before the scientific community arrives at a consensus. As noted earlier, what for many people is the conclusion of an investigation is but the beginning of a patient, methodical investigation for the scientific community.

The initial study on sex education could have many possible results. Our scientist might observe a higher incidence of VD for the sex education group, but a lower incidence of pregnancy, possibly implying promiscuity and the use of nonprophylactic contraception or only that teenagers having had sex education are more likely to report VD. Or we might find that there is a percentage difference in the two groups in the hypothesized direction, but that the difference is small. Would this confirm the hypothesis or not? How should we interpret the possible result of teenagers in the group without sex education having a higher incidence of pregnancies but a lower VD rate? What if there is a slight difference in the percentages, but in the opposite direction of the original hypothesis?

Although most scientists believe that the world is governed by simple truths, on the surface it is a complicated place, and any experiment or test of a hypothesis is actually a test of a complex web or *set* of hypotheses. In our example, our scientist focused on whether sex education causes teen sex problems, but much more than this is being tested. We are also testing the result of when students have sex education, and how they have it. Having a sex education class in the fifth grade may be very different from in the sixth or seventh grade. The type of sex education class could also be important. Was it taught by a woman or by a man? Did the teacher make use of visual aids and movies? Not only must we breathe some life into the logic of scientific method before it works but also we must do more than simply state that if our hypothesis is true, then such and such circumstances should be observed under controlled conditions. The logic of a scientific study is complicated. If the results do not occur as predicted, all we can conclude logically is that something is wrong somewhere in the total set of hypotheses and assumptions we used to deduce what should happen.

The game of science requires great patience. Like many fishermen on a shoreline, we throw out our net of ideas on the limitless sea of nature hoping to catch a few of her secrets.

Furthermore, facts are seldom just facts. Observations must be interpreted. For instance, if two teenagers are found to have produced a baby within a few years of having a sex education class, does this necessarily count as an instance supporting the hypothesis that sex education classes cause teen sex problems? Suppose upon examining this particular case carefully, we find that both teenagers are intelligent, are at the top of their class academically, and are in love, and because of their perceived understanding of the world situation and the amount of nuclear weapons, they made a conscious decision to have a baby before it is too late (in their opinion) to experience the joy of parenthood. Should this count as an observation of a teen sex problem?

For reasons such as these, the results of many scientific experiments and studies are inconclusive. Scientists have learned that nature does not reveal its secrets easily. They have learned to be shy suitors in their romance with the universe. Through countless experiences of having their ideas rejected by nature, they have learned to be very cautious in proclaiming what we know. The game of science requires great patience. Like many fishermen on a shoreline, we throw out our net of ideas on the limitless sea of nature hoping to catch a few of nature's secrets. Most of the time we pull in empty nets, and it requires a whole community of fishermen, cooperating and communicating, constantly critiquing and evaluating each other's fishing technique, to accumulate anything substantial. Let's look at one more example.

In the 1970s the warning label on a package of cigarettes was changed from "Caution, smoking *may* be hazardous to your health" to "Caution, smoking *is* hazardous to your health." This change shows that within the scientific community a consensus had been reached that, among other things, cigarette smoking is the principal factor in *causing* lung cancer. What exactly does this mean? It does not mean that every person who smokes cigarettes will get lung cancer. To say, as is believed today, that 75 percent of all lung cancer cases are caused by smoking is not the same as saying that 75 percent of the people who smoke will have, and eventually die of, lung cancer. Some may die of other causes, such as other forms of cancer, accidents, or old age. It is now also believed that cigarette smokers are about 140 times more likely than nonsmokers to get lung cancer. In terms of logic, this implies that it is possible to have lung cancer and not smoke and to not have lung cancer and smoke. The possibility of dying from lung cancer is very low if you do not smoke, and the possibility of dying from lung cancer is also somewhat low if you do smoke. But the possibility of dying of lung cancer if you do not smoke is so low that even though the possibility of dying of lung cancer as a smoker is also low, it is 140 times higher.

Several decades ago public health officials began to notice a significant increase in the number of lung cancer cases at the same time cigarette smoking was increasing. In the 1930s there were about 3 lung cancer deaths per 100,000 persons per year. By 1955 this had increased to 25 deaths per 100,000 persons per year. At first this was primarily a male disease when many more men smoked than women did; as more and more women began to smoke, the death rate for women began to increase. Of course, air pollution, the use of

Science is intelligence in action with no holds barred.

P. W. BRIDGMAN

automobiles, daily stress, and the pace of life had also increased. Thus, as in our sex education example, this was only a good place to start an investigation, not conclude one. At this point, the evidence linking smoking and lung cancer was only circumstantial.

One of the first systematic scientific studies was carried out by the medical doctors E. C. Hammond and D. Horn in the 1950s.[1] Over 180,000 American men between the ages of 50 and 69 were interviewed, targeting their smoking habits, general health, and living environment. At the beginning of the study none of the men had any serious illness. After a period of time, the doctors compared the death rate due to lung cancer between those men who smoked and those who did not.

To simplify the results, the statistics can be presented from the point of view of the proportion of deaths per 100,000 men over a 10-year period. In the group of men who had never smoked, the study indicated that 34 cases of lung cancer resulted per 100,000 men over 10 years. In the group who had smoked cigarettes, the study indicated that 4,719 cases of lung cancer resulted per 100,000 men over a decade. Furthermore, the study showed that the more a man smoked—from less than one-half a pack per day to more than two packs per day—the more likely he would be one of the 4,719 after 10 years. This was a significant difference (140 percent). Could the high percentage of lung cancer deaths among the men who smoked be a coincidence?

The Hammond and Horn study can be criticized on several points, something the tobacco companies were quick to do. Although a basic environmental variable was controlled—there was no significant difference in the rates of lung cancer for those who smoked and lived in the city and those who smoked and lived in the country—many other important factors were not controlled. For instance, it was possible that all of the men who died of lung cancer, including those who had never smoked, had a diet that was significantly different from those who did not die of the disease. Or they may have had unhealthy occupations, whether in the country or the city, that placed an unusual amount of stress on the lungs. Perhaps all the men with cancer had very stressful occupations involving a great deal of daily nervous tension. Finally, and this has been a point the tobacco companies have used consistently, the variable of heredity was not controlled. That is, it is possible that all the men who had cancer were prone to some form of cancer genetically, and this would be revealed by finding other incidents of cancer in the family. Parents, grandparents, and other relatives of the men who succumbed to lung cancer might have shown a large percentage of cancer, if this was investigated.

Such criticisms serve a positive function. Critiques that reveal weaknesses serve as a basis to make the next study better. Since the Hammond and Horn

If a man will begin with certainties, he shall end in doubts. But if he will be content to begin with doubts he shall end in certainties.

FRANCIS BACON

---

[1] E. C. Hammond and D. Horn, "Smoking and Death Rates," *Journal of the American Medical Association* 166 (March 8, 1958): 1159–1172, and 166 (March 15, 1958): 1294–1308.

study, dozens of similar studies have been conducted, each incorporating tighter and tighter controls based on possible oversights of the previous studies. By the early 1980s, studies included controls on all of the variables just mentioned. The results have not changed much. By 1985 over 140,000 people a year died of lung cancer, an impressive majority smoked cigarettes, and for the first time lung cancer rates for women exceeded those of breast cancer. Other diverse corroborating factors have also been identified—from the effects of secondhand smoke to chemical analysis of cigarette smoke revealing over 200 toxic substances, including radioactivity.

Does this prove that cigarette smoking is the principle cause of lung cancer? For logically technical reasons it does not. In fact, there is no such thing as a scientific proof, if proof means something that is known with absolute logical certainty. A controlled study can never be completely controlled. The number of ways people can differ, the number of possible variables, is infinite. In all the studies on cigarette smoking, it is possible that some obscure third factor was the real culprit. All of the people who had lung cancer could possibly have had some subtle factor in common that went unrecognized in every study. Perhaps the real cause was that all of the people who had lung cancer were given a piece of bubble gum on their fifth birthday, and it just so happened there was a chemical in the gum that reacted with their bodies in such a way that it activated a virus later in their lives, which then led to lung cancer 50 or so years later! This and millions of other "off the wall" possibilities have never been tested.[1] The link between smoking and lung cancer cannot be known in the sense of "known beyond any logical or conceivable doubt." The point is, however, can we say we know that cigarette smoking is a principal cause of lung cancer beyond a "reasonable doubt"? Is it rational if we claim to know something even if we are not absolutely sure that we know something? Can we distinguish between what is "conceivably" true and what is "reasonably" true?

Without room for doubt, there would be no room for self-correction.

A Little Logic

In the philosophy of science this issue is known as the *problem of induction:* No matter how much evidence we have for a conclusion, the conclusion could still conceivably be false. Unlike deductive reasoning, which preserves certainty, if all our evidence is true, then we can at best say it is "unlikely" that

---

[1] A more likely and interesting possibility is the role of radioactive radon gas prevalent in many homes. Radon gas is believed to also cause lung cancer and as of this writing has not been controlled for in studies linking smoking and lung cancer.

our conclusion is false when we are using inductive reasoning. To understand this, first consider the following *deductive* argument:

1. All American cars built after 1969 were equipped with seat belts at the factory.
2. John has a 1972 Ford.
3. Thus, John's car was equipped with seat belts at the factory.

If the first two statements are true (logicians call these statements premises), then the conclusion must be true. We may not know if the premises are true, or we may find out that the conclusion is false, but there is no doubt that the conclusion is true, if the premises are true. John could possibly have a 1968 Ford, or the year when all cars were equipped (by law) with seat belts might have been 1973. But if it were true that every American car after 1969 was indeed equipped with seat belts, and John has a car built after 1969, then there would be no doubt that he has one of the cars that was equipped with seat belts at the factory. Similarly, if we discover that he does not have a car that was equipped with seat belts at the factory, then we can conclude that at least one of the premises is false. Deductive reasoning provides certainty. Either the conclusion is known to be true, if the premises are true, or at least one of the premises is known to be false, if the conclusion is known to be false. With valid deductive reasoning, we cannot have true premises and a false conclusion.

Contrast this with the following example of inductive reasoning. Suppose we bring into a room a barrel full of apples. Suppose someone tells us that there are 100 apples in the barrel. I reach into the barrel, pull out one apple from the top, and place it on a desk. Upon inspection we can see that no one in his right mind would eat this apple. It is considerably overripe and is soft and full of maggots. Although few people would purchase the barrel of apples for human consumption, few also would be willing to wager from this single apple that we know all the apples in the barrel are rotten. On the basis of only a single instance, would it be wise to conclude that all the apples are rotten?

Small amounts of evidence need not always be weak. A biologist, for instance, might be willing on the basis of this one apple to wager that all of the apples are likely to be rotten, if other information were provided. Given the general knowledge of the existence of bacteria and their ability to spread rapidly, if it were known at what temperature the apples were stored, and for how long, a conclusion that all the apples are rotten might not be unreasonable. Often in science a few observations are made to do a lot of hypothetical work, especially when they are placed within a framework of a general consensus of belief of how the world works. This general consensus is sometimes referred to as the world view or *paradigm* of the time. In Chapter 1, we mentioned the work being done at CERN and other particle accelerators around the world. With just a very few facts, theoreticians follow long, excruciating mathematical trails until they arrive at statements about the first microseconds of the universe. In Chapter 5 we will see how the world view of the Renaissance played a crucial role in interpreting the observations of the motion of the planets.

But without anything else to go on, concluding that all the apples are rotten from a single positive case is a very weak inductive inference. To make the inductive inference stronger, more apples need to be sampled. If I pulled out four more, for a total of five, and all of them were just like the first one, we would now have a better basis for concluding that all the apples are rotten. This is called *induction by enumeration*. In general, the more positive cases in favor of a hypothesis, the stronger the hypothesis.

Yet if all five were simply pulled off the top, it is still possible that the apples on the bottom are not rotten. Thus, a stronger case could be made by choosing a representative sample, by selecting one from the top, one from the very bottom, one from each side of the barrel, and one from the middle. If all five are rotten, this would strengthen the hypothesis considerably. As noted previously, political pollsters know that we are much more likely to know the likelihood of the voting behavior of 50 million voters by examining a representative sample of 1,200 voters distributed throughout the country rather than 100,000 people living in New York. A small representative sample is much stronger logically than is a large unrepresentative one. Five representative apples are better than 20 just off the top.

Suppose we continued to sample the apples by pulling out 45 more. If all 50 sampled were rotten, then it would be much more likely that all the apples are rotten. But would you bet your life savings on the proposition that all the apples remaining are rotten? Probably not, because we know that it is still possible that some, even many, of the remaining apples are not rotten. Suppose we pulled out 49 more. Suppose that all 99 sampled are rotten. Do we know that the last remaining apple is rotten? Many people would probably bet their life savings at this point, but they would still have considerable anxiety as the last apple was pulled from the barrel, because it is still possible that this last apple is not rotten. It is still possible the hypothesis, "all the apples are rotten," is false, even though we have an overwhelming number of positive cases supporting it.

Much about the logic of science can be summarized with reference to this little example. Earlier we mentioned that science uses a hypothetical-deductive method. It is given this name because starting with a hypothesis, a prediction is deduced about what should be observed under particular conditions if the hypothesis is true. If the hypothesis is that all the apples in the barrel are rotten, then one selected from the middle should be rotten. The entire method is based on induction, however, because if an apple is pulled from the middle of the barrel and observed to be rotten, this confirms the hypothesis, but it does not logically prove it. Understand that it is possible to deduce true conclusions (the apple will be rotten in the middle of the barrel) from premises that may be false (all the apples are rotten).

We have to live today by what truth we can get today, and be ready tomorrow to call it falsehood.

WILLIAM JAMES

This logical situation typifies all scientific theories because they are based on inductive inference. Because we can deduce true predictions from a false theory, no matter how long a theory has been successful in making predictions, it cannot be known to be true absolutely. It could be found to be false tomorrow. In our relationship with nature, the logical situation is much worse than our apple example. In nature the situation is analogous to a barrel with an infinite number of apples. In the cigarette smoking case, millions of people smoke, and they differ in an infinite number of ways. Every scientific conclusion mentioned in this book could be wrong.

Critics of science will often attempt to use this logical window to repudiate many scientific conclusions. In the next chapter we will see that one of the main arguments offered by creationists against the theory of evolution amounts to no more than saying the theory is based on induction, arguing that because it cannot be known to be true, it must be based on a leap of faith. True, the theory of evolution could be proved to be wrong tomorrow, but it is one of the most factually supported and independently corroborated theories of all science. Critics of this theory also often commit the logical fallacy of appealing to ignorance, arguing that because the theory cannot be proved absolutely true, it must be false. But absence of evidence for absolute proof is not evidence of absence of truth. Critics of science fail to recognize the positive aspect of this logical doubt. Without room for doubt, there would be no room for self-correction, and we would be left with a cluttered clash of irrefutable beliefs. The process of criticism and doubt may not always be comfortable, but as noted by Xenophanes at the opening of this chapter, "in the course of time, through seeking, men find that which is better."

We cannot analyze the evidence for every scientific claim cited in this book, but keep in mind, as we survey the worldview of modern science and its historical development in the following chapters, that this patient, critical attitude lies behind all the scientific explanations discussed, no matter how strange. Science does not claim to know all the answers. It does, however, claim to provide us with a method of test and interaction by which we can become more and more intimate with the universe.

Because scientists are human beings, many aspects of our humanity also play a role in scientific discovery: artistic creation and imagination, political manipulation and personal exploitation, wishful thinking, bias, egocentricity, critical review, and premature skeptical rejection. At its best, however, there is only one absolute truth: that there are no absolute truths. Every solution to a mystery creates new mysteries; the universe is an eternal riddle for which there are no final solutions. Science is a game that never ends, a game whose completion would render, as the ancient Greek philosophers believed, life boring.

Science then involves a logical process that is fallible, and it involves much more than just a logical process. Every scientist and the science of a time are subject to the forces of human nature and culture. Scientists are forced to make many assumptions; some are conscious and some are not. The job

Scientists believe many strange things, but not without reason. For the vulnerability of testable ideas forces upon scientists a cooperative, critical process, whereby they must confront, observe, and be more intimate with the world, rather than veil it with ideas that seem philosophically satisfying or comforting.

of the philosopher of science is to reveal these assumptions, so that they can be publicly discussed and critically evaluated. With this in mind, let's look at a few basic terms involved in philosophical analysis.

Philosophical Terms

Many physicists today will claim they are not interested in "reality." Because they are interested in how things work, not why they work as they do, many will claim that a scientist's task is to conduct experiments, make observations, and find mathematical connections. We want to know what atoms will do, for instance, not what they are. These scientists, influenced by a philosophical tradition known as *positivism*, will claim that it is not their job to interpret the mathematics and explain what underlying reality is causing observed results. If you want to know what reality is, they will say, "Go ask a philosopher."

There is nothing so absurd or incredible that it has not been asserted by one philosopher or another.

RENÉ DESCARTES

Positivists are not interested in metaphysics. *Metaphysics* is a field in philosophy in which the main subject of interest is the question of what is most fundamentally real. Philosophers interested in metaphysics are particularly interested in distinguishing between what appears to be real and what is real. For instance, here is a metaphysical question: Can the table I am using to write on be considered real, or is it merely an appearance due to the way the human sense organs and brain interpret reality? Perhaps only the atoms are real. And what are the atoms? Subatomic particles, quarks, empty space, energy? What is energy? Sometimes a metaphysical question will be phrased, "What is the ontological status of atoms?" This is another way of asking if atoms are real.[1] Perhaps "atom" is only a useful human concept that has no corresponding concrete reality. What is the ontological status of a concept? An idea? Most physical scientists today believe there is a corresponding concrete reality, but profess not to be interested in what it is; they say that the role of a scientist is only the pragmatic task of finding concepts, especially mathematical ones, that work.

Although many scientists pretend not to address these questions, many have been accused of assuming, either consciously or unconsciously, a classical metaphysical position known as *materialism*. The scientist Carl Sagan, in his book *The Dragons of Eden* claims unashamedly that materialism is the most reasonable assumption for a scientist to make. Metaphysical materialsim

---

[1] Ontology is a branch of metaphysics that studies the question, What does "to exist" mean?

What is necessary "for the very existence of science," and what the characteristics of nature are, are not to be determined by pompous preconditions, they are determined always by the material with which we work, by nature herself.

RICHARD FEYNMAN,
THE CHARACTER OF PHYSICAL LAW

states that there is no evidence that anything called "mind" exists and that all that exists are concrete material things, forces, and empty space. Scientific method, however, does not depend necessarily upon making this assumption, and scientists themselves disagree on whether the scientific evidence supports one metaphysical assumption or another. John Eccles, a Nobel laureate and physiologist, argues in his book *Facing Reality* that recent developments in physics and neurophysiology warrant a reexamination of this question.[1] Some scientists have even held a position close to what is known as classical *idealism*, believing that the universe can be best understood by assuming that "thought" or "consciousness" is the most fundamental reality. During the Renaissance the major scientific figures—Copernicus, Kepler, and Galileo—all assumed that certain mathematical concepts had a special ontological status (they were ideas in the mind of God) and that any physical reality, such as the motion of a planet, must conform to these ideas. In spite of the claims of positivists, it is difficult to avoid all metaphysical assumptions. In scientific development, metaphysical assumptions can play a major background role insofar as they may govern the outlook of any individual scientist or culture within which a scientist is working.

Some philosophical critics of science will claim that science is absolute and dogmatic in terms of its epistemology. *Epistemology* is a field in philosophy concerned with the study of knowledge itself, particularly the best way of knowing. There are different ways people claim to know things. Much of our personal knowledge is based upon *testimony*. Someone may tell me that New Orleans has excellent restaurants or that Los Angeles has many freeways. If I believe this, but have never been to these cities, my belief is based on testimony. Sometimes the testimony is based upon *authority,* as would be the case if a chef told me about the restaurants in New Orleans. Many religions claim that a valid method of knowing involves *revelation,* whereby important truths about life, impossible to find out any other way, are disclosed to human beings by a divine being or God. Bill Maupin claimed to have had a *mystical vision.* Mystics, in general, claim that after years of special training it is possible to know some very important things about life and the universe "intuitively" while in a deep state of meditation. These intuitive visions should not be confused with revelation, because the visions not only involve personal effort and training but also do not necessarily involve divine aid or God.

Scientists, of course, do not believe it is possible to know the details of the dark side of our moon intuitively; they believe that observational experience is necessary, either indirectly via a robot with sensors and cameras or directly by putting human observers there. Science assumes the epistemological po-

---

[1] Eccles has also coauthored a book with the philosopher of science Karl Popper, *The Self and Its Brain* (New York: Springer International, 1977). As the title implies, these men believe that the human mind has an independent existence and is separate from the brain.

sition known as *empiricism*. Beliefs must be validated by public experiences. They must be objective.

Galileo Galilei is often cited in introductory science texts as being the first complete, empirical scientist. Instead of accepting ideas as true based upon authority or revelation, in 1609 he used a telescope to observe the Moon's surface. He saw that it was quite similar to the mountainous terrain of the Earth. He also observed that the planet Jupiter had at least four moons and that Venus had phases like that of the Moon; he even witnessed dark spots on the Sun. Such observations clearly contradicted both the authority of the ancient Aristotelian cosmology,[1] which taught that heavenly bodies such as the Moon and planets were not Earthlike, and what was thought to be the revealed word of Biblical scripture, that the Earth was the center of God's universe. Aristotelian science also taught that lighter bodies would tend to go upward and heavier ones downward. Thus, it was assumed that a heavier body would fall faster than a lighter one. Instead of accepting this popular and commonplace belief, Galileo showed that actual measurements of falling bodies did not support this idea. Thus, it is often claimed that Galileo's work represents an epistemological revolution, from which point on observational experience became the key method for attaining truth.

Mystics will claim that they too are empiricists. They claim that only through personal experience can the important truths of life be known and that anyone with the proper training can experience these truths. Furthermore, mystics claim that this training, although different in kind, is no different in degree than that necessary for a scientist to interpret complicated data from a robot sensor. Scientists generally scoff at this notion, because mystics will admit that these truths can only be experienced personally and cannot be communicated or described in a public language.

Epistemologists also discuss the degree to which beliefs can be certain. We have already noted that because scientific conclusions are based on inductive reasoning, they can never be absolutely certain. An epistemological position often opposed to empiricism, but historically greatly influenced by the discovery and development of mathematics, is called *rationalism*. The rationalist has a great faith in the logical power of the human mind and is skeptical about the universal validity of our observational perceptions. Do we not sometimes see things that are not really there? Do all people always see things the same way? The rationalist believes that the most important things about life, or at least the starting place for knowing the basic premises, can be known only through a logical intuition. Some things are so clear logically or mathematically that we just know they are true. For instance, don't we know that

---

[1] Cosmology is often considered a branch of metaphysics in which the main interest is the structure of reality, what overall shape the universe is in, and what forces or processes have caused it to be in its present state.

there are no round squares on the dark side of the Moon? We don't need robot sensors and observational experience to know that a round square is inconceivable. Don't we know that it is impossible for a woman to be born after her son has died? In short, the rationalist believes that we can know some things about life ahead of time, so to speak; we can know some things that no conceivable experience will contradict.[1]

Zeno, a fifth-century B.C. Greek philosopher, along with Parmenides, another famous Greek philosopher, believed that our senses cannot be trusted. They argued that often when something at first seems to be true it is, on closer inspection, not true. Sticks in water appear bent. Pulled out of the water, they are straight again. It is the critical ability of the human mind that enables us to know that the stick is not really bent. If we were limited to only observational experience, we would have to conclude that the stick bends when put into water and magically straightens itself out when pulled out.

One of the most famous examples of rationalism is known as Zeno's paradox. It makes use of an apparent inconsistency between what our experience indicates is true and what logic dictates must be true. One of the most basic experiences of life is that of motion. Everyday we witness ourselves and others moving from place to place. According to Zeno and Parmenides, this is just as much an illusion as the stick example. To move from any point A to another point B, a person must first travel half the distance. To move from this half-way point to point B, he must again travel half the distance. To move from this point to point B, again half the distance must be traversed, and so on. No matter how close point B is approached, there will still be some distance left, and the next half distance must be traveled again, and again, and again. Point B can never be reached. Since point B can be any distance, including a very short distance, it is thus logically impossible to move to even the shortest distance. We may experience ourselves and others moving about from point to point every day, but careful logical inspection, according to Zeno, reveals such movement to be an illusion. According to Zeno, and Parmenides, because our experience is often mistaken, we should follow the logical guide of our reason, no matter how strange the result.

This paradox may seem silly to people about to enter the twenty-first century. We move quite successfully every day between our homes, schools, and stores, and voyages by spacecraft that travel millions of miles are routine. We should keep in mind, however, that it was very difficult for people living dur-

---

[1] The empiricist also believes that some statements can be known with certainty. The difference is that the empiricist does not believe anything significant about the real world is conveyed by such statements. The only statements about the world that can be known ahead of time, prior to experiencing the world, are statements like "it is raining or not raining outside." Such statements are known to be true prior to checking the outside world only because they cover all possibilities.

ing the Middle Ages, and probably for many people today, to imagine that the Earth is moving and not the Sun. We do not experience ourselves moving at 1,000 miles per hour; instead we "observe" the Sun to move. That a belief is inconsistent with our common observational experience is not by itself a conclusive argument that it is false. The desk that I am writing this on appears to me to be very solid. Yet the atomic physicist informs us that the desk is 99.9 percent empty space. The "solidity" is another illusion due to the electromagnetic field generated by the electrons in the surface atoms of the desk and my hand. Because like electromagnetic forces repel each other, I cannot pass my hand through the table. It would be a very strange world without electromagnetic forces.

An empirical scientist would claim that we have used experience to learn these things. If this is true, it is one of the great ironies of science that it uses experience to prove other commonplace experiences to be illusions.

Empirical scientists also believe that motion is real. But because of the early Greeks' faith in the ability of the human mind to figure things out, to use reason to peel away layers of illusion and ignorance, they took Zeno's ideas seriously, even if they did not all agree with his conclusion that this proved motion and the physical world to be an illusion. For the early Greeks, such a fundamental inconsistency between common sense and reason, or any paradox for that matter, was nature's way of taunting us, of revealing one of her important secrets. As we will see in Chapter 4, when we discuss the contribution of the philosopher Plato to our scientific culture, the ancient Greeks had tremendous confidence in the logical and mathematical powers of human thinking. Such confidence was a key ingredient in the development of modern science.

The intellectual forces necessary to produce our modern scientific revolution were not in place until the Renaissance, when a synthesis of rationalism and empiricism were incorporated into one method. The hypothetical-deductive method combines the logic of the rationalist (deduction) and the observational experience of the empiricist (induction). Before continuing with this epistemological discussion, let's look at another example of this early faith in the power of reason. In the process we'll also discuss another indispensable part of scientific method, the creativity mill.

In trying to distinguish appearance from reality and lay bare the fundamental structure of the universe, science has had to transcend the "rabble of the senses."

LINCOLN BARNETT

Science and Creativity

Have we not all wondered at some time or another where the human race has gotten all its ideas? As a small child I remember watching an aunt make wheat bread. I remember bothering her with question after question about what she was doing: "How is it made? What are the ingredients? Wheat?

What is wheat? A plant? From the seed of a plant?" How could human beings ever figure out such a thing? How did we connect the seed of a common weed to make the bread for my peanut butter and jelly sandwich?

Jean-Paul Sartre, in his novel *Nausea*, tells the story of a man who has decided, "Science! It is up to us." This man has decided to read every book in the library so that he will know everything there is to know. He started with the first book on the first shelf, and by the time Sartre's hero Antoine Roquentin meets up with this "self-taught man," he is into the L's and reading *Peat Mosses and Where to Find Them* by Larbaletrier. According to Sartre:

He has read everything; he has stored up in his head most of what anyone knows about parthenogenesis, and half the arguments against vivisection. There is a universe behind and before him. And the day is approaching when closing the last book on the last shelf . . . he will say to himself, "Now what?"

I do not think it is possible really to understand the successes of science without understanding how hard it is—how easy it is to be led astray, how difficult it is to know at any time what is the next thing to be done.

STEVEN WEINBERG, *THE FIRST THREE MINUTES*

He will know many facts, but will he have made any connections? Will he be able to add to our knowledge of nature's secrets? During the third century B.C. in Alexandria, Egypt, another admirer of books by the name of Eratosthenes was reading about a curious fact. Eratosthenes was the head librarian of the famous library of Alexandria. The knowledge stored in this library was far ahead of its time, and he had access to many great thoughts and facts. The fact that preoccupied him this day was rather ordinary on the surface. He read that in the city of Syene (the site of modern-day Aswan), approximately 500 miles away, on the day of the summer solstice, June 21st, the longest day of the year, at exactly noon, the Sun shone directly over a particular water well casting no shadow. The average person would probably not even notice this. For the slightly more thoughtful, it was at least curious. Someone had to notice the event and connect it with the summer solstice. For the brilliant Eratosthenes, however, nature was taunting us with one of its secrets. There was a very important connection.

It is doubtful that it happened this way, but for dramatic effect imagine that Eratosthenes first came across this little item of knowledge a few minutes before noon, June 21st. We can imagine him looking up from the book with a passionate startled look and staring blankly into space. He then rushes outside to find a tall, straight pole and observes at exactly noon a shadow cast by the pole. He then jumps up and down in ecstasy at this great discovery: There was no shadow in Syene, but at exactly the same time there was still a shadow in Alexandria. What is the connection?

With a few reasonable assumptions, and a lot of rational deduction, Eratosthenes discovered not only that the Earth is spherical but also that its circumference is very large—approximately 25,000 miles. It was an ordinary fact, but a very big fact indeed.

He had a great deal of help in making this discovery. Not only did he have help in obtaining all the facts—someone needed to measure the distance from Alexandria to Syene and someone had to observe and note the connection of a shadowless well at Syene with the date and time—but, more importantly, the intellectual environment in Alexandria that made possible the connection of these facts owed its existence to the ancient Greek rationalists. The idea that a few facts could explain much with a little reason was already part of the Greek heritage.

The city of Alexandria was founded by the Greek conqueror Alexander the Great. With him came the intellectual tradition of faith in reason, the passion to explore the physical and biological world of his tutor Aristotle, the belief in natural law of the pre-Socratics, and the triumph of the mathematical teachings of Pythagoras, Plato, and Euclid—particularly geometry. It was geometry, and a belief, fostered by the rationalists, that with mathematics we possess a mystical power to transport our minds and see what our physical eyes may never see that allowed Eratosthenes, to a large extent, to connect these curious facts.

First of all, what could account for there being a shadow in one place, but not at another? One possibility was that the Sun was relatively close to the Earth. When the Sun was directly over the well in Syene, it would be at an angle in relation to the pole in Alexandria, thus casting a significant shadow. A more likely possibility was that the Sun was very far away, so far that by the time its light reached the Earth, the light rays would be virtually parallel to each other. If this were so, then one way to account for the shadow in one place, but not the other, would be to assume that the surface of the Earth was curved. And if it were curved, it was reasonable to assume that its curved surface met to form a sphere, a round Earth. This was not a new idea. Pythagoras had argued as early as the fifth century B.C. that the sphere was the most perfect object for the Earth to be shaped, and Aristotle and others had noted that during the eclipse of the Moon, the dark shape made on its face was that of a disk. (They assumed that the eclipse was caused by the Earth's shadow moving across it.)

As the master keeper of this tradition, Eratosthenes knew that circular objects can be measured and analyzed using a method of proportion. (For the purpose of illustration we will assume he knew about degrees. Actually, he used fractions.) He also knew from the geometry of Euclid that the alternate angles formed by a line intersecting two parallel lines would be equal. Thus, he reasoned, by measuring the angle of the shadow in Alexandria, we would automatically know the angle made by drawing a line from Alexandria to the center of the Earth and another line from Syene to the center of the Earth (Figure 2-1). This angle would then automatically tell us not only the proportion of space taken up by the angle compared to the entire 360° circumference but also the proportion of space taken up by the distance of Alexandria to Syene compared to the circumference of the entire Earth. Using a simple sundial, he measured the angle to be 7°, approximately $\frac{1}{50}$ of a circle. Thus,

FIGURE 2-1

By using basic geometry, a few facts, and a lot of insight, Eratosthenes was able to show that if the 500 miles from Syene to Alexandria was ¹⁄₅₀ of the entire circumference of the Earth, then the Earth must be approximately 25,000 miles in circumference (500 × 50 = 25,000). By measuring the angle *a* of the cast shadow in Alexandria to be 7°, he knew the angle *b* to be 7° or about ¹⁄₅₀ (⁷⁄₃₆₀) of a circle. Although the angles in this illustration are exaggerated, note that a crucial premise in Eratosthenes' conclusion is the geometric principle that a line *L* intersecting two parallel lines creates alternate angles that are equal (*a* = *b*).

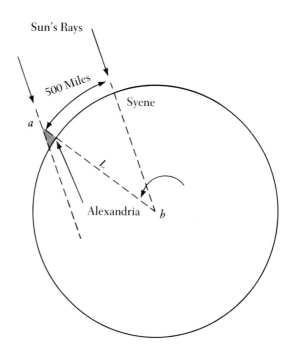

if the distance from Alexandria to Syene was ¹⁄₅₀ of the circumference of the entire Earth, then the Earth must be approximately 25,000 miles in circumference (500 × 50 = 25,000). At the dawn of Western civilization our ancestors were already beginning to understand the magnitude of the universe.

Can we say that Eratosthenes' conclusion was reasonable for his time? No one then could confirm this conclusion empirically. Although the observation of a lunar eclipse supplied empirical evidence supporting the belief in a spherical Earth, and it was reasonable to assume that the Sun was very far away, no one possessed the means to circumnavigate the globe in the third century B.C. There were no pictures from space showing a beautiful blue sphere. The factual evidence for a large spherical Earth was hardly overwhelming. At this time, whether one believed in the validity of Eratosthenes' picture of the Earth depended more on one's faith in the power of reason than on the facts.[1]

---

[1] Eratosthenes also had to assume that Syene and Alexandria were on the same meridian of longitude, on the same line connecting the North and South poles. This assumption is necessary for an accurate measurement of the Earth's circumference, but it is wrong. Hence, by today's standards of measurement, his result, although remarkable, is not accurate. Modern measurements, using Eratosthenes' methodology, give a 24,800-mile circumference.

Geometry and an intellectual tradition, curious facts noted by someone else, accident and serendipity,[1] all of these supplied the ingredients for Eratosthenes' creativity mill. The connections are easy to see after someone has made them. The number of potential pieces, however, is always infinite; you can possess all the facts in the world, as with Sartre's self-taught man, but unless you can connect them, all that results is "Now what?"

Eratosthenes had many biases, but these biases helped him make connections. He accepted the rationalism of the ancient Greeks and the idea that mathematics granted us a special power to transport our understanding to invisible places. Today, physicists in many ways continue this tradition. As Morris Kline has noted in his *Mathematics and the Search for Knowledge*, "adorned with a few physical facts," we have pushed our understanding to unimaginable places, places we will undoubtedly never visit.

There is an overwhelming consensus among contemporary philosophers of science that the scientific method is based on the epistemological position of empiricism. No matter how much logical deduction and mathematical analysis is used, at some point the world must be checked for the confirmation of a belief. Historically, however, spurred on by the power of mathematics and the tendency to conclude that we know something even though complete empirical observations are not available, rationalism has played both a constructive and creative role in the development of science. Philosophers are often criticized for being too rationalistic, using speculative logic in creating ivory-tower armchair fantasies. This criticism, however, overlooks the fact that many great scientific discoveries are made by scientists sitting at desks, following the elegant trails of mathematical equations. Creative ideas are the result of a complex web of influences. The key is to have ideas with which to make connections.

Science is, and must be, culturally embedded; what else could the product of human passion be? . . . Culture is not the enemy of objectivity but a matrix that can either aid or retard advancing knowledge.

STEPHEN JAY GOULD

Philosophical Issues

Of course, not all ideas are fruitful in making connections. Nor have great scientists been immune from detrimental rationalistic tendencies. The famous sixteenth-century astronomer Tycho Brahe was the best observational astronomer of his time. The data he recorded on the motion of the planets were crucial to our modern understanding of the solar system. Mathematically, he knew that one of the implications of this data was that the Sun was the center

---

[1] Luck has played a role in many scientific discoveries. Was Eratosthenes looking for facts about water wells and shadows? What if he had not read the particular text that described the shadowless well in Syene?

of the motion of all the planets, which further implied that the universe was very large and that the stars were an immense distance away. He could not bring himself to accept this radical conclusion, however, and accepted instead a more traditional view for his time. Why? Because it was inconceivable that God would "waste" this much space!

Johannes Kepler, who with Tycho's data finally solved the problem of planetary motion, was motivated by his certain belief that the Sun was the most appropriate object to be placed in the center of the universe because it was the material home or manifestation of God. Galileo, in spite of his brilliant astronomical observations and terrestrial experiments, failed to see the importance of Kepler's solution of planetary motion because it did not involve using perfect circles for the motion of the planets. More recently, consider the response to observations of primitive people in India who did not know about fire. The "world's foremost authority" on the ethnology of this area said that this is "inconceivable." True, this observation was inconsistent with the best accepted theories of the time, but do we know that the observations were inaccurate?

Many modern physicists reject literal interpretations of mathematical equations that imply that an event could take place before its cause. Could a bomb explode before a signal is sent to detonate it? Is this inconceivable? Do we know if it is impossible for any human being to ever experience this? Do we know if it is impossible for any form of consciousness to experience it? How do we know this? If we believe we know this on the basis of experience, then the nature of inductive reasoning requires us to say only that it is unlikely that such events will ever happen. If we believe we know absolutely that such events cannot happen, then we will be accepting the rationalist's basic claim that some things can be known without observational experience.

Must science assume some ideas dogmatically? Is science dogmatic in assuming a particular epistemology as absolute? Must we assume that the scientific method, a synthesis of reason and experience, is the only avenue to truth? Are there better ways of knowing than that allowed by inductive reasoning? How can science validate its own method?

The mystics claim that some simple acts of knowing cannot be described by an objective language. For instance, how does one really know how to drive a car? Is it not an intuitive feeling that one just gets after some practice? Someone else may help point the way with descriptions of what to do, but the final knowing cannot be described by any language. How about the experience of seeing a death on the highway? Does a cold scientific description, "the cause of the cessation of bodily function was due to a rapid deceleration," accurately convey the truth? What about our own deaths? That each of us will die someday is an objective fact. But there seems to be much more to this truth than can be described in the statement "I am mortal." Are there subjective truths, truths that cannot be described in an objective language? Is science being dogmatic in excluding the possibility of subjective truths?

An extremely healthy dose of skepticism about the reliability of science is an absolutely inevitable consequence of any scientific study of its track record.

MICHAEL SCRIVEN,
REASONING

What happens when metaphysical and epistemological assumptions are combined? Most scientists today accept an assumption that can be traced to the ancient Greeks: Whatever they are, the basic truths of the universe are "laws" that do *not* change—only our ideas about them do. Scientific objectivity presupposes that there is one truth, a collective truth, and our personal beliefs or the beliefs of scientists of a particular time either match these truths or they do not. The Earth is spherical, not because scientists believe it is, but because it is one way for everyone whether they know this or not. In other words, most scientists assume an epistemological-ontological assumption, which holds that beliefs about what is real do not affect what is real. Truth results only when our beliefs about what is real *correspond* to what is real.

This may not, however, be essential to science. In Chapter 8 we will see that some physicists have actually proposed that the points of view implied by our experiments can affect the nature of reality: that instead of assuming there is only one reality, and one set of true descriptions about it, these physicists have proposed that there can be "complementary" realities. And reputable scientists in physics and medicine are not only reexamining this traditional scientific assumption but also wondering candidly if a person's state of mind may have a bearing on whether he or she is prone to diseases such as cancer and whether cures and remissions are possible using a mental therapy. The belief that there is only one reality can itself be subjected to scientific scrutiny. There could be multiple realities or none at all! Even if these ideas are controversial, they are at least discussed.

Although scientists at any given time may be caught within a web of many assumptions, science at its best does not rely on many assumptions. Science also assumes, as did the ancient Greeks, that the more we think critically about our beliefs, the more likely we are to know the truth. There are cynics, however, who believe that critical thinking is not a marvelous human characteristic at all. Critical thinking, they argue, makes life more complicated, uncovers distracting details, and makes it less likely that human beings will discover the simple solutions to life's problems. There are also nihilists who argue that our so-called intelligence and our ability to be aware of the details of the universe are an evolutionary dead end, that far from producing the good life, our awareness and rationality are the cause of our craziness.

Defenders of science will often argue that even if some assumptions are necessary in the application of scientific method, these assumptions are validated by the record of success. However, there is a major logical problem with this justification. It simply raises the problem of induction again. It is circular reasoning to attempt to vindicate inductive reasoning by asserting that so far inductive reasoning has worked, because this vindication itself is an inductive argument. The history of science shows many beliefs that have been "successful" for a time, only later to fail. No matter how much evidence there is for an inductive conclusion, it is always logically possible for the conclusion to be false. Thus, it is logically possible for the scientific method to completely fail tomorrow, even if it is true that it has been successful for centuries. And if it could fail tomorrow, the question remains whether it is reasonable to continue to believe in the scientific method as helpful for our future.

Has science been successful? No other creature in the universe that we know of is potentially as violent as the human race. At any moment our Earth could be destroyed by nuclear holocaust. No matter what one believes is responsible for this situation, it is indisputable that the scientific method has made the means of this destruction possible.

The vantage point of philosophy makes discussion on these topics possible. Philosophers believe these abstract questions are important because they are intimately related to our more personal concerns about who we are, where we have come from, and what may be in store for us in terms of the survival of our species on this fragile fragment of the universe. It is important, however, to tie these discussions down with as much concrete science as possible. Thus, with a few terms and issues under our belt, let's continue our tour of the cosmological development that has made it possible for us to ask such questions, by discussing what scientists have discovered to be our biological roots.

Concept Summary

People believe many things. Anything is possible, but how can ideas that are reasonable be separated from those that are merely conceivable? The use of the scientific method presumes that the testing of ideas is the first step in this process; ideas must be vulnerable to disproof. To succeed, scientific ideas must be able to fail by checking the logical implications of each idea with observations of what takes place in the world. Scientists believe many strange things, but not without reason. For the vulnerability of testable ideas forces scientists to be cooperative and critical; they must confront, observe, and be more intimate with the world, rather than obscure it with ideas that seem philosophically satisfying or comforting.

Although the critical process of science involves many patient, disciplinary techniques that promote objectivity (such as controlled studies, statistical analysis, standards of replication and corroboration), our interaction with nature involves creativity and a complex web of ideas, assumptions, perspectives, influences, interpretations, and paradigms or world views. Ultimately, all scientific explanations are based on inductive reasoning; no matter how successful an explanation is in making predictions, it is always possible for true predictions to be deduced from false theories, so no scientific explanation can be known to be true absolutely. But the essence of science is self-correction and the obtainment of beliefs that have a reasonable chance of being true, not absolute truth.

Philosophers study science for many reasons. Positivism, materialism, and idealism are traditional philosophical concepts that can be debated in terms of scientific results. But because science involves first and foremost a way of knowing, the field of epistemology, the study of knowledge, is most often associated with philosophical interest in science. The modern scientific method synthesizes two epistemological traditions: the rationalist, who placed great emphasis on reason, logic, and mathematics in the knowing process, and the empiricist, who believed that only observational experience validates knowledge. Although philosophers of science generally agree that scientific method is ultimately based on the epistemological postulate of empiricism, modern scientific ideas, especially cosmological ones, are often the result of the contemplation of abstract mathematical trails.

If the results of science are based on inductive reasoning, and hence can never be certain, then the question arises as to whether science can be self-corrective. That an idea works by being tested many times is no guarantee that it will work in the future. Furthermore, correction implies that something is better. Can one idea be said to be better than another if it cannot be shown to be absolutely true? This philosophical question and many others often discussed in relation to science in turn have a bearing on the great questions of who we are, where we have come from, and what may be in store for us.

Suggested Readings

Dismantling the Universe: The Nature of Scientific Discovery, by Richard Morris (New York: Simon and Schuster, 1983).

A book intended for the informed general reader as well as scientists and philosophers. Includes a discussion of the critical evaluation process of scientific method, plus the importance of the nonrational and illogical creativity in scientific discovery. Also emphasizes the prominence of theories in observing the world, and the dismantling of the conceptual universes implied by our theories as a necessary condition for scientific revolutions.

How We Know: An Exploration of the Scientific Process, by Martin Goldstein and Inge F. Goldstein (New York: Plenum Press, 1978).

The Goldsteins explain the scientific method and attitude through case studies with commentary. They discuss extended examples from biology (cholera), physics (heat and atomic theory), and psychology (schizophrenia), experimentation, measurement, and the nature of theories, logic and mathematics, and probability and statistics.

The Rise of Scientific Philosophy, by Hans Reichenbach (Berkeley: University of California Press, 1951).

Written by an important philosopher of science, who studied with Einstein and made major contributions to the philosophical understanding of the theory of relativity, this book was one of the first attempts to reach a wider audience in explaining the importance of the philosophy of science.

Conjectures and Refutations: The Growth of Scientific Knowledge, by Karl Raimund Popper (New York: Harper & Row, 1965).

For anyone interested in the philosophy of science, this is a must-read book by this century's most influential philosopher of science. Emphasizing the significance of the refutability and vulnerability of scientific theories, Popper discusses induction and other matters related to the problem of scientific justification from a broad historical perspective.

The Limits of Science, by Nicholas Rescher (Berkeley: University of California Press, 1984).

Although this is a valuable recent appraisal of traditional issues in the philosophy of science, the reader will need to labor through some "professional" terminology—"intrasystemic" explanation, "ideational" versus "pragmatic commensurability," "question propagation." Beneath such convoluted locutions is a provocative reflection on many of the problems posed in our philosophical interlude. Why does each generation of scientists select a new version of what is reasonable from the conceivable? Because, according to Rescher, there is an infinite texture to nature that constantly eludes the grasp of finite human beings and their scientific explanations. But the limit of science, condemned forever as it is to being an open-ended enterprise, gives it an open-ended creativity to explore the infinite textures. What then justifies our relationship with nature as progressive? "Pragmatic commensurability"—modern views are technologically superior to previous ones.

Science and Subjectivity, 2nd ed., by Israel Scheffler (Indianapolis: Bobbs Merrill, 1982).

This book is concerned with the problem of objectivity in science. If inductive reasoning does not offer a means to separate the reasonable from the conceivable, and our theories inevitably distort observations and mask reality, how can we say science is an objective enterprise? Scheffler attempts to answer this question and in the process provides a good discussion of the issues and the major philosophers of science.

The Structure of Science: Problems in the Logic of Scientific Explanation, 2nd ed., by Ernest Nagel (Indianapolis: Hackett, 1979).

A popular upper-division college textbook for introductory philosophy of science courses. Other good introductory textbooks include *Philosophy of Science,* by Arthur Coleman Danto and Sidney Morgenbesser (New York; Meridian Books, 1960), *Philosophy of Science: The Link Between Science and Philosophy,* by Philipp Frank (Englewood Cliffs, N.J.: Prentice-Hall, 1957), and *The Anatomy of Inquiry: Philosophical Studies in the Theory of Science,* by Israel Scheffler (New York: Knopf, 1963).

CHAPTER 3

Our Biological Roots:
Evolution and Philosophical Issues

There is grandeur in this view of life, with its several powers, having been originally breathed into a few forms or into one; and that, whilst this planet has gone cycling on according to the fixed law of gravity, from so simple a beginning endless forms most beautiful and most wonderful have been, and are being, evolved.

CHARLES DARWIN

To concede that evolutionary biology is a theory is not to suppose that there are alternatives to it that are equally worthy of a place in our curriculum. All theories are revisable, but not all theories are equal.

PHILIP KITCHER

An abandonment of the hope that we might read a meaning for our lives passively in nature compels us to seek answers within ourselves.

STEPHEN JAY GOULD

The library of Alexandria, mentioned in Chapter 2, held many texts on subjects remarkably ahead of their time. Some writings said that the Sun was the center of our planetary system and that the stars are very far away; others claimed that the human race had developed from fish. Others dealt with neurology and medicine. The thoughts scratched out on papyrus and parchment in this Egyptian library represented a culmination of an inquisitive attitude with cultural roots to the ancient Greeks. By the so-called Dark Ages this attitude was no longer considered a virtue, and the library and most of its contents had been destroyed. The intellectual leaders of the time had become tired of thinking about the secrets of nature. According to St. Augustine, a principal figure of the Dark Ages:

There is another form of temptation, even more fraught with danger. This is the *disease of curiosity*. . . . [emphasis added] It is this which drives us on to try to discover the secrets of nature, those secrets which are beyond our understanding, which can avail us nothing and which men should not wish to learn. . . . In this immense forest, full of pitfalls and perils, I have drawn myself back, and pulled myself away from these thorns. In the midst of all these things which float unceasingly around me in everyday life, I am never surprised at any of them, and never captivated by my genuine desire to study them. . . . I no longer dream of the stars.[1]

Why did it end? What changed curiosity from a virtue to a disease? Perhaps as the freshness of childhood inevitably changes to the insecurity of adolescence and the fearful responsibility of adulthood, societies too go through stages. Perhaps societies change philosophical perspectives in the same way fads and clothing styles change. Or, perhaps, as Carl Sagan has claimed, it is because the scientists and philosophers of Alexandria did not communicate with the common people, and when a scapegoat was needed for political purposes the destruction of what people do not understand was a natural consequence.

Today, opinion polls reveal that almost half of the people in the United States believe in the literal interpretation of the story of Adam and Eve in

---

[1] St. Augustine, *Confessions*.

the book of Genesis. Some fundamentalists even believe that life did not evolve, but was created all at once as recently as 4004 B.C. Yet Carl Sagan has claimed that "evolution is a fact, not a theory," and there is almost unanimous agreement among scientists that life on Earth is at least 3.5 billion years old. Obviously, there is a substantial communication problem, not unlike the days of Alexandria. In this chapter, our purpose will be to explain the theory of evolution, discuss why scientists are so sure it is true, examine its epistemological foundation, and explore its important philosophical implications for an understanding of human nature and the human prospect.

Evolution, as a complete belief system about how nature has produced the infinite variety of plant and animal life visible on Earth today, is a theory, not a fact. What Sagan means, and what most scientists do not dispute, is that the record is clear that some sort of evolution has taken place. The theoretical issue concerns what sort of process is at work. When people discuss the theory of evolution today, often with controversy, they are referring to the mechanism of evolution, first explained by Charles Darwin and Alfred Russell Wallace. This is the theory of *natural selection*. Most religious critics of the scientific theory of evolution refer to it as "only a theory," implying that it is only a speculation. As a refutable theory, it may not be true, but in terms of our discussion in Chapter 2, the theory is not just a speculation but also one of the most confirmed and corroborated inductive generalizations of our time; this generalization accounts for such an extensive amount of data, it serves as a guiding mental framework for research in many sciences.

Darwin and Natural Selection

Although not true evolutionists by modern standards, as early as the sixth century B.C. the ancient Greek philosopher Anaximander had proposed that the human species had arisen from fishlike creatures, and Xenophanes, a contemporary of Anaximander, had noted that layers of rock showed a gradation of animal existence in fossilized pictures. Thus, the concept of evolution was well known by the time of Charles Darwin. Darwin's unique contribution was his rigorous methodology and persuasive advocacy of natural selection as the primary, but not sole, determinant or mechanism of evolution. According to Darwin, physical characteristics of the individuals of a species[1] naturally vary, and when the variation enhances survival value and contributes a reproductive advantage, the environment selects those that are better

---

[1] A species is defined in terms of what is called "reproductive isolation," by similar organisms that are capable of interbreeding and producing fertile offspring.

adapted to prevailing conditions. In other words, because the characteristics of every plant and animal differ to some extent, when the differences between individuals, major or minor, begin to affect the chances of their living successfully in their environment and producing more offspring, a genetic trend will be established, resulting in a new species. An older genetic trend may branch off and be reproductively isolated from a new trend—the two groups will not be able to mate—or the old trend may be eliminated entirely, becoming an extinct species.

As a playful analogy for how this process works, suppose that the cultural environment of the human race changed drastically. Suppose that the popularity of the sport of basketball increased to the point that participation became absolutely essential. Everyone must play to survive, and, of course, the more successfully one played, the more successful one would likely be at obtaining the fruits of life, being able to afford a family and reproducing.

In accordance with Darwin's theory, there would be no mystery if we took off on a very long space voyage and upon returning found that the average height of the human race had increased dramatically. The environment would have selected height as a favorable characteristic.

According to Darwin and his modern followers, tallness initially would be a random accident. Early mothers would be just as likely to give birth to short children or to tall children. As children we have perhaps all at one time or another been enthralled by looking at the pictures in books such as *Ripley's Believe It or Not* or the *Guinness Book of Records,* which revealed some unusual event. In the back of my mind I can still dimly see the startling picture of a gigantic young boy standing next to his very short parents. According to evolutionary biologists, such unusual births are the stuff of evolution. In a sense every species alive today is the result of the accumulation of freakish chance events. This does not mean that a mother has an equal chance of giving birth to a mass of flesh shaped like a wheel. According to evolutionary biologists, "random" means only "in no preferred direction." There is no plan directing births in a special direction.

We must be careful here, because there are many misconceptions of the implications of Darwin's theory—both among critics and supporters alike. Some say that this is a "survival of the fittest" doctrine, but Darwin did not originate this phrase, and the theory does not necessarily imply a survival of the most physically capable. The environment 65 million years ago on Earth favored not the powerful dinosaur but an ancestor to the lowly tree shrew, which began to flourish and evolve into many new creatures, among them, eventually, primates and human beings. Prior to this, for millions of years the ancestors to the mammals were small insignificant powerless creatures, hiding during the day and venturing forth only at night.

Fitness does not mean that there are inherently "better" physical characteristics. In Chapter 1 we mentioned the cassowary, mole, and sloth as examples of animals that have lost what seem like valuable survival characteristics. As with cave-dwelling fish that are blind, what is advantageous in one environment is not necessarily a plus in another. The same can be true with

We are not here concerned with hopes and fears, only with the truth as far as our reason allows us to discover it. I have given the evidence to the best of my ability; and we must acknowledge . . . that man with all his noble qualities, with sympathy which feels for the most debased, with benevolence which extends not only to other men but to the humblest living creature, with his godlike intellect which has penetrated into the movements and constitution of the solar system. . . . Man still bears in his bodily frame the indelible stamp of his lowly origin.

CHARLES DARWIN

the characteristic of size. The fossil record reveals that one of the first land creatures was an ancestor to the modern tiny millipede. Because it had little competition, this ancient version attained the length of a cow. Today, 400 million years later, the descendants of the first millipedes are less than an inch long for a good reason: It is much better to be small when you have a lot of creatures looking for you for dinner. In our playful human example above, imagine what would happen if the force of gravity were to gradually increase on Earth. Eventually, the force would reach a critical level where it would be intolerable for large, tall humans. The result would be a world full of short, squat basketball players. Gaining an apparent advantage at one time can also lead to extinction at another time. The male of the now extinct Irish deer had antlers that extended horizontally for up to 12 feet with huge palmlike spiked lobes at each end. Undoubtedly this made for a very impressive display for the female of the species and gave certain males a reproductive advantage, but ultimately it was as functional as a tall basketball player on a world where the force of gravity is too strong. The modern peacock is another example of this process.

Darwinism implies a built-in equality of treatment and preference in nature. There are no special creatures, characteristics, or directions. Environments exist in which a flimsy mosquito has a much better chance of survival than the most powerful lion. Perhaps this is why there are over 100 different species of mosquito. Some bacteria can survive in environments where no human being could, in temperatures close to the boiling point of water. Evolutionary biologists speak of "the tree of life" because no special branches are perceived in the development of life on Earth. Every modern animal and plant at the end of each branch, including the human species, is the result of millions of lucky events. And not one, including human beings, has any special guarantee of a long future.

Fossil Evidence:
"A Million Facts"

But is it true? When pressed by doubters or religious fundamentalists, supporters of Darwin reply that there are a million facts to support evolution, in terms of fossil representations in museums all over the world. A fossil is a rock picture of an animal or plant. We know that the overwhelming majority of animals and plants that have lived on Earth have perished leaving no trace of their existence. Flesh and foliage decay, bone and wood eventually turn to dust. Occasionally, a few out of many thousands die under just the right circumstances to produce a fossil, a picture for us to read and contemplate.

Fossils occur in what geologists call sedimentary rock, rock that is the result of an accumulation of sediments, solidified by great pressure from the weight of accumulating sediments above. Shale and limestone are two examples. The best circumstance for fossilization is when an animal or plant dies in or falls into a body of water. As the centuries pass, dead vegetation and sediment accumulate and cover it. Eventually, in the case of bone, the pressure is great enough to cause the calcium phosphate to undergo a chemical change and turn the bone to stone. In the case of plants, and even soft-bodied, boneless creatures such as jellyfish, some may maintain their shape just long enough for fossilization to occur. Through our knowledge gained in this century on radioactivity and a simple logic of rock layering, a history of life on Earth can be reconstructed. David Attenborough described this process:

Since the discovery of radioactivity scientists have realized that rocks have a geological clock within them. Several chemical elements decay with age, producing radioactivity in the process. Potassium turns into argon, uranium into lead, rubidium into strontium. The rate at which this happens can be estimated. So if the proportion of the secondary element to the primary one in a rock is measured, the time at which the original mineral was formed can be calculated. Since there are several such pairs of elements decaying at different speeds, it is possible to make cross-checks. . . . [Thus] anyone can date many rocks in a relative way by simple logic and by doing so put into order the major events of fossil history. If rocks lie in layers . . . then the lower layer must be older than the upper. So we can follow the history of life through the strata and trace the lineages of animals back to their beginnings by going deeper and deeper into the earth's crust.[1]

The mathematical probability, however, of the generalizations made from the fossil evidence and these million facts is less than the 1 rotten apple out of a barrel of 100 discussed in Chapter 2. It is an exceptional fate for an animal or plant to die under just the right circumstances for fossilization. Of those that have died under just the right conditions, there is little guarantee that we have found a representative sample of them. Our planet has many inaccessible areas of sedimentary rock.

We are ignorant, terribly, immensely ignorant. And our work is, to learn. To observe, to experiment, to tabulate, to induce, to deduce. Biology was never a clearer or more inviting field for fascinating, joyful, hopeful work.

VERNON KELLOGG

---

[1] David Attenborough, *Life on Earth: A Natural History* (Boston: Little, Brown, 1979), p. 15. Copyright © 1979 by David Attenborough Productions Ltd. Used by permission.

But scientists look at this situation in much the same way as we did earlier when we discussed the scientific connection between cigarette smoking and lung cancer. Recall that each individual in these studies could vary in countless ways, making it impossible to rule out some hidden factor as the real cause. Similarly, because a fossil is the result of an exceptional death, we could have a very distorted view of the history of life on this planet. But just as in the case of the lung cancer studies, when a consistent pattern begins to emerge and is repeated many times, we ought not to be timid in proclaiming that some beliefs are more reasonable than others. When rock formations in Australia have identical dates and fossils as those of the Grand Canyon, when all over the world a consistent pattern of evidence reveals an evolution from common descent—from bacteria to blue-green algae, to protozoa, to colonies of microscopic animals such as the volvox and sponges, to jellyfish and corals, to flatworms, tubular worms, and segmented worms—with branches to the segmented insects and shelled worms such as lingulella and clams, to sea slugs and the nautilus, with land versions of snails and slugs, to the squid, the octopus, to crabs, shrimps, and lobsters, to fish, amphibians, reptiles, and to mammals—then despite the problem of induction and the element of uncertainty in reading the fossil record, a reasonable person begins to realize that nature is revealing one of her many secrets.

Tomorrow we could find the undisputed fossil remains of human bones in a rock layer where heretofore only primitive trilobites have been found. But we have not, and it must be remembered that the depth of our inductive uncertainty is what makes evolutionary biology such an exciting science. Debates rage constantly among scientists about the details, but few doubt that the basic scenario is sound.

Representatives of an estimated 250,000 species have been found in the fossil record. From this, approximately 500 million species of animal and plant life are estimated to have existed on Earth. Of these, 99.75 percent are extinct, and of those still alive (1.25 million) two-thirds are animals and one-third are plants. To those trained in anatomy, geology, and paleontology, there is impressive evidence of common ancestry, that each of the species at the tips of the evolutionary tree developed from earlier ones.

Contemporary Observations of Evolution

Critics of natural selection will often claim that no one has ever observed an instance of evolution. Evolutionists will counter that there are examples of at least *artificial selection* all around us. With human beings playing the key role in the environment, we have selected out certain physical characteristics from nature that we have found desirable for our survival.

The ancestor to the modern corn plant was a scrawny weed with only a few seeds. By carefully taking care of the few "freaks" with many seeds and letting the others die, the modern ear of corn has evolved. At the turn of this century the Illinois Agricultural Experiment Station began an experiment that lasted 24 years. An initial crop of corn was grown in which the average height of each ear was between 3½ to 4½ feet from the ground. By selecting year after year the ears of corn both lowest and highest from the ground, after 24 years 2 separate strains were developed. The ears of the low variety averaged only 8 inches and those of the high variety were 10 feet!

A similar explanation accounts for cows with large udders full of milk, seedless fruit, and the extinction of many species. In this way the human species has sculpted nature in much the same way the natural environment does by accident. Furthermore, almost daily we read in the newspaper that another microorganism has become resistant to the drugs we have been using to eradicate diseases caused by these creatures. By chance a few members of a bacterial species will be resistant to our drugs. As the others die, a new resistant strain emerges in much the same way a new species emerges when a radical change in the environment kills most of the members of a species. A few just happen to have the right characteristics to live in the new environment. Much to the discouragement of farmers, the same process works with insects when, after a chemical spray is used for many years for protection, a new resistant version of the insect emerges to devastate their crops.

Another example of observational evidence for natural selection is a phenomenon known as *industrial melanism*. In the midnineteenth century, a peppered moth flourished in Great Britain. Its color made it almost invisible when it landed on a tree covered with grayish lichen, a type of fungus, making it difficult for birds, the moths' natural predator, to spot them. As the industrial revolution intensified, smoke and pollution killed the lichen and darkened the trees, thus making the moths easy targets for birds. A "black sheep" version of this moth had appeared earlier, but its numbers were suppressed because birds spotted them easily. As the environment changed favorably for their color, they flourished, and the grayish variety almost became extinct. In the present century, a series of Clean Air Acts were passed to control pollution emissions. As the lichen returned and the trees again lightened in color, the original peppered variety began to flourish again, and the darker variety was suppressed.

Relative to the vast span of time that evolutionary biologists and geologists believe has taken place on Earth, the amount of time consumed by the changes just cited is very small. One of the most obvious deficiencies in the arguments by critics of natural selection is a failure of imagination—a failure to extend the changes capable within a short span of time to the countless possibilities that could develop given millions and even billions of years of variation and environmental change.

All of these examples are independent of the fossil evidence for natural selection, but they point to the same conclusion: Life has evolved on this

If superior creatures from space ever visit Earth, the first question they will ask in order to assess the level of our civilization, is: "Have they discovered evolution yet?"

RICHARD DAWKINS

planet through a process of chance variation and environmental determinism. Once the physical characteristics emerge by chance, the environment then determines whether or not this characteristic is detrimental, neutral, or advantageous to an animal or plant.

Embryology:
Ontogeny Recapitulates Phylogeny

If you ever feel a need to intimidate or impress someone with your intellectual depth, you might find a way to mention that you have studied the phenomenon of ontogeny recapitulating phylogeny. Translated, it means that the embryological growth or development (ontogeny) of a complex organism tends to repeat (recapitulate) characteristics of its evolutionary ancestry (phylogeny).

All complex plant and animal life begins from a single cell—a fertilized egg or seed—and develops into a colony of cells and then a multicellular organism of great complexity. The embryos of all animals also display structural similarities. A human fetus at various stages of its development has gill arches like a fish, a tail indistinguishable from other primates, and a divergent big toe like that of a monkey or ape. In most humans the big toe realigns parallel to the other four toes before birth. For every 100,000 human infants, on average one will be born with a tail, much to the dismay of the parents. The amniotic fluid surrounding all embryos of reptiles, birds, and mammals has the same salt content as the sea, the birthplace of the first forms of life. The brain of a human being shows a clear pattern of structuring, whereby reptilelike, mammallike, and primatelike sections appear one on top of the other.

Fossil evidence indicates that whales and dolphins evolved from a land ancestor with a doglike body. The embryos of both whales and dolphins show a recapitulation of this phylogeny in that they have four limb buds, the two hindmost usually being resorbed before birth and the two front developing into fins. Occasionally, just as in the case of a child born with a small tail, a baby dolphin or whale can be observed with the hind limb buds still in place.

Many forms of animal life that do not give their young a head start within a womb reveal the same process of recapitulation in a larval stage. For instance, fossil evidence indicates that flatworms evolved from corallike creatures, and as corroborating evidence of this, the larval stage of some flatworm species (there are some 3,000) resemble tiny free-swimming coral organisms. For many millions of years one of the most successful creatures on Earth was the ancient trilobite. Some 250 million years ago it became extinct. Its closest living relative is the horseshoe crab. The adult horseshoe crab shows little sign of the segmentation of its ancestor, the circular grooves that run perpendicular to the length of the trilobite's body—a common physiological mechanism found in worms and insects. However, the newly hatched offspring of the horseshoe crab show this segmentation before developing the shelled armor of the adult. For this reason, they are called trilobite larvae.

Universality of the Cell:
Deoxyribonucleic Acid

The most recent development in support of the theory of natural selection, and to most scientists the most striking corroboration, is the discovery that just as the endless variety and arrangement of words are different sequences of the same letters, the physical development of all life on Earth, and its diversity, is directed by different sequencing of the same chemicals. The same basic percentage of composition of water, salts, carbohydrates, amino acids, nucleotides, and fats exists in the lowly bacteria as in the cells of a human being. Six basic elements—carbon, hydrogen, oxygen, nitrogen, phosphorus, and sulfur—make up 99.1 percent of the composition of all life on Earth.

The basis of all life is deoxyribonucleic acid—DNA. With the exception of a few viruses, in the deepest recesses of the cells of every form of life is a molecule 30 billion atoms long. This molecule is the coded instructions for the overall directional development and form for every living thing on Earth. The basic pieces of this extremely long molecule are the same for me as those of a tree. There are only four "letters" for the language of life; the alphabet may be small, but the words are very long. Each living thing is like a different play or drama written in the same language. With the same chemical alphabet, different words are constructed, and the different words are arranged to form different meanings and works of art. The difference between a human being and a tree is the arrangement of letters and words and how these letters in turn form what biologists call *genes* and different arrangements of genes. Although this is not completely understood in terms of every chemical and physiological step, we know that the genes convey a direction to the development of physical structures, functions, and behavioral tendencies because we have learned how to tinker with genes and produce different physical characteristics and functions in various forms of life.

Darwin was not aware of the mechanism of genetic variation, DNA replication, and radiation-induced mutation. We have learned since Darwin's time that just as mistakes can be made in the retyping of a play, subtle copy mistakes can be made in DNA. Also, just as outside electromagnetic forces can cause drastic changes in the information encoded on electromagnetic media, such as a computer disk, so minute outside forces, such as cosmic radiation, can alter a small segment of the 30-billion-atom spiral staircase of

Nothing in biology makes sense except in the light of evolution.

THEODOSIUS DOBZHANSKY

DNA and produce mutations. From this potential disorder, an infinite variety of change is possible. The majority of the changes are detrimental to the individual, because chances are these changes will not match what is needed to live successfully in the environment. At least some are no doubt neutral. From infinite diversity, however, a few lucky matches are possible given enough time.

The science of microbiology, which has corroborated so much of Darwin's original theory, is highly complex. To help the nonspecialist understand what the microbiologist has discovered, and how it relates to the mechanism of natural selection, science authors Gerald Feinberg and Robert Shapiro have offered the following analogy:

> Imagine a process in which you select a place in the text of *Hamlet* at random. You then decide arbitrarily whether one or more words are to be removed, added, or exchanged for new ones. If new words are to be added or used to replace existing ones, you select them, again at random from a dictionary. Usually this process will not increase the literary value of the play. . . . In very unlucky cases, the meaning of an entire section may be damaged. In rare instances, however, the text will be improved. The play may have been optimal for its original audience, and at that time any changes made by the random process would have been for the worse. But changes in society and language since the play was written have made *Hamlet* increasingly less understandable to modern audiences. (This is analogous to a need for adaptation of an organism due to changes in its environment.) It is a tribute to the genius of its creator that *Hamlet* has retained so much meaning after the passage of centuries. One can see, however, that if the random replacement process were carried out a great many times and if the errors were discarded and improvements preserved, an updated *Hamlet* would eventually be produced.[1]

The giraffe is an example of an updated *Hamlet*. Although there is no direct paleontological evidence for this particular case, the following is a likely Darwinian scenario. Millions of years ago the environment of the ancestor to the modern-day giraffe (the Pre-okapi) was becoming drier. The first foliage to die was the lower bushes and grasses. The last was the large trees. As this process

---

[1]Gerald Feinberg and Robert Shapiro, *Life Beyond Earth: The Intelligent Earthling's Guide to Life in the Universe* (New York: William Morrow, 1980), p. 102. Used by permission.

continued gradually over thousands of years, any animal that had a feeding advantage would be more likely to survive and reproduce. Within this time a mutation, or a series of mutations, coupled with normal genetic variation, produced a creature with a longer neck and legs. Today, evolutionary biologists are debating whether or not this would be a gradual process emerging over many generations or would be relatively sudden (geologically speaking), and whether or not the long neck served no useful purpose at first or was useful for other purposes. What they agree on is that through genetic variation and mutation a much different animal emerged, a new play in the drama of life, one that now has an advantage in being able to eat the topmost leaves of acacia trees. If the environment had not changed at the time of the mutation, the ancestor of the giraffe might still be with us. If the mutation had not occurred, there would be no long-necked giraffes, and the entire lineage may have died out completely. Just as plays do not survive if the response of the audience and reviewers is negative, so species do not survive unless they play to rave reviews of the environment.

Philosophical Considerations

As noted previously, a remarkably humbling philosophical implication follows from this theory: Evolution has no direction and no purpose. There is no inevitable, unswerving progress to "higher" creatures. There is only the spectacular result of happenstance. We have learned that given enough time, chance can produce incredible works of adaptation. There is no superior adaptation; there is only adaptation. There may be more complex adaptations, but there is no guarantee that complexity is better than simplicity.

When human beings survey the biosphere, our art, our history, and our technological achievements appear superior to that of other creatures. An objective observer, however, from another world might well conclude that insects are the dominant creatures and that the insect body is the most adaptable for living on this planet. As David Attenborough has pointed out:

If the historical development of science has indeed sometimes pricked our vanity, it has not plunged us into an abyss of immorality . . . it has liberated us from misconceptions, and thereby aided us in our moral progress.

PHILIP KITCHER, *ABUSING SCIENCE: THE CASE AGAINST CREATIONISM*

Insects swarm in deserts as well as forests; they swim below water and crawl in deep caves in perpetual darkness. They fly over the high peaks of the Himalayas and exist in surprising numbers on permanent ice caps of the Poles. One fly makes its home in pools of crude oil welling up from the ground; another lives in steaming hot volcanic springs. Some deliberately seek high concentrations of brine and others regularly withstand being frozen solid. They excavate homes for themselves in the skins of animals and burrow long winding tunnels within the thickness of a leaf.[1]

Insects have existed on Earth far longer than modern humans have, and they could have a longer future. We may have landed men on the Moon and built air-conditioned skyscrapers, but consider the flea, which can initiate a jump onto a human or animal body within a millionth of a second, a distance of over 300 times its own height—a comparable distance of a person being able to jump over a 70-story building. Or, consider the termite that can construct a building comparable, given its size, to the Empire State Building complete with effective natural air-conditioning, in spite of an outside temperature of over 100°.

Human beings are purposeful creatures, and it is difficult for us to think of anything that does not have a purpose, just as it is difficult for us to conceive any event taking place that was not caused by something. Could a building fall down for no reason—not an unknown reason, but no reason whatsoever? Similarly, it is much easier for us to believe that our species is special, not only because there appears to be evidence for it—our art, science, technology—but also because we want to be part of a universe that is meaningful and goal directed. But "purpose" is a human concept, and we must guard against interpreting everything through human filters, as if these were the only filters possible.

We may or may not be the only intelligent creature on Earth, or the entire universe, but there is no guarantee that the use of intelligence is a good idea, or that it is part of any plan. For all we know, intelligence is a characteristic that will soon go the way of the saber-toothed tiger.[2] What we call intelligence could be nothing more than another brief experiment, an experiment that has created nuclear weapons.

There is an important message in this view of life. Our consistent philosophical bias of our "specialness," which permeates most of our religions and politics, as well as part of science—this bias that somehow we are guaranteed a necessary survival and justice—has led us repeatedly to indulge in frivolous destructive behavior. The perspective that natural selection provides, added to the cosmological perspective discussed in Chapter 1, can serve as a foundation for an authentic attitude of preciousness and preservation.

I once lived next door to a very nice family who ran a small truck farm. They were apparently very devout, never missing a Sunday at church. In addition to vegetables and fruits, they raised pigs. The neighborhood could

Our climb to the top has been a get-rich-quick story, and, like all nouveaux riches, we are very sensitive about our background.

DESMOND MORRIS

---

[1] David Attenborough, *Life on Earth: A Natural History* (Boston: Little, Brown, 1979), p. 87. Copyright © 1979 by David Attenborough Productions Ltd. Used by permission.

[2] This is perhaps unfair to the saber-toothed tiger. The saber-toothed tiger is often thought of as a brief freak experiment of nature. Actually it survived as a species for hundreds of thousands of years.

always tell when there was a special occasion, because we would all wake up to the high-pitched, desperate screams of a pig being killed for a barbecue. The screams would sometimes last for over a half an hour. It finally occurred to me that this was a very long time and that surely there was a faster and more humane way to kill the pig. When I inquired about this one day, the matriarch of the family replied that they killed the pig in a special slow way, by deliberately stabbing the neck of the pig so as not to sever the jugular vein, which would result in instant death. She explained that in this way the blood from the heart of the pig is pumped efficiently to all the muscles in the pig's body, making the meat taste better. Horrified, I asked her if she did not feel a little guilty in doing this. She replied with a laugh, "Oh, you know what it says in the Bible, man has dominion over the animals."

We continuously take pride in our potential for empathy, compassion, and a superior value system. Yet until very recently manufacturers of women's makeup routinely used the eyes of rabbits to test for toxicity of new inventions. When the new chemical was dropped into their eyes, if the rabbits became blind, it was discarded, along with the now handicapped rabbits. We continuously keep cows impregnated to produce milk and then kill the majority of the baby calves born, producing veal cutlets and sweetbreads (the thymus gland) for fancy French restaurants. And there is our steak from steers, made possible by castrating millions of young male cattle per year.

Even a casual study of historical and contemporary national conflicts reveals people who believe that their existence is so special that death in a "just" war guarantees a reservation in heaven. No doubt some people in the United States do not fear nuclear war with the Soviet Union because, after all, the United States is a Christian nation. We will win, no matter the outcome, because our place in heaven is guaranteed.

Most people go about their daily lives taking for granted the everyday stability of human commerce. Many people are terrified at the thought that we are just another lucky finite creature. But as Darwin noted, there is "grandeur in this view" of life as the result of an equality of chance. And as Stephen Jay Gould has pointed out:

The average species of fossil invertebrate [such as insects] lives five to ten million years. . . . Vertebrate species [such as fish, amphibians, reptiles, and mammals] tend to live for shorter times. If we are still here to witness the destruction of our planet some five billion years or more hence [when scientists believe our sun will die], then we will have achieved something so unprecedented in the history of life that we should be willing to sing our swan song with joy.[1]

---

[1] Stephen Jay Gould, *The Panda's Thumb: More Reflections in Natural History* (New York: W. W. Norton, 1980), p. 142.

Lamarck and the Inheritance of Acquired Characteristics

As we will see, the lack of apparent purpose and direction implied by the theory of evolution disturbs those who hold traditional religious views on creation. Before we look at the philosophical and religious objections to evolution, we need to examine another purported scientific theory. This is the view first proposed by the French biologist Jean-Baptiste Lamarck, a half century before Darwin, and changed somewhat over the years; it is now called the *inheritance of acquired characteristics*. Unlike the orthodox Darwinian view, this theory states that the variation observed in nature and the amazing adaptability of life are *not* the result of chance genetic variations. For Lamarck, variation is the result of an improved or adaptive response to a changing environment. For Darwin, variation comes first, and selection or rejection by the environment afterward. For Lamarck, changes in the environment have priority; these changes cause improved creatures. Following our play analogy, an updated *Hamlet* is produced naturally and inevitably as people and their use of language change.

According to Lamarck's theory, there is an inherent wisdom in evolution. Instead of random unplanned genetic changes, there are "purposeful" genetic changes. Thus, a direct communication link is assumed to exist between the changing environment, the animal's response to the environment, and DNA in the germ or sex cells. An animal that experiences a radically changing environment will begin to behave in a certain way, either using or disusing its inherited characteristics, in attempting to adapt to the new environment. This in turn will produce acquired characteristics, which will then result in the production of offspring that are more adaptable to the environment. In the case of our giraffe, long-necked giraffes would tend to predominate in a much more directed way than in the Darwinian scenario. The elevating of the food source of the ancestral giraffe would cause a stretching of the muscles of the legs and neck, eventually resulting in offspring with longer legs and necks.

Modern advocates of this theory of evolution will point to the ostrich and note that it just happens to have calluses on its rump, breast, and other body parts that come into contact with the ground when it sits down. The same calluses exist in the unhatched chick. Lamarckians argue that it "takes a lot of believing" to accept the orthodox view that genes developed purely by chance for the calluses to develop only for the required spots. They argue it is much more reasonable to believe that the calluses developed first in the adult ostrich, just as calluses develop on the hands of a laborer or a baseball player, and then "somehow" the code for this trait is passed on to the sex cells. In this way the calluses become a *preadaptation*. Other examples are the tough skin of the human foot and the elephant's trunk.

Modern evolutionary biologists, following Darwin's emphasis on natural selection as the main cause of evolution,[1] claim that although this view may be more psychologically and philosophically satisfying, it conflicts substantially with the observed facts. First of all, there are many examples of environmentally imposed physical changes that are never communicated to the sex cells of an animal, and hence their offspring. For instance, take the practice of circumcision, the now widespread practice of removing the outer covering of skin from a baby boy's penis. Although of religious origin, this practice is now routine in Western culture to prevent infection. Here, evolutionary biologists will argue, is a substantial physical change imposed by the environment that has a definite advantage for survival, yet no one has ever witnessed the birth of a male child already circumcised. Lamarckians have yet to explain the "somehow" of a communication link between an acquired characteristic and the DNA of a parent's sex cells. The majority of microbiologists believe that the inheritance of an acquired characteristic is a chemical impossibility, that information can pass in only one direction at the level of DNA. Coded messages can be sent from DNA via a chemical messenger to develop a particular physical characteristic, but not vice versa.

Second, the entire concept behind Lamarckism is inconsistent with the paleontological evidence, the fossil record that indicates that the vast majority of species are extinct, implying a much messier view of evolution than that implied by the notion that acquired characteristics can be inherited just when they are needed. In other words, if there is an inherent wisdom in nature, it is very difficult to reconcile this with the apparent massive amount of trial and error reflected in the fossil record. (Incidentally, the ostrich has calluses in places where they serve no apparent function in protecting it from the ground.)

One of the most striking facts about both creatures living today and those extinct is the infinite variety of forms of life. Over 500 million years ago an animal existed that was so strange that scientists have named it hallucigenia. It had seven pairs of legs and seven tentacles each ending with a mouth. When we consider such creatures, life may be amazing, but it does not appear to be planned. There are many different environments, and there are many different variations adaptable to those environments. Darwin's theory implies a large amount of extinction; Lamarck's implies no extinction at all. For Lamarck, on the tree of life each branch is a ladder, where each lower animal simply changes to another higher form; Darwin's tree branches again and

[1] Darwin, as was customary for his time, also gave some credence to use and disuse as a cause of variation. Although this is not at all accepted by modern evolutionary biologists, even Darwin emphasized that use and disuse must be "largely combined with" or "overmatched by" natural selection. See page 177 of the sixth edition of *On the Origin of Species by Means of Natural Selection or the Preservation of Favoured Races in the Struggle for Life*: John Murray, (London 1900).

again with some of the branches dying and dropping off completely. Hallucigenia is one example of many of a tree branch that has dropped off completely; it did not evolve into any "higher" form of life.

Finally, supporters of Darwinism will point out that what often may appear to be superficially a case of an inheritance of an acquired characteristic is easily explained in terms of natural variety and selection. In the past half century behavioral scientists have used thousands of white rats in laboratory experiments. These researchers have noticed that the rats have adapted to their captivity for the most part by becoming naturally more docile compared to their wild relatives. Is this an example of an inherited acquired characteristic? Has a wild creature, confined to a small space, become less active and then passed this characteristic on to its offspring? No, say the Darwinians. This is actually a classic case of selection. Given the original population of wild rats, there will be a distribution of rat "personalities" from the very hyperactive to docile. Being hyperactive or active in the wild would be an advantage. In a captive laboratory environment it would not, and it is unlikely that the active rats could mate in such an environment. Hence, after several generations, laboratory rats are almost all docile.

What is intriguing about Lamarckism, and undoubtedly one of the reasons for its staying power, is that it incorporates in scientific form, or at least attempts to, the major philosophical objection mentioned earlier. Life appears to be too intricately adapted to every environmental niche to be an accident. Human beings appear superior to other creatures, and Lamarckism does imply that humankind is the end product of evolution. When Darwin first proposed the theory of evolution, the religiously minded rejected it out of hand as impossible, not on scientific grounds but on philosophical grounds. It implied that the Bible was wrong and that God, if He existed at all, was a bungler. Today a new approach has emerged that also rejects the orthodox Darwinian position and claims not only to be consistent with the Bible but also to be a "better" scientific theory. This view has come to be known as scientific creationism. The adherents of this view have lobbied throughout the United States for equal treatment in science texts and biology classes. In a democracy should not students be informed of competing scientific theories?

Scientific Creationism

For the most part, the argument of scientific creationism can be divided into two main categories: (1) attacks on the apparent weaknesses in the orthodox theory of evolution and (2) a number of traditional philosophical arguments.

In attacking what can be called the epistemological weaknesses of evolution theory, creationists argue that it requires a gigantic leap of faith to reconstruct the orthodox evolutionary scenario from the fossil record. A paleontologist or an evolutionary anthropologist will find a tooth and conclude from this tooth that the creature who possessed it was apelike, lived 7 million years ago, did not walk upright, and was a vegetarian, eating primarily fruit and coarse vegetation. Isn't this a bit much to conclude from just a tooth? As we have already noted, a fossil is the result of an extraordinary death. Do the fossils that we have found represent the whole? Millions of fossils representing 250,000 species is a very small sample compared to the over 500 million species that we think have existed on Earth.

There are also acknowledged gaps in the fossil record. Even Darwin noted that the fossil record does not seem to show a gradual evolution. And then there is the biggest gap of all, a gap from the time of the very first microorganisms, 3½ billion years ago, to the Cambrian explosion, 600 million years ago and the beginning of 99 percent of the evolutionary development on Earth. It is imagination that fills in these gaps, not evidence, so the creationists maintain. If the dates are wrong, the fossil record supports an abrupt beginning of all life on Earth, one consistent with the Bible's idea of creation by a supreme being.

Is this argument a criticism of evolution theory or of the entire scientific method? In Chapter 2 we learned that imagination and creativity always fill in the gaps in evidence. The major issue is always whether the numerically scanty evidence supports one conceivable picture better than another. Many scientists have noted that the energy received from outer space used to figure out our cosmological roots—from the Big Bang theory to galaxies, pulsars, and black holes—amounts to no more than the energy of a few snowflakes striking the ground! Recall also how little factual evidence was available to Eratosthenes. The history of science has demonstrated repeatedly that the critical process inherent in the scientific method allows a few facts to go a long way.

The truth may not be helpful, but the concealment of it cannot be.

MELVIN KONNER, *THE TANGLED WING*

Epistemologically, the status of the evidence for the modern theory of evolution is no different than that of political polls and the cigarette smoking studies discussed in the previous chapter. If there were no questions raised by gaps in evidence, there would be nothing to discuss and debate, and there would be no room for further growth and understanding. All of the questions raised by the creationists were raised by scientists themselves long ago. Debates and discussions on evolution take place all the time, and more often than not, new ideas and levels of understanding emerge from these debates.

For many years scientists pondered the gap of 3½ billion years between the emergence of the first bacteria and the Cambrian explosion. A recent explanation is that because the first bacteria were anaerobic—oxygen is a poison to them and is excreted—and because the primitive atmosphere of

Earth had no oxygen, it required billions of years for the oxygen level to build up in the atmosphere. This accumulation then allowed for the creation of both a protective ozone layer to screen out harmful ultraviolet light and a level of ordinary oxygen capable of sustaining the more familiar life on Earth today. Furthermore, the puzzling gaps in the fossil record have produced a suggested modification of a gradualist interpretation of evolution, known technically as *punctuated equilibrium*. Rather than seeing the fossil record as incomplete, this view suggests we interpret the data in many cases literally: Evolution often is marked by long periods of stability, punctuated by rapid change. (Keep in mind that "rapid" for a geologist and paleontologist means hundreds of thousands of years.) If the environment of a particular niche is stable for a long time, there would be no favorable selection of novel variations. According to this view, the life of any species is marked by "long periods of boredom and short periods of terror."

On the question of bone deductions and the interpretation of fossils, we also find scientists debating, critiquing, and discussing the interpretations of any individual scientist. Usually the interpretation of what a tooth means is the result of a cooperative community effort: An anatomist will confirm that the tooth is indeed similar to that of present-day apes; a physicist will take an electron microscope or a computerized tomographic picture of the tooth, revealing microscopic scratches and a pattern of wear identical to the teeth of fruit- and vegetable-eating animals living today; a geologist will analyze the rock strata and date it both by an analysis of the layers of rock and the radioactive elements in the rock; another geologist may be called in to confirm that millions of years ago the location's environment was a lush jungle similar to what modern apes live in; a paleobotanist, a specialist in extinct plant biology, will analyze the plant fossils, sometimes even the fossilized pollen grains, to confirm the jungle plant life of the area; and so on. Could they all be wrong? Yes.

It is conceivable that the creature in question was a strange type of cow with a deformed tooth, or that the creature was indeed apelike, and this particular one had developed a bad habit of chewing on sticks. Or perhaps a hurricane blew the tooth, the pollen grains, and the remains of plant life many hundreds of miles and deposited them in the middle of a desert. Many scenarios are conceivable, just as it is conceivable that the cause of lung cancer is not cigarette smoking, but exposure to a strange chemical substance that activated a virus many years later. The question is which of all the possible scenarios is the most reasonable. Inductive evidence from one point of view is always scanty; the constant challenge is to be able to separate the most reasonable view from all the conceivable ones.

The evolutionary scenarios that scientists have put together could conceivably be wrong, but they are not "gigantic leaps of faith." It is hardly epistemologically fair to criticize the theory of natural selection on the basis of the problem of induction, a generic problem for all science, and then conclude there is a more scientific approach. For the most part, the only thing scientific about this aspect of creationism is the willingness to now discuss the evidence, rather than object to evolution on philosophical grounds alone.

The Design Argument

Creationists often refer to the theory of evolution as an "animal fairy tale." Objective observations and a little common sense, they say, show it is much more "logical" to believe that God purposely created all at once the beautiful adaptations that we see today. The phrase "much more logical" might, however, only mean what seems more comfortable. But could a new version of *Hamlet* be written by randomly replacing words, as Feinberg and Shapiro claimed earlier? Mathematically, this is possible given the opportunity for billions of word replacements, but is it reasonable to believe this could happen in a reasonable amount of time? Could a monkey banging the keys of a typewriter haphazardly produce by chance a *Hamlet*?

A major argument of the creationist is actually an old philosophical position called the design argument. This argument proposes that it is much more reasonable to assume that the adaptable intricacies of nature were designed and could not possibly be the result of accident and chance. Consider just a few of nature's artworks.

In South America there exists a moth that spreads its wings to scare predators away. When it spreads its wings, it reveals a design that looks very much like a ferocious monkey. There is also a passion fruit vine that is covered with yellow spots strategically placed on tender new shoots. To a butterfly looking for a place to deposit its eggs—eggs that would turn into worms and eat the leaves—the yellow spots look like some other butterfly has already staked out this territory. Because it is disadvantageous for the butterfly to lay its eggs where others already exist, it looks for a better place to reproduce. Also in South America, there is a ventriloquistic lizard, which is capable of throwing its voice, making it seem to a predator looking for a meal that the sounds of the lizard are originating in another location. Could these marvelous adaptations be accidents?

Consider also that there are insects that can withstand incredibly low temperatures because they have antifreeze in their blood; four-eyed fish, two eyes to see out of the water and two eyes for seeing under water simultaneously; fish with flashlight eyes containing phosphorescent bacteria to see in dark ocean depths; a bottom-feeding shark that looks like a rock; flowers that stink to attract flies and ensure pollination; fish, frogs, and squirrels that fly (glide); a "Jesus lizard" that can run across lakes and streams; and the panda with a sixth finger (a thumblike appendage) that enables it to eat bamboo shoots. Finally, consider the monarch butterfly and its color. The monarch's orange

The celestial order and the beauty of the universe compel me to admit that there is some excellent and eternal Being, who deserves the respect and homage of men.

CICERO

hues are a message to all possible predators (especially birds) that sickness and possibly death will result if the butterfly is used for a meal. Consider also that this same sequence of colors is seen in such diverse creatures as snakes, frogs, and even sea slugs with the same warning implied and in other species that are not actually poisonous but nevertheless "mimic" the real thing for protection.

Is it not much more reasonable to assume that such unique features are the result of intelligent design? How could an anglerfish develop a little fishing pole by accident, one complete with a fishlike appendage of bait at the end? One species of trilobite had an eye with a double lens conforming to a sophisticated mathematical principle that human beings discovered only 350 years ago in developing the telescope. How could such a thing develop by chance?

The design argument usually employs the following analogy. Suppose you were shipwrecked on an apparently deserted island. Your ship had been blown off course by a violent storm—so far off course that the island was not on any map you had ever seen. There was, at first, no sign of human life, past or present. One day, while walking along the beach, you found a wristwatch. Which would be more reasonable to conclude? (1) That nature, through a process of a number of coincidental events, had produced the watch; that bolts of lightning had somehow struck various rocks and produced at various times pieces of metal and glass, which just happened to resemble the intricate parts of a watch; that the wind just happened to assemble these pieces and they all just happened to fit! Or (2) that somewhere, at some time, an intelligent human designer crafted the watch, and it somehow ended up on the beach. Surely, even if we allow for a time span of millions and millions of years, it is much more reasonable to conclude that the watch is the result of the craftsmanship of an intelligent creature, capable of purposeful planning.

The physical universe and any item of life are much more intricate and complicated than a simple watch, so the argument goes. Which is more reasonable then: that this is all the result of blind chance or that it all has been crafted by an intelligent being? Could moths just happen to grow monkey faces on their wings? Could insects evolve by chance with antifreeze in their blood? If all things are possible, why didn't the little fishing pole of the angler fish develop backward, away from the fish's mouth? Is it merely a coincidence that it faces forward so that the fish can catch a meal? Or, consider a type of sea slug that lives off the Great Barrier Reef. It eats stinging jellyfish. The stinging cells of the jellyfish do not affect the sea slug; instead, the slug uses the cells for its own protection. After the jellyfish is consumed, the stinging cells are not digested, but migrate intact to the back of the sea slug. Here they offer formidable protection from any animal wishing to make the sea slug a meal. Is it not just as unlikely that such coordination is the result of chance as lightning producing the parts of a watch?

Consider also the amazing complexity of a human being. The human body consists of trillions of cells. In each cell there is a nucleus and DNA. It is estimated that human DNA has between 50,000 to 500,000 genes. It is because of the genes that heart cells coordinate their activities to pump blood, that pancreas cells make insulin, that hair cells form hair, that fertilized egg cells multiply and become babies, and that thousands of other things essential to life occur. That eyes see and brains think is the result of a process much more complicated than a watch. It could not have happened by accident, so the authors of the design argument argue. There must exist a designer.

There is a force to this argument that has appealed to many intelligent men and women through the ages, not just to the philosophically and religiously minded but to great scientists as well. From the beginnings of Western culture, the roots of the scientific outlook have been infused with a sense of wonder at the marvelous "order" around us. Many great scientists have been impressed not only by the ecological beauty of the biological world but also by the apparent mathematical harmony that we can find everywhere we look in the physical world.

We must be careful, however, with our sense of order, logic, and reasonableness. If we are sincerely interested in the truth, we must consider where our sense of order might be coming from. Could it be more of a projection of the human mind than an actual independent reality? Might not our inability to fathom how an insect could end up with antifreeze in its blood reflect the limitations of our imagination, especially the difficulty of maintaining a sense of the time that has elapsed on Earth, and the many possibilities during this time compared to our short life spans? Perhaps we think too much of ourselves when we refuse to believe that accident, given enough time, could write a better *Hamlet*.

The eighteenth-century British philosopher David Hume was the first to point out the following argument. Let us assume for a moment that our observational experience of nature does indeed point clearly to a creator, a master craftsman of all the intricate beauty around us. The fact that we now know that a creator exists does not tell us much about the important characteristics of this creator. How would we know that this creator is good, all-powerful, all-knowing, and eternal? How would we know that the human species is a special creature in the grand scheme of things? There is also a great deal of suffering and chaos in the world. The data are consistent with not only a good God but an evil one as well, a sadist, a bungler, whose "watch" sometimes falters or is used for evil purposes. If we are to be honest about the watch analogy, the fact that we find a watch on a beach would only indicate that a finite craftsman existed somewhere, an ordinary person who lives and dies, sometimes does not get along with others, and has a potential for good and evil. For all we know, our master craftsman could be a temperamental artist capable of great art, but not at all a nice person to be around.

Hume's point is that the argument from design is logically invalid, if one attempts to infer a God with all the Judeo-Christian qualities from the notion that nature has apparent design features. The argument needs to be supplemented by a great deal of faith in the Bible. Additional premises would need to be added about the life of Jesus, his mission, his resurrection. We would need to discuss the evidence that the Bible was divinely inspired and the

[The brain is] an enchanted loom where millions of flashing shuttles weave a dissolving pattern.

SIR CHARLES SHERRINGTON

This world is only the first rude essay of some infant deity, who afterwards abandoned it, ashamed of his lame performance.

DAVID HUME

epistemological issue of whether revelation is a valid means of gaining knowledge. These are not silly issues and should be discussed in democratic schools. They are not, however, ideas that need to be discussed in a biology class.

It does not prove much to say the universe has to be ordered in some way or it would not work. Could a universe work that was not ordered in some way? If the anglerfish had grown its little fishing pole the wrong way, it would die. If insects in frigid climates did not have some kind of antifreeze in their blood, they would not be around for us to wonder about. The fossil record shows that many exotic animals have existed, that natural selection has experimented with many different designs, and that most of these designs survived and reproduced for a time, but when the environment changed they ultimately failed.

Observational experience, of both human and animal life, demonstrates that mutations are born regularly, but for the most part these mutations are detrimental. The apparent design we see today is the result of a relentless rolling of the dice and a vast amount of death.[1] Observational experience shows a messiness to evolution, and this is best accounted for so far by the theory of natural selection.

Nature is not always pretty. From a human point of view, the sea cucumber has a rather ugly way of defending itself. If a fish or crab gets too close, it defecates its internal organs, leaving the surprised predator entangled in a sticky mess. The sea cucumber remains alive and grows back its intestines in a few weeks. As many readers of Hume have pointed out, it may be consoling to see how perfect the shape of the banana is for the human mouth, but then one must wonder about God's purpose for the watermelon.

Finally, the analogy of a monkey randomly banging away at a typewriter to characterize natural selection is a false one. It is crucial to remember that a key to natural selection is that the environment over time responds. In other words, a better analogy would involve a monkey that randomly strikes a key and then receives positive, negative, or neutral feedback of some sort. In this way we can imagine how a *Hamlet* could be produced.

If there is anything divine in man, it is that his front paws were liberated at some remote time to the freedom which lets them clutch, carry, feel, measure, compare!

HOMER SMITH

The Cosmological Argument

Another argument borrowed by the creationist from the past is known as the cosmological argument. Is it conceivable that the universe could have been

---

[1] During the early 1980s, A. R. Thatcher, Britain's registrar general of population, estimated that nearly 60 billion humans died between 40,000 B.C., the time of modern *Homo sapiens,* and A.D. 1980, 12 times the present population.

created from nothing? Or that the universe is here for no reason? Is it not more reasonable to believe that there is a purpose to all this, rather than absolutely no purpose? Because evolution implies that there is no direction or purpose to life, and it is more logical to believe that there is a purpose to the universe, evolution must be false.

Proponents of the cosmological argument claim that the universe had to have a beginning, otherwise it is infinite. Since infinity minus infinity is still infinity, this makes it impossible to explain why we are here now, rather than at some other time. Why didn't the human species exist an infinite time ago? So assuming that the universe had a beginning, could it have arisen from nothing, without a cause, and for no reason?

Although the theory of evolution does not imply any claim about the origin of the universe—it is primarily a scientific explanation for the development of life on Earth—the question of purpose has provoked much philosophical debate. It is difficult for human beings to imagine a universe emerging from pure nothingness. Advances in science, however, have repeatedly taught us that a failure of human imagination to grasp or make sense of a possible reality is more indicative of a clash between the truth and our common sense, rather than a proof that our common conceptions of space and time are more logical. It is also very difficult for a human being to imagine what the world looks like to an insect using ultraviolet light rather than the visible spectrum, or what a tree would look like in infrared light only. Yet we know that both of these perspectives exist.

Science has demonstrated that our common sense is based upon a built-in perceptual filtering system, valid for our existence, but only a fraction of all the possible windows to reality. Using the powerful tool of abstract mathematics, science has been able, again and again, to point to unimaginable realities. The Big Bang theory, which involves a mathematical description of matter, space, and time emerging from a single point from "nowhere" and "nowhen," is no longer thought of as unusual in modern physics. As we have seen, this theory is so well accepted that large sums of money are being spent on creating laboratories the size of small cities to understand the details. Paul Davies points out in his book *God and the New Physics*, "Failure of the human imagination to grasp certain crucial features of reality is a warning that we cannot expect to base great religious truths (such as the nature of the creation) on simple-minded ideas of space, time, and matter, gleaned from daily experience."

The universe as a single point? With space and time inside? But what is outside the point? Don't space and time have to be outside the point? Our minds balk. It is inconceivable that a point can exist without being "in" space and time. For scientists during the time of Isaac Newton this was indeed inconceivable, but for physical scientists after Albert Einstein it is not only conceivable but the evidence also demonstrates it is reasonable. At the end of the Middle Ages many intelligent people understood the mathematical implications of the Sun being the center of our system of planets and the move-

Failure of the human imagination to grasp certain crucial features of reality is a warning that we cannot expect to base great religious truths . . . on simple-minded ideas of space, time, and matter, gleaned from daily experience.

PAUL DAVIES

ment of the Earth. But they rejected these ideas because of the implication that there was a huge, unimaginable space between the Earth and the stars. So much space was inconceivable. Today, the concept and plausibility of the Big Bang theory undermine much of the intuitive force of the cosmological argument. If it is conceivable that the entire universe could emerge from a single point, then there is a natural explanation for the universe emerging from a kind of nothingness. These incredible theories are not something scientists have made up but the result of a long struggle, being the best explanations that we have for modern astronomical observations.

But what about the intuitive feeling that there must be a reason and purpose for the universe? Many scientists, after all the natural explanations have been stated, have stood back from their equations and technical data and have been swept away emotionally by the wonder of it all. Any sensitive human being may experience this same feeling by simply immersing himself or herself in a dark starlit night and contemplating the fact that the points of light above are trillions of miles away. Why is all this here rather than not here?

It is admirable that human beings have this emotion. Science has not destroyed this emotion; it has enhanced it. Scientists are no different from the rest of humanity. Few, however, would be willing to assert that there is a scientific connection between this feeling and the existence of a particular conception of God or that this feeling proves that there is a better scientific explanation for life than the theory of evolution. For some, if anything at all can be made of this feeling, it makes more sense to believe in evolution as being more consistent with a truly glorious God. If there is no direction, purpose, or value system easily revealed by studying the natural universe, this then places a much greater responsibility on each human being to make authentic decisions about right and wrong. A truly exalted being, if it exists, would create a universe where we are truly free to choose. There would be no obvious purpose so that we would be responsible for creating our own. We would then do unto others as we would want them to do unto us, not out of fear of punishment but because we have decided that it is the right thing to do.

The Moral Argument

But how do we know what is the right thing to do? Consider the following hypothetical situation. Suppose you are in a car alone, driving on an isolated road, far from any town or houses. It is a stormy night with rain and lightning, thunder and wind. The road is bordered by tall trees, and the wind is breaking off large branches. The lightning occasionally strikes close by. You feel relatively secure, however, because your car has plenty of gas, is well made, and has a set of good tires. You have only a little way to go to connect with a much safer road, so the one thing you do not want to do is stop. Suddenly, just ahead, on the side of the road, you see two young children. One is about five years old and the other is only three. They are both drenched from the heavy rain and very scared. They are holding hands and crying, but the five-year-old is trying to be brave as she pulls the three-year-old along. At any moment a falling tree branch could crush them. What would you do? Would you stop? What is the right thing to do? How would you judge the person in a similar situation who does not stop, who concludes that they are not his children, and that whatever their reason for being alone in this isolated place is none of his business?

Creationists will argue that any compassionate, moral human being will know what to do, and the only way that we can account for this knowledge and the rightness of stopping for the children is that there is a moral sentiment and conscience given to every human being by a higher Being, a moral sentiment reflecting an absolute set of values revealed to us by this Being. In the purposeless, directionless universe of the supporters of Darwin, there would be no sanction for such an act. Some people would stop and some would not, but there would be less motivation to stop and no objective criterion for deciding the rightness or wrongness of stopping.

As in the case of the cosmological argument, the creationist's argument here takes aim at engaging our intuition. There is a strong feeling that stopping is the right response. This feeling is so strong that it is impossible to reconcile with the explanation that we have such a feeling only because of our cultural conditioning. Would someone stop simply because someone told him it would be the right thing to do? The sense of rightness here makes the way a person was brought up, or the cultural background, irrelevant. There might be an explanation for why a person did not stop (poor childhood, for example), and we might even understand in some intellectual sense why a person did not stop, but not stopping would still be wrong.

We possess then, argue the creationists, an innate potential for a strong internal sense of moral sentiment—a sense of empathy and compassion for others that could lead to acts inconsistent with our own well-being—plus a strong internal sense of right judgment. In other words, human beings possess the potential to develop and express a superior set of values, a value system that can be termed other-directed: values such as a concern for universal justice, charity, and love. How could such sentiments be possible in a purposeless universe of blind chance? If the goal of evolution is individual survival and reproductive success, how can we account for the intuitive feeling that acts of compassion and love convey a sense of universal rightness that seems to transcend the physical dimensions of space and time? Or how can natural selection account for apparently crazy acts of altruism?

In 1985 millions of TV viewers around the world witnessed firsthand a tragic fire at a British soccer stadium. Only minutes after the fire started, the heat from it had become so intense that the clothing and hair of people

If God indeed exists, then one of his greatest gifts to us was our reason. To deny science—to deny evolution—is not to be truly religious or truly moral. It is indeed the opposite. There is nothing in modern evolutionary theory which stands in the way of a deep sense of religion or of a morally worthwhile life.

MICHAEL RUSE

in the middle of the soccer field, a considerable distance away from the burning stands, began to burst into flames. With the announcer moaning, "Oh, look at this," the cameras showed a man in a daze walking slowly across the field, his hair and clothes ablaze. Suddenly, with little thought for their personal safety, people turned back, running to the man and removing their jackets. They threw the man to the ground and attempted to smother the flames as quickly as possible. The heat was so intense where they were that at any moment their own clothing could burst into flames. As the flames engulfing the man continued to resist extinction—some of the jackets used to smother the flames began to burn—more and more people returned, ripping their jackets off and surrounding the man.[1]

Such actions are almost always spontaneous and *noncognitive*. That is, acts of altruism do not result from a carefully thought out cost-benefit analysis of what the individual may gain or lose and what dangers exist to one's personal well-being. They are thoughtless, and from the standpoint of individual survival, clearly irrational. If the driving force for evolution is individual survival and reproductive success, what possible reason exists for risking one's life for a complete stranger?

Because of considerations such as these, creationists note, even the great British naturalist, Alfred Russell Wallace, the cooriginator with Darwin of the theory of natural selection, balked at applying this theory completely to the evolution of the human species. There is, thought Wallace, such an obvious moral distinction between the valueless world implied by natural selection and the potential moral sensibility of man that God must have stepped in and supplied man with a soul. For the creationist, Wallace's inconsistency is a symptom of the major problem the theory of natural selection has in explaining meaning in life and humankind's potential for noble values.

The Evolution of Emotion

There are three possible responses to the creationist's argument. First, the intuitive sense of the rightness of certain values and actions could be a cultural illusion. *Relativism* holds that a distinction between right and wrong is simply a matter of cultural conditioning, that what is right in one society may not be right in another. This applies not only to matters of taste, such as food and clothing but also to how to treat the children in the storm. In another culture it might be appropriate to keep on driving, because this would

---

[1] It is also worth noting that a few weeks later, TV cameras captured another soccer phenomenon. This time, however, millions of viewers were shocked to witness a riot where many people were killed.

"toughen up" the children, and it would be bad for the children for anyone to interfere with this process.

Second, one could argue that stopping is indeed the right thing to do, but an appeal to God and a sense of universal purpose are unnecessary. Such actions can be given a purely rational basis. Insofar as human beings are rational animals, we can learn that actions of compassion and actions based on other higher values are reasonable. Just as we are capable of a greater understanding of the natural world, so we are capable of learning a rational value system. (But could any extension of this same explanation account for apparently noncognitive acts of altruism?)

Whether or not moral values must have an absolute or rational basis, or whether or not values are objective or relative, is much too involved an issue to deal with here. What is more appropriate for a discussion of evolution theory, and this is the third response, is whether or not the theory can account for human feelings of compassion and apparent acts of altruistic behavior.

Note that acts of altruism in general, and acts of emotional bonding in particular, are not exclusively human behaviors. There are many examples of individual sacrifice in the animal world. There is not only the well-known action of a mother bird drawing attention to herself to save her offspring, but there are examples where the individual sacrifices itself for the good of the species and in some cases even other species. The field of study that compares animal and human behavior is *ethology*. Since the time of Darwin, ethological studies have been combined with evolution theory to produce a remarkable scientific explanation for the evolution of emotion and human values.

Until approximately 60 million years ago, the primary reproductive strategy on Earth, as exhibited by fish, amphibians, and reptiles, was for the female parent to produce as many eggs and offspring as physically possible. In this way, although most would die, a few would likely survive to continue the parents' genetic line. In this strategy, there was no need for much of a relationship between the parents and offspring. With few exceptions[1] the pattern continues for these animal groups: The parents completely remove themselves from any responsibility of care, and the offspring are on their own in the game of survival. For this practice to be successful, the offspring must emerge in a biologically mature form ready for survival, complete with instinctual behavioral patterns for gathering food and defending themselves against predators. There is no time for significant biological development and learning after birth.

With the evolution of the mammals, a revolution in reproductive strategy took place. Instead of the expensive production of many eggs, only a relative few are produced. For this system to work, a motivation to stay around must

It must not be forgotten that although a high standard of morality gives a slight or no advantage to each individual man and his children over the other men of the same tribe, yet an advancement in the standard of morality will certainly give an immense advantage to one tribe over another.

CHARLES DARWIN

---

[1] Some species of alligators and frogs take care of their young to some extent, and at least some dinosaurs are now believed to have practiced some child care. Recent fossil evidence has focused on the nesting behavior of Maiasaura, which means "good mother lizard."

exist, and the parents must display a large investment of protective time. Thus, for the first time on a heretofore, for the most part, uncaring Earth an emotional bond became common between mothers and their offspring.[1] From this point, adopting a very broad perspective on the evolution of mammals and the human species, we can see a pattern of generalization of what might be called "companion feeling" emerging with time and circumstance.

At first there is only maternal love. For the male of many early mammalian species, the old pattern is still workable. They do not stick around to care for their offspring, and their time is much better spent finding and mating with as many partners as possible. Eventually though, a new strategy emerges, a minirevolution within a revolution: paternal love and the family. Just as it was useful to invest a large amount of time on a few offspring rather than to waste that same energy producing many that would die, so we find that it is expedient for the males to invest their time partaking of a protective family arrangement, strengthening the probability of the survival of their offspring. Given the proper environment, not only is this an advantage in terms of efficiency but it also increases the probability that the male will find a mate, because females have begun to select mates on the basis of whether or not they will stay around and help with child care.

The next functional relationship is the clan or herd. Just as it is a personal advantage for the mother to care for her offspring (so part of her genes will survive), so it is a personal advantage to care about the members of one's extended family and eventually even unrelated members of one's own species. The more the species prospers, the more likely one's offspring will prosper. By the time we reach the human species and human civilization, forms of companion feeling have become potentially very abstract. Why should I care about children that are not mine? Why should I risk personal injury for strangers? Why should I stop? Because my own children have a much better chance of survival in a world where people do such things.[2]

Finally, although an ecological consciousness can hardly be attributed to contemporary humans alone, today we see an ever-increasing concern for the well-being of the entire natural environment. Some have called this *biophilia,* literally "love of the biosphere." Why should the members of organizations such as Greenpeace risk their lives in little boats attempting to halt the

---

[1] Mammals first evolved some 200 million years ago, but they remained primarily small nocturnal creatures hiding from the dominant dinosaurs. As the dinosaurs disappeared 60 million years ago, the mammals began to radiate over the entire face of the Earth.

[2] Much overlooked by popular understanding of Darwin's famous work, *On the Origin of Species by Means of Natural Selection or the Preservation of Favoured Races in the Struggle for Life,* 6th ed. (London: John Murray, 1900), is his statement (p. 78), "I use the term [struggle for existence] in a large and metaphorical sense including dependence of one being on another, and including (which is more important) not only the life of the individual, but success in leaving progeny."

slaughter of the majestic whale? Why should we be concerned about the extinction of the snail darter?

Moorpark College in Ventura County, California, is a small community college famous throughout the world for its Exotic Animal Training and Management Program. This two-and-a-half-year associate degree program is for students interested in careers in caring for animals. Every year over 1,000 applications from around the world and all walks of life are received. Of this, only 100 are granted an interview, and eventually only 40 students are accepted. Each student is expected to dedicate two to three years of his life as a minimum necessary condition for being part of this program. There are no holidays or vacations, and each student is expected to be on call 24 hours a day. Everything must be sacrificed for the well-being of the animals, and even though there are substantial living expenses and program fees, no outside employment is permitted.

We are learning that our Earth is a fragile, contingent place, and the smallest disruption to the smallest creature can have major consequences to all life on Earth. From the standpoint of evolutionary biology, however, this awareness is not totally one of learning. The instinctual potential to touch the world sparingly has its roots in the relationship of companion feeling between the mammalian mother and baby. In general, we find ourselves reacting positively and protectively, not only to babylike features within our own species, but to the puppylike and juvenile features of other species as well. Perhaps we are "fooled" by the evolved response to our own babies, an inappropriate psychological transfer, or perhaps it is a genuine adaptive feature of human nature. Whatever the case, the fact remains, people do these things.

Although much debated, this ethological-evolutionary explanation for the development of the human potential for higher values is an intriguing possibility with major philosophical ramifications. Suppose love is an accident, a result of a fortuitous change in reproductive strategy? Should it be valued any less?

In support of this explanation, recently there has been a related interest in a developing subfield of ethology called neuroethology, the comparative study of the physiology of the brains and behavior patterns of different animals. One result of these studies is the theory of the "triune brain," authored by Paul MacLean.[1] Briefly, it maintains that human beings and other recently evolved mammals show in their brains at least three major evolutionary steps.

---

[1] See Paul D. MacLean and Kral Adalbert Vojtech, *A Triune Concept of the Brain and Behavior* (Toronto: University of Toronto Press, 1973); Paul D. MacLean, "The Triune Brain Evolving," *American Journal of Physical Anthropology* 52, no. 2 (1980): 215; "Evolutionary Psychiatry and the Triune Brain," *Psychological Medicine* 15 (1985): 219–221; "Brain Evolution: The Origin of Social and Cognitive Behavior," *Journal of Children in Contemporary Society* 16 (1983): 9–21; "Brain Roots of the Will-to-Power," *Zygon: Journal of Religion and Science* 18 (1983): 359–374; "Family Feeling in the Triune Brain," *Psychology Today* 15, no. 2 (1981): 100.

At the base of the human brain is a neurological structure known as the R-complex. The "R" is for "reptilian brain." Whether in reptiles, birds, or mammals, this structure is concerned with controlling instinctive behavior patterns that are routine and fixed, especially those related to self-preservation, such as defensive aggression and defense of territory. Above the R-complex is the "old mammalian brain," referred to more commonly today as the limbic system. This structure is associated with emotion. In the light of the preceding discussion on reproductive strategies of reptiles and mammals, it is significant that this neurological structure is found only in mammals. Finally, on top of the old mammalian brain, in human beings, primates, and other recently evolved mammals such as whales and dolphins, is the cerebral cortex—the site of advanced information processing and learning.

A female hamster that has her cortex removed at birth will develop almost normally, able to carry out all the basic routines of food gathering, defense, and reproduction. She will even be capable of being a mother. Only the precision of her behavior seems to be affected. But further removal of the limbic system damages maternal and play behavior significantly. Human beings with Parkinson's disease or Huntington's chorea often have a partial or total inability to initiate and carry out routines. In one well-known case, a woman who could consciously decide that she should start dinner found herself incapable of carrying out this routine task. As would be expected, such diseases have been traced to defects in the R-complex.

Thus, there is general agreement that the neurological structures in animals and humans can be roughly correlated with general behavior patterns. Reptiles lay many eggs and do not care for their young, thus necessitating immediate, instinctual routines necessary for survival. Mammals lay few eggs and care for their young, allowing for slow development and learning, and requiring a neurological structure, not only to mediate an emotional response to the world but to process the complexities of child rearing as well. At least on Earth, emotion is associated with bigger brains, slower development to adulthood, and greater awareness and learning.

The Naturalistic Fallacy

Do these three responses answer the creationist? Of course not, if by an answer we mean a justification for the rightness of moral actions. The fact that there may be a natural explanation for human behavior, including those rare moments of altruism and moral action, does not prove that this behavior is the best or is right. The creationists want much more than the how of love and justice. They are not interested in what goes on in a person's brain when compassion is experienced. They want our lives to have meaning and direction. They want a clear, absolute perspective that says there is a difference between being kind or cruel to a child. In response to such a need, the theory of natural selection gives us nothing.

Of the few things that philosophers agree on, most agree that it is not possible to logically connect what *ought* to be the case with what *is* the case. To think otherwise is to commit the *naturalistic fallacy*. From the standpoint of natural selection, the mammalian reproductive strategy, which serves as the basis for the development of emotion and companion feeling, is just another successful strategy, adaptable and efficient given the proper environment. If there is no direction and purpose in evolution, then the mammalian strategy is not necessarily better. From a description of reproduction strategies, a morality cannot be derived logically. In birds, for instance, there are both monogamous and polygamous species. For species whose food source is abundant it is not necessary for the males to help rear the young. It is also a relative disadvantage for more than one parent to frequent the nest. It calls too much attention to the nest. Natural selection does not sanction one value system as better than another. The system that works well for one species may not work well for another.

But is it necessary to prove that humankind's so-called higher value system is better? As in the cases of the other arguments for the existence of a God-like purpose in the natural world, an appeal to an intuitive sense of rightness of human action could well be more of an indication of anthropocentrism than a crucial proof that something is wrong with the theory of evolution. Natural selection can at least offer an alternative view of how it is possible for human beings to base some of their behavior on higher values. Evolution can account for how these values are possible in a directionless universe. Evolution may not be able to account for the rightness of such behaviors in an absolute sense, but this demand may be unnecessary in the first place. Whether a biological illusion or a fortuitous abstraction, the potential to act altruistically and for others exists.[1] Do we need to justify these values, or choose them? Like our very existence, our potential for higher values may be a result of happenstance. Natural selection is messy, and this is reflected in the fact that human beings are morally messy. Sometimes we care for our children and sometimes we do not. Our awareness of our messy evolutionary background can help us choose who we want to be. At the present crossroads of our tenuous existence, we do not need to justify life—we need to choose life. It is a matter of choice, not logic.

The issues raised by the creationists are not silly. They are important, but are they scientific issues? The design, cosmological, and moral arguments are all traditional philosophical arguments. Aside from these, the objections raised

Some men who call themselves pessimists because they cannot read good into the operations of nature forget that they cannot read evil. In morals the law of competition no more justifies personal, official, or national selfishness or brutality than the law of gravitation justifies the shooting of a bird.

VERNON KELLOGG

[1]We speak of the "possibility" and "potentiality" of other-directed values having a biological foundation. This foundation is an "open program" only; the actualization depends upon culture.

by the creationists against evolution theory consist primarily of attacking the apparent weaknesses of the theory. It is a basic epistemological tenet of science that extraordinary claims require extraordinary evidence. That the universe and all life on Earth were created by a benevolent intelligent higher being is an extraordinary claim. As such, to gain the title of a scientific hypothesis, it must be refutable and have direct positive evidence for it. Vague connections to established scientific fact are not sufficient.[1]

In terms of logic, the argument of today's creationists can be reduced to a questionable dilemma fallacy. They have given us only two choices: either evolution is true or creationism is true; intuitively something feels wrong about evolution theory and there are epistemological problems with its being "totally" certain; thus, creationism must be true. But aside from other theories of evolution, there are many other views of creation. Even if it could be demonstrated that something is seriously wrong with the Darwinian theory of natural selection, it would not follow that science has thus proved that a traditional Western religious conception of God and creation is true.

Still the emotional response beckons: 100 billion to 400 billion stars per galaxy, at least 10 billion galaxies, 500 million chances at life on Earth and 99 percent extinct. Out of this relentless shuffling, there is now at least one creature capable of loving and thinking. On Earth the universe seems to have been experimenting with different forms of awareness. Love and emotion are associated with bigger brains and greater awareness. Is this a unique situation? In this immensity of space and time, are we the only creatures that have evolved to ponder such questions?

There is no consensus yet whether, given the right conditions, the formation of DNA is inevitable or a rare occurrence. Some scientists believe that it is so rare that it could not have happened on Earth first. The probability is too low. The possible chemical interactions of an entire galaxy are needed. Others believe that given the chemical conditions that existed on Earth billions of years ago, the formation of a self-replicating molecule is almost guaranteed. There is agreement, however, that the basic chemical compounds needed for DNA formation not only existed on Earth approximately 4 billion years ago but also are common throughout the universe. Thus, the probability of life evolving in many parts of each galaxy may be high. The universe could well be teeming with life and different forms of awareness. We will address this question further in Chapter 9.

If the Darwinian view is true, however, the message is clear that there will be no human beings anywhere else. The likelihood of exactly the same

---

[1] For instance, some creationists will point to the Cambrian explosion as evidence that life started on Earth all at once. But the Cambrian explosion started over 600 million years ago and extended for over 100 million years.

sequence of events that produced mammals and human beings on Earth ever being repeated is infinitesimally small. On our own Earth there are countless events that could have easily occurred otherwise, and we would not be here to contemplate these questions. Approximately 60 million years ago the evolutionary spread of the mammals became possible with the extinction of the dinosaurs. Although there is no consensus yet on how this happened, recent thinking has converged on the possibility of a total climatic disruption caused by an impact of a large comet. Such an impact would have been equivalent to many nuclear explosions. Tons of dust and debris would have been thrown into the atmosphere, thus causing a drastic change in the climate. The dinosaurs, the dominant creatures on Earth for over 100 million years, could not adapt. If this comet had missed the Earth, the development of the mammals would likely have remained restricted to small, lowly, nighttime, ratlike creatures. There would be no monkeys, apes, and human beings.

The fact that the number of contingent factors necessary for human beings to evolve are too numerous to be repeated and that highly successful creatures can vanish in a geological instant are sobering, humbling thoughts. Why there is something rather than nothing and why we are here are not as important as that there is a universe and we are a fortuitous part of it. We are here now, a complex creature with a complex evolutionary past: a "hopeful monster," a mixed bag of evolutionary survival strategies, capable of both great compassion and great destruction.

Concept Summary

We should understand before we judge. One of the most misunderstood and misjudged theories by the layperson, among both supporters and nonsupporters alike, is Darwin's theory of evolution, or natural selection. Although the theory itself has evolved since the time of Darwin, the basic principle remains the same: Natural variation of physical characteristics results in the selection of those characteristics better suited to prevailing conditions of the environment; individuals possessing these characteristics have a better chance of survival and are more likely to reproduce and pass these characteristics on to their offspring.

Because there is no preferred direction to initial variations, and because the environment inevitably changes over time, a superior or adaptable characteristic is a relative phenomenon: Gaining an advantage at one time can lead to a disadvantage at another time. Evolutionary biologists speak of "the tree of life," because there are no special branches produced by natural selection; the branches of this tree, its twists and turns, are shaped by contingency, and each modern plant and animal at the tips of each branch is the result of a long, mixed process of determinism and happenstance.

Although Darwin's theory implies the possibility of exemplary adaptation, it also implies a "messiness" to the development of life on Earth. Both adaptive branching and an unprejudiced awkwardness are reflected in the fossil record. So is a consistent pattern of evidence of evolution from common descent. Human-initiated examples of artificial selection, embryological similarities in animal life, and especially modern advances in genetics and the understanding of universal processes within the cellular life of all species have further produced a synthesis of understanding and corroboration of the evolutionary process.

Because natural selection is neutral in terms of purpose or direction, many people find more philosophical satisfaction in views that imply more design to life. Lamarck's theory of evolution implies that variation is a planned, purposeful, or directed response to a changing environment and that the changing environment shapes life on Earth in a more focused, less awkward, and perhaps more compassionate way. Animals improve naturally, in time for and in tune with the changing environment, by acquiring useful characteristics and passing these characteristics on to their offspring. Although perhaps more conceptually reassuring, neither the amount of extinction shown in the fossil record nor the survival of neutral or useless characteristics seems consistent with this view. Lamarckians have also failed to explain how information can be transmitted from an acquired useful characteristic to genes and automatically passed on to offspring.

Like Lamarckism, scientific creationism postulates a more purposeful explanation of life on Earth. Arguing that a correct understanding of the scientific evidence demonstrates that life began all at once in an act of creation and that Darwin's theory, based only on inductive evidence, rests on imaginative leaps of faith, creationists further argue that a little common, philosophical sense shows that evolution from common descent is illogical and impossible. Using the traditional philosophical arguments from design, cosmology, and morality, creationists argue that the primary facts of life—the beauty and order of nature, humankind's ability for ethical judgment, and our potential to express other-directed values—can only be understood by believing in an intelligent supreme being who initiated the universe with direction and purpose.

Placed within the proper context, the issues creationists raise are important. Science, however, based on a critical process that forces us to confront and be more intimate with the world, also reasonably compels us to accept ideas that transcend philosophical satisfaction. Our difficulty in imagining how chance and contingency could produce elements of apparent design may be more a reflection of the limitations of our imagination. That it seems counterintuitive to the human mind that the universe could emerge from nothing does not mean that it is impossible. The theory of natural selection is also consistent with the so-called facts of life, and implies or at least is further consistent with (because it is primarily a descriptive scientific theory, not a philosophical one) an even more striking possibility: Our potential for higher values originates in a fortuitous change in reproductive strategy.

Finally, natural selection is consistent with just as grand a view of life as that of creationism: Because it does not clearly endorse one value system as better than another, it leaves us truly free to choose what is right. Such a view is not inconsistent with a broad-minded view of a supreme being. Thus, the exclusive choices given to us by creationism, either evolution and sense-lessness or God and creation, are questionably too limited, and although its conclusion could be true, such a method of reasoning is not acceptable scientific practice.

Suggested Readings

Life on Earth: A Natural History, by David Attenborough (Boston: Little, Brown, 1979).

A persuasive portrayal of the story of evolution and its infinite diversity, using pictures and discussion of living species of animals and plants resembling the extinct species that gave rise to the present life on Earth.

Extinction, by Steven M. Stanley (New York: Scientific American Books, 1987).

With drawings and photographs, Stanley takes the reader on a guided tour of the geological and biological history of the Earth. An excellent overview of the paleontological and geological evidence for mass extinctions.

The Flamingo's Smile: Reflections in Natural History, by Stephen Jay Gould (New York: W. W. Norton, 1985).

Perhaps Gould's best book. As with his previous books *Ever Since Darwin* and *The Panda's Thumb,* Gould continues exploring the fascinating quirkiness and messiness of evolution and the implications for meaning in life. Includes interpretations of Darwinism, the politics of science, mind in the universe, and the issue of extraterrestrial life.

Biophilia, by Edward Osborne Wilson (Cambridge, Mass.: Harvard University Press, 1984).

Unlike Wilson's *Sociobiology* and *On Human Nature,* two previous controversial books, *Biophilia* is a more personal and less technical presentation of the idea that the perplexities of human behavior can only be fully understood within the context of our biological history and its connection with the rest of the Earth's species. It is also a poetic and scientific appeal to see the rationality inherent in human concern for nonhuman species and their environments.

Evolution and Human Nature, by Richard Morris (New York: Seaview/Putnam, 1983).

Although many stimulating books have attempted to relate the human condition to evolutionary biology and the modern study of ethology (Robert Ardrey's *African Genesis,* Konrad Lorenz's *On Aggression,* Desmond Morris's *The Naked Ape,* and Melvin Konner's *The Tangled Wing,* for example), this book is a good overview of the many issues and controversies.

The Blind Watchmaker, by Richard Dawkins (New York: W. W. Norton, 1986).

A provocative defense of Darwinism and a forceful antidote to recent books (see *The Neck of the Giraffe: Where Darwin Went Wrong* by Francis Hitching and *The Great Evolution Mystery* by Gordon Rattray Taylor) that claim there must be something more than natural selection to explain the intricate functional adaptive complexity of life on Earth. Author also of *The Selfish Gene,* Dawkins shows how apparent design features of nature can be accounted for by natural selection and how this is the best theory to account for these features.

Abusing Science: The Case Against Creationism, by Philip Kitcher (Cambridge, Mass.: MIT Press, 1982).

Another antidote to the critics of natural selection, focusing on creationism, but at a deeper and more inclusive philosophical level. Kitcher, a professional philosopher of science, integrates the educational controversy between the teaching of evolution and creationism in public schools with a philosophical analysis of the nature of science. The book is also informative in discussing fossil details and evolutionary transitions and thought-provoking in defending the notion that modern evolution theory is not inconsistent with a sense of religion or morality.

The Growth of Biological Thought: Diversity, Evolution, and Inheritance, by Ernst Mayr (Cambridge, Mass.: Belknap Press, 1982).

A must-read book for anyone interested in a complete academic understanding of the history and content of evolution theory. Mayr, formerly a Harvard University professor and one of the major players of what is called the "Neo-Darwinian synthesis," has written the definitive text on evolutionary problems and their historical background.

The Origin of Species, By Means of Natural Selection or the Preservation of Favoured Races in the Struggle for Life, by Charles Darwin, 6th ed. (London: John Murray, 1900).

This edition of the primary source is recommended, because it contains additions and corrections resulting from Darwin's reflections on further evidence for natural selection and criticism of his theory. For an introductory presentation of Darwin's writings, see *The Essential Darwin,* Robert Jastrow, general editor, with selections and commentary by Kenneth Korey (Boston: Little, Brown, 1984); this book contains excerpts from Darwin's *Autobiography, The Voyage of the Beagle, Origin of Species,* and *The Descent of Man.*

CHAPTER 4

Cultural Roots:
1. The Ancient Greeks

In a part of the world that had for centuries been civilized . . . there gradually emerged a people, not very numerous, not very powerful, not very well organized, who had a totally new conception of what human life was for, and showed for the first time what the human mind was for.

H.D.F. KITTO, *THE GREEKS*

I would not be confident in everything I say about the argument: but one thing I would fight for to the end, both in word and deed if I were able—that if we believe we should try to find out what is not known, we would be better and braver and less idle than if we believed that what we do not know is impossible to find out and that we need not even try.

SOCRATES

The entire intellectual history of western culture is but a footnote to Plato.

ALFRED NORTH WHITEHEAD

Nothing. Fifteen billion years ago only a single point, at nowhere and nowhen. Approximately 5 billion years ago, the formation of the Earth and the Sun. Billions of years later, perhaps just 15 million years ago, less than one-third of 1 percent of Earth's history, the lush jungle habitats of primitive apelike creatures began to shrink because of the moving of continents. Eventually the vast grasslands that resulted selected a new type of apelike creature, that stood upright and perhaps could run like a track star and carry things with its hands. Finally, just a few minutes ago on the Cosmic Calendar (see Chapter 1), modern *Homo sapiens* emerged and perhaps began to ask why there is something rather than nothing.

Belief in Natural Law

For generation upon generation after this, our ancestors had only the wonder of the stars to entertain and to hint at an answer. Most of the human race answered the questions of why and how in terms of the idea that the universe is like a large puppet whose strings are pulled by a god or gods, that unexpected events, or even expected ones, are subject to the whims of a supernatural creature who has the same emotional stability problems as a human being. At a time of multicultural awareness, it is perhaps not fashionable to attribute so much to one culture, but before science could exist a radical idea had to be accepted.

In the northeastern Mediterranean area—Ancient Greece, the west coast of Asia Minor, and the various islands in between—from the seventh century to the second century B.C., a mere 2,500 years ago, the last five seconds or so on the Cosmic Calendar, a revolution in thought took place. Some people began to think that the universe is a rational place, with an internal order and governed by universal, natural laws. Most important, they believed that this order is knowable, that human beings can figure out these laws. These people also thought that the cosmos[1] is a good place, that our knowing about it is good, that knowing takes place not through divine revelation and obe-

As regards the first [principle of science], "that nature can be understood" . . . the most astonishing thing about it is that it had to be invented, that it was at all necessary to invent it.

ERWIN SCHRÖDINGER,
MIND AND MATTER

---

[1] *Cosmos* is a Greek word for "all that is." In this book we will use *cosmos* and *universe* interchangeably.

dience to authority but through open inquiry and critical evaluation of competing ideas, and that knowing is good not only for practical utility but for its own sake as well. Knowledge is an essential ingredient of happiness and the good life. As the astronomer Kepler would echo centuries later, the ancient Greeks believed that as the birds sing and the grass grows, human beings were meant to know. In short, faith in the gods was replaced with a faith in the cosmos, and the human species discovered that reason was a tool with which to romance the universe.

In light of our discussion of creationism in Chapter 3, note that an aspect of Greek religion was instrumental in the development of this outlook. Of the many creation myths circulating in the Mediterranean area, one maintained that the present state of the cosmos evolved from chaos. From this, it was thought, evolved human beings and the gods. In this creation story an ordered universe exists before the gods, rather than after. The gods did not create the universe; they are simply another part of it. The human species and the gods were considered both to be children of the cosmos, implying an important consequence for the relationship between human beings and their gods: Human nature and the gods were on a somewhat equal footing. Although the gods were granted immortality and carefree living and were more powerful, human beings were potentially morally superior and more intelligent, and human existence in general was more challenging and potentially happier. The Greeks prayed to their gods standing upright. Devotion was limited to a superficial devotion. The gods were kept happy to avoid trouble, but there was no love lost between humankind and the gods. The human family had to pay its taxes, but did not have to enjoy doing it.

The Greek word *philosophy* means literally "love of wisdom." The word *philos*, however, does not mean a sexual love (eros) or a charitable love (agape) but a passionate never-ending striving, a loving contemplation of the cosmos. In essence, devotion to a personalized divinity began to be replaced by a devotion to wisdom, a seeking of knowledge, and the correct use of that knowledge.

An important consequence of this new attitude was that happiness and the good life no longer required "divine aid," or the help of the gods. Human beings, as part of a purposeful nature and a rational plan, had everything they needed to be happy. Happiness was simply a matter of developing the potential the cosmos gave us—especially our ability to know. Actualizing the potential to know was considered fun, because in doing so human beings fulfilled nature's purpose. Continual discovery and exploration were thought of as joyful, even though there is never a completion to this process, a final resting place of understanding everything. Heaven, immortality, and the afterlife for these early contemplators of our cosmological roots were boring in comparison.

Philosophers and psychologists refer to such a theory of happiness or the good life as a *self-actualization theory*. The emphasis is on the process of moving toward the achievement of a goal, rather than the actual achievement of the goal. Consider the mind of a child. Try to remember the fresh joy of every new experience. We are all born curious, full of questions, with an open mind and a built-in sense of awe and wonder about life. Each new object is a new universe, each day a new beginning. A child would not be disappointed if the daily ecstasy of experiencing and learning new things were to last forever. The Greeks in many ways were the children of our scientific culture.

From the standpoint of eternity, all of our pursuits are futile. There will be no end to learning; there will be no final day when everything is known. Moreover, at any second the universe could suffer a cosmic "burp," a star close to us could supernova, and we, and all our progress and knowledge, would be gone in an instant.

Many people become disenchanted with science when they learn that science does not provide absolutely certain truths, a perfectly safe harbor in the buzzing world of daily turmoil and change. They fail to realize that if it did this, science could not grow. Adults and cultures can become tired of learning; like a child, a true scientist must relish the freshness of every new day. We are all born with this feeling of freshness. Most of us lose it. But it is a necessary part of the scientific attitude.

To be philosophically inclined toward the scientific attitude, however, a price must be paid: One must be able to live with a cosmic insecurity. One must be able to use this insecurity as a creative force to keep traveling and remember that it is better to travel than to arrive. This concept is not easy either for a culture or a human being. Like a romantic affair, our quest for understanding is an uneasy mixture of rapture and incompleteness.

In ancient Greece, and the surrounding Mediterranean area influenced by Greek philosophy, the fruits of thinking in natural terms ripened quickly. Thousands of years before modern times there were people who believed that humankind had emerged from fish and that many more forms of life had existed on Earth than we see now. Some people believed that the Earth was a sphere unsupported by any physical thing, that the Moon was a place with hills and valleys like the Earth, not a god or an original source of light, but reflected light from the Sun, a glowing stone. Eventually, some people believed that the Earth moved and was not the center of the cosmos. There were also people who believed that everything consisted of atoms and empty space and that light had a finite, but very great, speed. Some of these people intently studied the physical and biological world, conducting physical experiments and making detailed astronomical observations. Some people even attempted to estimate the absolute dimensions of the Moon and Sun and their distances from the Earth, and some were successful in deducing the existence of geological change from observing seashells on mountains and fossils of seaweed and fish in stone quarries. Others studied embryos, dissected animals, and even examined the anatomy of the brain (see boxes on pages 110–111). A world view emerged for a brief time not unlike our modern view of a vast universe of change, atoms, and empty space; of a Milky Way composed of millions of unresolved stars; of worlds that evolve and decay.

The analogy between childhood wonder and adult creativity is biology, not metaphor.

STEPHEN JAY GOULD

Learning is ever in the freshness of its youth, even for the old.

AESCHYLUS

Some Ancient Mediterranean Philosopher-Scientists

Thales (625?–547 B.C.): One of the first to conceive of the universe in natural terms—that nature has laws that can be known through a process of rational criticism. Is known to have predicted an eclipse; fell in a water well while looking at the stars. Also known for applying his knowledge of astronomy to weather forecasting, predicting a bumper olive crop, buying up all the olive presses, and making a financial killing. He taught that everything was made of water and that the Earth was a flat disk floating on water.

Anaximander (610–ca. 547 B.C.): An early believer that the human species arose from lower animals, specifically from fishlike creatures. Made a sundial, determined accurately the length of the year and seasons, made a map of the known world, and taught that the Earth is not supported and sits at the center of the cosmos. He also believed that the Sun, Moon, and the stars are made of fire seen through moving holes in the celestial globe of the sky and that all reality evolved from a boundless, nondefinable stuff.

Anaximenes (?): Among the first to suggest that qualitative changes (from a liquid to a gaseous state, for example) can be explained as changes in the density of one stuff. He thought this ultimate stuff to be air. He also believed that the Earth and the planets float and that the Moon shines by reflecting light from the Sun.

Anaxagoras (ca. 500–ca. 428 B.C.): Held that the Sun and Moon were not gods, that the Moon was a place made of ordinary matter with hills and valleys, and that the Sun was a red-hot stone. For this he was imprisoned for impiety. He dissected animals, studied the anatomy of the brain, and discovered that fish breathe through their gills. However, he thought the cosmos could be better understood qualitatively rather than quantitatively.

Pythagoras (ca. 580–ca. 500 B.C.): Argued that the earth was a sphere and moved around an invisible central fire (the Altar of Zeus). Is most noted for teaching that mathematical harmonies are the basis for all natural events.

Empedocles (ca. 490–430 B.C.): Taught that the basic elements were earth, water, air, and fire. Believed that light had a finite speed, but a very great one, and that many more life forms once existed on Earth than we see now. Conducted a famous experiment with a water clock (clepsydra), indicating that air must be made of something. Some accounts claim he thought of himself as a god and died by leaping into hot lava of the volcano Aetna.

Democritus (ca. 460–ca. 370 B.C.): Nicknamed the laughing philosopher. Taught that atoms and empty space made up all existing matter, that worlds such as the Earth evolved and decayed, that there were many others, some inhabited, others not, and that the Milky Way was composed of millions of unresolved stars.

Socrates (470–399 B.C.): Held few doctrines, but believed passionately in the fruitfulness of the pursuit of truth, the unlimited growth potential of humankind, and the goodness of the cosmos. Although he was more interested in human behavior than the natural world, he is quoted as saying, "Man must rise above the Earth—to the top of the clouds and beyond—for only thus will he fully understand the world in which he lives." He was condemned to death for impiety and corrupting the youth of Athens.

Plato (427–347 B.C.): A student of Socrates. Defended the Greek rationalist tradition against the attack of the sophists. Emphasized the supreme importance of mathematics in gaining knowledge, recognized the limitations of experiment and empirical knowledge, and accepted an operationalist, or instrumentalist, view of physical science. Founded the first school dedicated to the pursuit of universal knowledge, the Academy, and suggested that women have the same intellectual potential as men.

Aristotle (384–322 B.C.): A student of Plato. The world's first systematic biologist. He emphasized the importance of empirical observation, tutored Alexander the Great, and developed a physics of motion that had a great influence until the sixteenth century. As with Plato, his primary interest was to reaffirm the existence of a public and knowable reality, which the sophists had questioned. Founded a school, the Lyceum.

For myself, I like a
universe that includes
much that is unknown
and, at the same time,
much that is knowable.
A universe in which
everything is known
would be static and
dull, as boring as the
heaven of some weak-
minded theologians.

CARL SAGAN, "CAN WE
KNOW THE UNIVERSE?
REFLECTIONS ON A GRAIN
OF SALT"

Most important of all, there were people who took for granted the notion of "objectivity," believing in an independent truth that could be discovered through a critical exchange of ideas by a community of knowledge seekers. There was one world and one truth about this world. A person could be free to believe that the world is round or flat, but if the world is round, then those who believe that it is flat were just plain wrong. The cosmos was a harmonious place, and behind the apparent complexity of daily life, the show was run by universal natural laws.

There was, however, also much disagreement. Some thought that all physical matter was actually water in disguise; others that the ultimate stuff was air; still others that there were four basic elements: earth, water, air, and fire. Some even believed that physical matter, change, and motion were an illusion and only thought, or consciousness, existed. There were wars and political turmoil as well, and some people were executed because of their beliefs. Within a few centuries, for various reasons, a crisis in confidence developed, and people soon sought safer harbors of thinking or perhaps excuses not to think at all.

Protagoras and the Sophists

Confronted by confusion and uncertainty, people usually choose one of two extremes: (1) to believe what one believes adamantly and absolutely, rationalizing one's beliefs into an irrefutable fortress, or (2) completely reject the notion of universal truth and accept all beliefs as true for those who believe them. The former we can call *absolutists* and the latter *relativists*. Both pride themselves in having discovered something profound, but both positions result in the same end—the cessation of critical thinking.

A consequence of the ancient Greek crisis in confidence was a movement known as sophism. The sophists claimed to be neither scientists nor philosophers but "educators." They traveled about the Greek city-states claiming to teach the key to a successful life. One of the most famous was Protagoras (ca. 485–410? B.C.), the author of the statement "Man is the measure of all things, of all things that are, that they are, of all things that are not, that they are not." In this cryptic statement rested the complete rejection of the old Greek ideal. We cannot know what The Truth is, according to Protagoras. Objectivity is a myth. Each individual sees the world through his own filters; we are each locked within a cage of our own appearances, and there are no criteria to select some filters or appearances as better than others. What is true for you is thus true for you, and what is true for me is true for me.

Protagoras was perhaps one of the world's first anthropologists; he was also a lawyer, a diplomat, and a teacher of rhetoric. As a cultural anthropologist he traveled widely about the Mediterranean, coming into contact with many different cultures and life-styles. He saw that people, societies, and cultures differ. Protagoras concluded that it was presumptuous of the Greeks to think of Greek virtues as the best way to live and other societies as barbaric. As a lawyer and rhetorician, Protagoras taught that the wise man learns to be ideologically flexible. In cases of disagreement the wise man knows that there is no objective solution, no "right" answer. The goal of resolving disagreements is not to find out the truth, for there is no such thing. Rather, the goal is to cure such disagreements by having people agree. As a diplomat and rhetorician, he taught that the truly wise man is able to persuade others to accept his view of things and "cure disagreements in his favor." This persuasion must be subtle, so that the wise man can make his position appear right or good to anyone who formerly did not think so. Life and language have an infinite texture, and the wise man learns to use the freedom this implies. The skilled rhetorician learns how to fill the many empty spaces between the thoughts of opposing views.

There is no distortion or con-artistry in the behavior of the wise man, because this would imply that some right view is being distorted, and for Protagoras, there is no such thing. The goal of life is not the result of a scientific endeavor—to discover the truth. The goal is an artistic one—to mold reality from a particular perspective and persuade others to see this perspective. Not that the perspective of the wise man is better—it is just more expedient to have others see things your way. Agreement through diplomacy is what we seek, not truth through science and logic.

For Protagoras the relationship between the human mind and reality is like the well-known story of the four blind men attempting to "know" what an elephant is. One of the blind men holds on to the trunk and proclaims this is the true elephant. Another has hold of a leg, another the tail, and the other the back—all describing what they are experiencing accurately and proclaiming that they have discovered the real elephant. Our experience of reality is always indirect and always a point of view of the whole, a perspective, not the whole itself.

Of course it is not possible always to persuade others to one's particular way of thinking. In such cases, according to Protagoras, we should *conform*. Because one idea is no better than another, the wise thing to do would be to conform to the practices of one's community most of the time. Above all, chaos and anarchy should be avoided. Because no idea is better or worse than any other, it does not matter which ideas are accepted. As long as the majority conform, there will be less chaos and confusion in the world. So we should accept the universals of our culture, not because we have discovered that they are the best way to live, but because life is easier, more predictable, and safer if everyone agrees to live by the same truths and values.

Although Protagoras was highly respected and was not interested in destroying the traditional Greek virtues and institutions, the epistemological relativism of his philosophy soon gave birth, by Greek standards, to more

The wise man learns that the game of life does not involve uncovering an objective truth that is already there for all to see, but instead an expansion of a perspective by persuading others to live in that perspective.

No human being is constituted to know the truth, the whole truth, and nothing but the truth; and even the best of men must be content with fragments, with partial glimpses, never the full fruition.

WILLIAM OSLER

reprehensible philosophical fruits. Callicles, another sophist, taught that the key to a successful life is to realize that the conventional ideas of right and wrong of a society are designed by inferior people as a trick to control the truly strong. Ideas such as "just treatment" are actually part of a strategy to keep the strong from taking what would be rightfully theirs, if the course of nature were followed instead. In nature the superior creatures always rule over, and have more than, the inferior. Conventional ideas of right and wrong are not discoveries of reason, but rather rationalizations that the fearful inferior have devised to protect themselves.

Finally, another sophist, Thrasymachus, argued that all discussions of right and wrong, true and false, are just a lot of hot air. All that really matters in life is power. Beliefs are true or false, actions are considered right or wrong, depending upon the best interests of those in power. Political legislation is not the result of a painstaking objective analysis into what is the truth or what is best. Rather, it is the other way around: What is considered the truth and the best are dictated by the political exploitations of those in power. For Thrasymachus, the most important thing to learn about life is that "might is right."

> *What, then, is truth? A mobile army of metaphors, metonyms, and anthropomorphisms . . . truths are illusions about which one has forgotten that this is what they are.*
>
> FRIEDRICH NIETZSCHE, "ON TRUTH AND LIE IN AN EXTRA-MORAL SENSE"

Socrates and Plato

> *The best of the joke is that Protagoras acknowledges the truth of their opinion who believes his opinion to be false; for in admitting that the opinions of all men are true, in effect he grants that the opinion of his opponents is true.*
>
> PLATO, the *THEAETETUS*

Although there is much that is sobering about this sophist philosophy, and although the scientific attitude shares with Protagoras the rejection of absolutism, science would be impossible if such a philosophy were taken to its logical conclusion. If the human race had followed Protagoras, we would not know what we know today about the planets, stars, and galaxies, or our cosmological and biological roots. There would have been no incentive to search for better ideas. Relativism is not so much a philosophy as it is an epistemological ghost that materializes whenever people become tired of seeking. Historically this has happened many times. Fortunately, there has thus far always been those who defend the old Greek ideal.

Protagoras recognized, in his own way, the problem of induction. As we saw in Chapter 2, "evidence" needs to be interpreted, "facts" are a matter of perspective, and no matter how much evidence one has for a generalization or an abstract representation of nature, there is no guarantee that it is true. In addition, given any set of agreed-upon facts, there are always, in principle, an infinite number of possible explanations. Because there is never an absolute guarantee, Protagoras concluded that it was impossible to separate the reasonable from the conceivable.

Science takes the more burdensome epistemological path: It assumes that even though there is no guarantee, it is possible through great effort to separate the reasonable from the merely conceivable. We genuinely think we know it is more reasonable to believe that smoking is the cause of most lung cancers rather than some inconsequential happening on someone's birthday. But how do we know this when admittedly we can never be absolutely sure that smoking is the cause of most lung cancers? Can we say that we have knowledge, if we do not have certainty? Unless there is a method with a set of criteria to judge some ideas as being better than others, each human being is the measure of all things. This challenge has always been one of the major epistemological problems of Western thought. It is not an exaggeration to say that much of the intellectual history of Western culture can be understood by watching how each age answers this question.

Socrates was very impressed by the philosophy of Protagoras, but he could not accept his final conclusion that truth was a myth or the idea that there is no better way to live. In his youth and middle years Socrates was a soldier. He witnessed firsthand the violence and cruelty that human beings are capable of. There may be many different good ways to live, but surely there are also many bad ways to live. If so, he reasoned, what general idea about "good" enables us to judge some particular way of life as good or bad?

Protagoras had argued that each of us is limited, that most people claim to know the whole truth, even though they believe what they do because of the force of their cultural tradition. He is correct, according to Socrates, about how most people claim to know the whole truth, and how, when analyzed, we find that the reasons for their beliefs are very weak, usually based only on appeals to tradition, authority, and popularity. But our limitations and unanalyzed beliefs do not prove the sophists' epistemological relativism. To use the same analogy of the elephant and the blind men, if the four blind men could realize their limited perspectives and then communicate, they could at least construct a better image of the real elephant, according to Socrates. Each could realize that the whole truth could be approximated through collaborating. What we first must do then is know how little we really know. Then we must be willing to critically examine beliefs and communicate our different perspectives. For Socrates this was an urgent task—the future happiness of humankind was at stake.

Socrates believed it was his purpose in life to make people think, to make people realize their limitations as Protagoras had taught. Truth, as a community effort, was impossible unless people critically examined their beliefs and communicated their limited perspectives. He lived what he taught. In his later years he could be seen daily walking the streets of Athens, probing all who would listen on their opinions of the ultimate questions of life. Almost always the conversation would draw a crowd of young men looking for an adventure in ideas. And almost always the result would be the same: someone who initially thought he knew the answers to the great questions of life would leave embarrassed, his pretensions of knowledge destroyed by the logical questioning of Socrates:

The real purpose of scientific method is to make sure Nature hasn't misled you into thinking you know something you don't actually know.

ROBERT M. PIRSIG, ZEN AND THE ART OF MOTORCYCLE MAINTENANCE

It is literally true, even if it sounds rather comical, that God has specially appointed me to this city, as though it were a large thoroughbred horse which because of its great size is inclined to be lazy and needs stimulation of some stinging fly. It seems to me that God has attached me to this city to perform the office of such a fly, and all day long I never cease to settle here, there, and everywhere, rousing, persuading, reproving every one of you.

SOCRATES, the APOLOGY

—Why is the death sentence the appropriate punishment
  for treason?
—Because everyone knows this is right. (*Popularity*)

—How does everyone know this?
—Because it has always been this way. (*Tradition*)

—Why has it always been this way?
—Because it was decreed by the gods. (*Authority*)

—Has it been decreed by the gods because it is right or only
  because the gods have said so?
—Because it is right.

—Why is the death sentence the appropriate punishment
  for treason? Why is it right?

*One thing only I know,
and that is that I know
nothing.*

SOCRATES

*For man, the
unexamined life is
not worth living.*

SOCRATES

Socrates would demonstrate again and again to young, inquiring minds why appeals to popularity, tradition, and authority were not the result of thinking but an excuse not to think. If a behavior was considered right because it has always been approved, then why has it always been approved? What was the initial reason? Does this reason still apply today? If a belief is considered true because most people have this belief, then what are the reasons for this belief? What evidence has convinced so many people? More often than not, there were no answers to these questions from those who pretended to know.

Eventually Socrates embarrassed one too many famous citizens. He was arrested and tried for impiety (questioning the existence of the gods) and corrupting the minds of the youth of Athens. He was convicted and put to death. Of the many young men who had followed Socrates around Athens, one was Plato. The injustice of silencing this sincere, unpretentious man had a profound effect on Plato. Some things are just wrong, and essentially Plato dedicated the rest of his life to righting this terrible injustice. Plato's "revenge" was to ensure that the world and all future generations would know of Socrates, whose death was an injustice not only for Plato but also for all.

We could discuss the philosophy of Plato from many perspectives. So vast was his influence on succeeding generations of Western culture that no discussion of Western religion, science, or political theory is complete without documenting his influence. In this light the twentieth-century philosopher Alfred North Whitehead said that all of Western culture is but a footnote to Plato. Often those of a more limited perspective will find in Plato only a mystic who impeded the development of Western science, who alone somehow put a stop to the great initial development of ancient Greek science. From a modern perspective there is a small amount of truth to this, but the larger truth is that Plato answered the sophists and preserved the old Greek ideal of objectivity and universal truth.

We see in Plato someone who is desperately searching for a "hook of certainty," some sure guide through the world of conflicting opinions of truth and justice. The death of Socrates and the philosophy of Protagoras were not unrelated events for him. At the trial of Socrates the jury was persuaded by those skilled in the tricks of rhetoric to see the unjust as the just. If objectivity was a myth, if there were no universal laws that all rational human beings ought to believe, then even the death of Socrates was an unjust act only for him.

Plato was much too intelligent to dogmatically sweep away the arguments of Protagoras as silly. He saw that Protagoras's arguments were very powerful and that Socrates had not really answered Protagoras, other than to restate a faith in the existence of better ideas and a method for achieving them. Much more was needed.

To begin with, Plato concluded that the problems raised by the sophists concerning an absolute understanding of the physical world were insurmountable. Not only is every perception a matter of interpretation, not only is there the problem of induction, but also false ideas work. Consider some examples. Suppose I believe that a particular politician is a communist. As far as I am concerned, everything he has ever done as a public figure is consistent with him being a communist. On the basis of this I am able to predict how he will vote on the various legislative bills this year. My predictions are accurate and this convinces me further that he is a communist. I could continue to make predictions about his behavior. My theory has power. It is practical. It could, however, still be false.

As we will see in the next chapter, prior to the sixteenth century the most accepted model of planetary motion was that of the Earth-centered universe. The Earth was stationary at the center, and the Sun, Moon, planets, and stars circled around it. With an appropriate arrangement of circles, the nightly motion of the stars, the monthly motion of the Moon, the yearly motion of the Sun, and even the strange motion of the planets could be qualitatively understood and quantitatively predicted to a considerable degree of accuracy. It was an explanatory scheme that had power and great utility. Farmers could keep track of the seasons, navigators using principles derived from it could sail dangerous distances accurately, and astronomers could predict the appearance of the night sky of places on the Earth where no civilized person had set foot. Yet today we know this picture is wrong.

Plato knew that even though the Earth-centered system worked, this did not prove that the real physical situation corresponded to it. One could always have many workable opinions about the physical reality behind appearances. Plato believed that to answer the sophists some truths must be found that were beyond all doubt. There must be ideas that are so clear, so self-evident just by thinking them, that everyone would know that there are no alternatives. These truths must not be dependent on or relative to time, place, or person. Are there such truths? Plato believed that there were and we could

In questions of just and unjust . . . ought we to follow the opinion of the many and to fear them; or the opinion of the one man who has understanding? Ought we not to . . . reverence him more than all the rest of the world; and if we desert him shall we not destroy and injure that principle in us which may be assumed to be improved by justice and deteriorated by injustice?

PLATO

begin by finding them in mathematics. Consider the difference between the following statements:

1. 2 + 2 = 4.
2. There are nine planets in our solar system.

Plato argued that these statements have a different epistemological status. The second statement is considered true today, but it was considered false during Plato's time (when only five planets were known). Similarly, this statement could be false in the future in several ways. First of all, we could be wrong about there being only nine planets. There could be an undetected tenth planet. In fact, from time to time claims have been made for a tenth planet orbiting the Sun outside the orbit of Pluto. As viewed from Earth, Pluto's motion through the backdrop of the stars does not totally conform to the expected motion, assuming there are only nine planets. There are what astronomers call perturbations. A tenth planet has not been confirmed, but this could either be because there is no tenth planet (there being another as yet unknown cause of the perturbations) or because our technology is not sufficiently developed to detect it yet.

Secondly, we may be correct about there being only nine planets now, but at some future date a catastrophe could occur to one of the planets and there would then be only eight. In fact, astronomers predict that approximately 5 billion years from now our Sun's lifespan as a dependable life-sustaining body will end. It will become a red giant and engulf all of the inner planets, perhaps as far out as Jupiter. There will then no longer be nine planets— maybe only three, maybe none. The point is, some future observation could make a belief that we consider true today false tomorrow. Plato saw that any statement about the physical world was relative to time, place, and the subjective perspective of the observer, and hence, the epistemological skepticism of the sophists.

On the other hand, the first statement seems to be true independent of any future observation. Is it possible that 2 + 2 will equal 5 in the future? Could we some day travel this vast universe and find a strange planet where in counting two rocks and two more rocks the result is five? If the solar system were entirely destroyed, 2 + 2 would still be 4. If the entire physical universe ceased to exist, the thought "2 + 2 = 4" may be no longer useful, but it would surely not be false.

Consider another example. Suppose you are sitting in a room with no windows. Outside the building is a parking lot in the shape of a precise rectangle. There are 25 rows in this parking lot and each row has exactly 20 parking stalls. Suppose you are told that at this particular time every single slot contains a car and there are no cases of double parking. Also, the parking lot is closed and no one is driving around or waiting inside the parking lot for someone to leave. How many cars are there in the parking lot? We know instantly there are 500 cars in the parking lot ($25 \times 20 = 500$). Consider the strangeness of this knowledge. You cannot see the 500 cars. Nor do you need to go outside and observe the parking lot to make sure there are 500 cars. It would be a silly waste of time to count each car to make sure that there are 500 cars. You know there are 500 cars, even though you have no direct physical contact with the parking lot. Furthermore, you would possess this knowledge whether the parking lot was in the next town, in Paris, or on Mars. If the information is true—no double parking, the parking lot is full—then you know there are 500 cars in the parking lot.

Mathematics and Reality

Although for Plato the examples would be from geometry, the preceding examples should enable us to understand why mathematics was so important to Plato and other Greeks. For Plato the problem was to get his foot out of the cage of relativism, so to speak. Mathematics was the key. It seemed to provide the stability and certainty needed to defend objectivity. It was also powerful, rational, and mystical. To summarize some of the key characteristics the Greeks would see in the preceding examples, consider the following.

First, there is certainty. If all the information given to us is correct, and our calculations are correct, then we cannot be wrong about the conclusion. Second, the use of mathematics seems to give the user an unusual power of perception. With mathematics as a means we are able to transport our minds to places where we may never set foot physically. Think of the achievement of Eratosthenes in calculating the size of the Earth. In his day there was no physical way for him to travel and experience the Earth's 25,000-mile circumference. Yet by sitting at a desk in the library at Alexandria, he knew. In a sense he could transport his mind around the world while his physical body stayed in one place. Consider also the achievement of the third century B.C. astronomer Aristarchus of Samos, who devised a geometrically correct method for estimating the relative distances of the Sun and the Moon from the Earth. Although his actual results were incorrect, because of the primitive state of measurement, through geometry he "knew" these distances were very large. For the Greeks such achievements meant that one could contact a realm of truth by thinking. They were rationalists, to use the language introduced in Chapter 2. (But modern science, epistemologically based on empiricist philosophy, would not interpret the results of Eratosthenes and Aristarchus in this way.)

Third, although the physical circumstances to which one applied mathematical thoughts could still change, the thoughts themselves, to the Greek mind, seemed to be eternal. The physical universe could cease to exist, but the geometrical theorems used by Eratosthenes and Aristarchus would con-

All things are numbers.

PYTHAGORAS

Let no one without geometry enter here.

Inscription over entrance to Plato's Academy

tinue to exist. The Earth could explode, but the geometric theorem used by Eratosthenes could be applied to another physical sphere. There are an infinite number of physical things that "2 + 2 = 4" can be applied to, and even when the physical things change or disappear, the concept remains the same. Thus, and this is the fourth point, mathematical principles are not things. They are thoughts. Finally, mathematical principles are public entities. The geometry used by Eratosthenes and Aristarchus was not a private collection of truths. Mathematics consists of public ideas, discoverable by, and applicable for, any person, from any culture, at any time, past, present, or future.

But what kind of existence does an idea or thought have? Consider how difficult it must be initially for a child to grasp the concept of "two." A mother may point to a piece of chalk and say repeatedly "chalk . . . chalk." Eventually the child will begin to mimic the mother and associate the verbal expression with the object. But imagine the potential confusion when the mother picks up another piece of chalk and says "two"! The objects were just referred to as "chalk"—now the name is "two"? Where is the object "two"? Imagine the further confusion when she points to two tables or chairs and says "two." Eventually we understand that abstraction and others, and if we are fortunate to have the right teachers, we will understand the beauty of the achievement of Eratosthenes, Aristarchus, and modern scientists who with a few facts can calculate backward in time to the physical circumstances of the first billionth of a second of the existence of a universe 15 billion years old.

So important was mathematics for Plato that it served as a basis for the rest of his philosophy. This was not an idle game. For Plato the entire future of the human race was at stake. Truth and justice were not myths, games, or rationalizations as the sophists implied. He thus founded in the fourth century B.C. the first university of public knowledge, where educated men and women from all over the Mediterranean came for many centuries to study and exchange ideas. It was called the Academy and over its main entrance was an inscription that specified a background in mathematics was required before one could enter.

Plato's main concern, however, was ethical judgment and a political system that would maximize correct ethical judgment. He concluded that if there is a realm of truth that we have access to through thought, then there must also be an objective basis for ethical standards and justice. There must be eternal standards of right and wrong, just as there are eternal mathematical principles. Similarly, knowledge of these standards must proceed in the same way as knowledge of mathematical truths. For me to know "2 + 2 = 4," I must, of course, first experience particular examples of counting. But at some point I do not need to examine any further examples of "2 and 2 is 4." I understand the concept and know that it will always be true. Similarly, to know that any particular action is a just action, there must be a common property of just actions that we are capable of knowing as in the proposition "2 + 2 = 4." By examining particular cases and communicating our opinions, we can obtain a better idea of what justice really is. For Plato, real justice was not a matter of personal opinion, or relative to a culture or time, or a matter of power, but the result of a rational insight. As in the words of the Declaration of Independence, justice was a "self-evident" truth.

Plato's next step was to use the insights gained from mathematics to establish a metaphysics, a theory of reality. Mathematical truths are eternal, applicable at all times and places and true for all people. Yet they are thoughts, ideas, or concepts. We can apply them to the physical world, but they are not physical things. Because they are not physical, they do not change and are not subject to the doubts of experience. Yet they can be known with certainty through thinking and they can be communicated. But if they are not in the physical world, where are they?

According to Plato, the culmination of these insights is the realization that there is another realm of existence, a dimensionless, timeless, nonmaterial reality of pure thought, a realm of ideas. For Plato, the skepticism of the sophists and other paradoxes, such as Zeno's paradox of motion, are simply nature's way of revealing one of her secrets: a realm of true *being*, a realm of self-evident formal truth, one with which we can participate through thought. This realm is contrasted with the world of our normal experience, which Plato called *becoming*. Any statement we make about the physical world is subject to doubt and change, because the physical world is an illusion or shadow of a higher reality. A strange conclusion no doubt, conflicting with common sense, but according to Plato, it is what reason leads us to step by step.

Mind, for anything perception can compass, goes . . . in our spatial world more ghostly than a ghost. Invisible, intangible, it is a thing not even of outline; it is not a "thing." It remains without sensual confirmation and remains without it forever.

SIR CHARLES SHERRINGTON

Common Sense and Reality:
The Allegory of the Cave

In one of his most famous writings, *The Republic,* Plato warns the people of his time, and all future generations, that a true seeker of the truth must deal with the fact that new discoveries will always conflict uncomfortably with common sense. In what has come to be called the "Allegory of the Cave," he asks us to imagine a race of people imprisoned deep in a dark cave, all chained together facing a wall of the cave. Behind them is a large fire and another race of people moving about, such that shadows are cast on the cave's walls. The chained people do not realize that they are captives of the other race. The only reality they have ever experienced is the shadows on the wall. By observing the shadows carefully, they have been able to recognize generalities in the movements of the shadows and have been able to develop a practical science: They can predict what shadows will appear at different times and

Philosophy subverts man's satisfaction with himself, exposes custom as a questionable dream, and offers not so much solutions as a different life.

WALTER KAUFMAN

when food and water will become available, for example. Because their science works, because there is a regularity to their experience, there is little reason for them to believe that this science is false, that what they are seeing is an illusion, an appearance only, reflected from a more fundamental reality.

Suppose now that by accident one person escapes from the grip of his chains enough to turn around and see for the first time what is actually causing the shadows. His first reaction is uncertainty, confusion, and fear. He quickly turns back to what is familiar. All his life he has known only the shadows as real. He must be going crazy. It must be a bad dream. He concludes that he must never turn around again. Eventually though his natural curiosity becomes too much to suppress and he turns around once more. This time, because he is more familiar with the sight, he does not quickly turn back to the shadows. He realizes that he has made a great discovery. With great excitement he turns back to his companions, announces his discovery, and attempts to get them to turn around. At first they laugh at him. Then they reason with him, pointing out that the shadows are what everyone knows to be real and practical. As he persists, they ignore him, concluding that he has lost his mind. At first the force of this community rejection causes him to doubt what he has experienced. Maybe he is crazy. But it seemed so real.

Eventually though he turns for the final time, and with the resolve of an explorer, he frees himself completely from his chains and leaves his people to investigate this new reality. He discovers that he is in a cave, finds the passage out of the cave, and slowly and tentatively makes his way out, eventually reaching the outside world. At first it is too much to bear—so many new things, and the light of the Sun blinds him. When he finally adjusts, he is overjoyed by this simple, clear, beautiful new reality compared to the dim, confining reality of the cave's shadows. He must share this discovery. So back into the cave he goes to try once more to convince his former companions, most likely to no avail.

Mathematics Today:
Why Does It Work so Well?

Today physicists routinely use higher mathematical functions to explore strange dimensions of space and time, dimensions that conflict radically with common sense. Whether these dimensions are real or not is much debated by physicists and philosophers of science. That physicists can conduct such investigations, however, without fear of ridicule is just one of the many leg-

If the doors of perception were cleansed, everything would appear to man as it is, infinite. For man has closed himself up till he sees all things through the narrow chinks of his cavern.

WILLIAM BLAKE

Socrates said he was not an Athenian or a Greek, but a citizen of the world.

PLUTARCH

acies we owe to Plato. Although this legacy is of critical importance, modern philosophers of science believe Plato was wrong about the nature of mathematics and the impossibility of an objective physical science. Because of his need for certainty, Plato, as a rationalist, believed too much about mathematics and not enough about the possibility of a science of the physical world. Advances in mathematics have demonstrated that contradictory mathematical systems can be created and applied to the physical world. The geometry used by Eratosthenes works in mapping the circumference of the Earth, but a different geometry is needed to understand the large regions of space and time as analyzed by Einstein.[1] Plato recognized that mathematical principles could be applied to different hypothetical situations, even contradictory ones, but the mathematical principles themselves were always consistent. As we will see in the next chapter, the geometry of a circle can be used to construct an Earth-centered model and a Sun-centered model of the motion of the planets. Because only the geometry was stable, Plato concluded that only its principles were real. Today the mathematical situation is analogous to having one mathematical system in which "2 + 2 = 4" is true and another where "2 + 2 = 5" is true. It is doubtful that Plato would have had room in his perfect realm of being for contradictory ideas.

For Plato, mathematical ideas were eternal thoughts independent of human minds. The goal was to discover these thoughts and use them to organize the world of physical appearances, or to "save the appearances" to use his own words. Today, most philosophers of science believe that mathematical ideas are the result of free creations of the human mind. They are tools that we create to map the world. Just as a hammer is needed in one situation, a saw is needed in another. Experience is the key. Without some input of initial data derived from experience, our mathematical tools are useless. In the previous example of the parking lot, someone must first experience the parking lot and see that it is a rectangle, that it has 25 rows with 20 parking stalls each, and that it is full with no double parking. If any of this information is incorrect, then the conclusion that there are 500 cars in the parking lot could be wrong. Influenced very much by empiricism, philosophers of science today believe that mathematical equations do not convey any more certainty than is already contained in the initial data. In a sense, the conclusion that there are 500 cars in the parking lot is already contained in our knowledge of the initial data. The mathematics is simply a way of making explicit what

Philosophy [the universe] is written in that great book which ever lies before our eyes. . . . We cannot understand it if we do not first learn the language and grasp the symbols in which it is written. The book is written in the mathematical language . . . without whose help it is humanly impossible to comprehend a single word of it, and without which one wanders in vain through a dark labyrinth.

GALILEO, *THE ASSAYER*

---

[1] In the Euclidean geometry Eratosthenes used, the sum of the interior angles created by the line intersecting two parallel lines must equal 180°. The axioms that are used to prove this also lead to the conclusion that the sum of the angles of a triangle are always 180°. In the non-Euclidean geometry used by Einstein, the sum of the angles of a triangle can exceed 180°.

is implicit in the data. (In this regard, recall that the conclusions of both Eratosthenes and Aristarchus were technically incorrect because of errors of observation.)

What we believe today is not as important for the history of science as what Plato believed. Although perhaps wrong, Plato's conception of mathematics was one of the most influential philosophical positions ever taken in Western culture. It had a profound and lasting influence on both the development of Christianity and modern science. As we will see in the next chapter, by the time of the Renaissance, Plato's eternal realm of being full of harmonious mathematical ideas became the eternal mind of God, and as Plato believed that in doing mathematics we read the realm of eternity, so the Renaissance scientists believed that in doing mathematics we read the mind of God.

Few scientists today will profess that in doing mathematics they feel as if they are reading the mind of God. But when we find them talking about mathematics, and especially its application to understanding the physical world, the legacy of Plato is clear. There is something bewildering still about using mathematics to fly a robot spacecraft to the planet Saturn, at a distance of a billion miles from Earth, and be only 43 miles off upon arrival, or to the planet Uranus, close to 2 billion miles away, arriving on course and one minute ahead of schedule. Why should our mathematical thoughts work? Why should the free creations of the human mind work so well? According to Einstein, this is the greatest cosmological mystery, that our ideas should work at all.

The philosopher Aristotle, a student of Plato's at the Academy, saw too many problems with Plato's other-worldly interpretation of mathematics. According to Aristotle, universal principles such as those of mathematics work because they are built-in *formal* structures, and only human beings are capable of reading these formal structures. They are inseparably linked with the physical universe, and once we understand how they work on Earth, we immediately know something about the entire universe. By the Middle Ages these formal structures had become the floor plan by which God had created the universe, and as God's special creature, we were endowed with the special potential to read this floor plan. Scientists today profess philosophical detachment from such speculations. Nevertheless, indecision and debate continue. Consider the following statements in *Mathematics and the Search for Knowledge*, from some of the best minds of our time.[1]

There is inherent in nature a hidden harmony that reflects itself in our minds under the image of simple mathematical

When the bagel is eaten, the hole does not remain to be reincarnated in a doughnut.

GREGORY BATESON

---

[1] Morris Kline, *Mathematics and the Search for Knowledge* (New York: Oxford University Press, 1985).

laws. That then is the reason why events in nature are predictable by a combination of observation and mathematical analysis. Again and again in the history of physics this conviction, or should I say this dream, of harmony in nature has found fulfillments beyond our expectations.

HERMANN WEYL

The essential fact is simply that all the pictures which science now draws of nature . . . are mathematical pictures. . . . It can hardly be disputed that nature and our conscious mathematical minds work according to the same laws.

SIR JAMES JEANS

Here arises a puzzle that has disturbed scientists of all periods. How is it possible that mathematics, a product of human thought that is independent of experience, fits so excellently the objects of physical reality? Can human reason without experience discover by pure thinking properties of real things?

ALBERT EINSTEIN

Is there perhaps some magical power in the subject [mathematics] that, although it had fought under the invincible banner of truth, has actually achieved its victories through some inner mysterious strength?

MORRIS KLINE

Philosophically, there may well be serious difficulties in Plato's defense of objectivity against Protagoras and the sophists. Assuming the principles of mathematics are universal does not demonstrate necessarily the existence of universal ethical principles. Historically, however, it does not matter. Wrong or right, Plato was believed. He won the intellectual battle with the sophists. Ideas can be discussed and ranked by a community of knowledge seekers willing to acknowledge their limited perspectives. Our modern view of the universe is the result of believing in better ideas.

But how do better ideas come about? How do revolutions in thought take place? How do people break out of the intellectual caves of previous generations? If absolute proof is not possible, how are more reasonable ideas separated from merely conceivable ones? This is the subject of the next chapter.

Concept Summary

If evolution has shaped us, so have the thoughts of others who have come before us. Western culture and its principal analytic tool, science, have presupposed that the universe is a rational, harmonious place with an internal order, knowable by human beings through a process of open inquiry and critical evaluation of competing ideas. With childlike wonder, the ancient Greeks were the intellectual source of such thoughts and the idea that happiness can involve seeking knowledge for its own sake.

Because such a view of life is open-ended and necessitates a philosophical insecurity, its acceptance is often punctuated by views that involve less risk. One reaction to the philosophy of the ancient Greeks was that of Protagoras and the sophists. For Protagoras, open-endedness to the process of seeking knowledge implies a lack of certainty, and lack of certainty implies a lack of knowledge. Hence we do not discover truth, we make it. Each of us is the measure of reality; each of us molds reality from a particular perspective. Everything is relative, and the wise man or woman learns that the game of life does not involve uncovering an objective truth that is already there for all to see, but instead involves an expansion of a perspective by persuading others to live in that perspective. Truth is a myth, and if persuasion, in a given situation, is ineffective, conformity to the prevailing viewpoint is a diplomatic alternative.

Socrates saw such a view as stimulating and sobering, but ultimately questionable, because it implied that any way of life was conceivably good. There must be universal truths and standards, otherwise there would be no way to distinguish good ways of living from bad ones. Protagoras is correct, however, according to Socrates, in that each of us is not only limited—perceiving the world from our own cage of appearances—but also we fail to acknowledge the ignorance of our limited perspectives. Most often, when challenged, we fall back on modes of reasoning (authority, popularity, and tradition) that are actually excuses not to think. Truth exists but it is indeed difficult to obtain, according to Socrates. This is all the more reason we must acknowledge our personal limitations and share our thoughts critically with others.

Plato, a student of Socrates, dedicated his intellectual life to the ideals inherent in Socratic philosophy. Penetrating the problem of knowledge at a deeper level, Plato concluded that the challenge offered by Protagoras was insurmountable unless standards and truths could be found that were "self-evident." Plato thought he saw the key to the whole problem in the essence of a mathematical truth. Mathematical truths seem to be durable principles that remain true regardless of what happens in the physical world; our learning of these principles may be stimulated by our interaction with the physical world, but their self-evident truth is ultimately realized intellectually. At some point we just know that $2 + 2 = 4$ cannot be false.

This premise led Plato to the conclusion that beliefs about the physical world, even our most widely held ones, could be no more than practical, approximate statements, and thus agreeing with Protagoras, that a science of the physical world could never constitute true knowledge. But because truth and knowledge must exist, this in turn led Plato to believe in another realm of existence—a realm of pure thought comprehendible by the human mind, a realm of true being distinct from the shadowy, changing world of physical appearances.

Although Plato's legacy is great, especially his emphasis on mathematics, modern science has adopted a more complex epistemological path. Whereas Protagoras concludes that certainty, and hence knowledge, is impossible, and Plato concludes that certainty, and hence knowledge, is possible, modern science assumes that although certainty is not possible, knowledge is possible nevertheless. Reasonable beliefs about the nature of the physical world are possible, even though we cannot have absolute assurance that our beliefs are true.

Suggested Readings

The Greeks, by Humphrey David Findley Kitto, rev. ed. (New York: Penguin Books, 1957).

A classic brief introduction to the Greek culture and its significance for the development of Western civilization.

Nature and the Greeks, by Erwin Schrödinger (Cambridge, England: Cambridge University Press, 1954).

A short investigation of the scientific world-picture inherited from ancient Greek thinkers. Schrödinger, one of the main contributors to modern physics (see Chapter 8), searches for some insight into the epistemological roots of the present perplexities of modern science.

A History of Philosophy, vol. 1, *Greece and Rome,* by Frederick Charles Copleston, new rev. ed. (Garden City, N.Y.: Image Books, 1962).

Part of an eight-volume history of Western philosophy from the ancient Greeks to the twentieth century. Although there are many good introductions to the history of Western philosophy, this time-honored work is still one of the best in terms of completeness and objectivity.

For a somewhat more readable account, see *A History of Western Philosophy,* vol. 1, *The Classical Mind,* 2nd ed., by William Thomas Jones (New York: Harcourt Brace Jovanovich, 1969–75). This book has a stimulating chapter on the sophists, "Education Through Violence." Throughout, Jones emphasizes that one of the ways of understanding the classical Greek philosophers is from the point of view of their struggle with defending objectivity and a public truth.

Greek Philosophy: Thales to Plato, by John Burnet (New York: Macmillan, 1964).

Although the author died in 1928, the eight reprintings of this book since then attest to the authority it still commands. Packed with scholarly twists and turns, it is still quite readable for the highly motivated student.

The Collected Dialogues of Plato, Including the Letters, ed. by Edith Hamilton and Huntington Cairns (New York: Pantheon Books, 1964).

A must-read primary source for anyone interested in the cultural roots of Western civilization. Out of the barely penetrable mists of history comes the voice of a great intellect, asking the questions and defining the parameters of the answers that have guided our civilization for centuries. Most recommended are the following dialogues: *The Apology, The Republic, The Theaetetus,* and *The Protagoras.* Also see the *7th Letter* for a surprising conclusion.

CHAPTER 5

Cultural Roots:
2. Science and Religion— The Copernican Revolution

The story of the Copernican Revolution is not . . . simply a story of astronomers and the skies. . . . No fundamental astronomical discovery, no new sort of astronomical observation, persuaded Copernicus of ancient astronomy's inadequacy or of the necessity for change.

THOMAS S. KUHN, *THE COPERNICAN REVOLUTION*[1]

No experience whatsoever could prove that the heavens rotate daily and not the earth.

NICHOLAS ORESME, BISHOP OF LISIEUX, 1377

Whether a man is on the earth, or the sun, or some other star, it will always seem to him that the position that he occupies is the motionless center, and that all other things are in motion.

NICOLAS DE CUSA, BISHOP OF BRIXEN, 1450

When we encounter abstract intellectual discussions on the conflict between science and religion, or between science and philosophy, or between the scientific and artistic temperament, "science" is nothing more than an abstract category. But science, we should remember, is a human activity, and as such it cannot be divorced from other aspects of culture. In introductory texts science is often idealized, as if it were some huge computerized, noncreative machine marching relentlessly along grabbing up facts and truth, always untouched by prejudice, dogma, religion, and philosophy. Authors of other popular books on science in their excitement to emphasize the joy of growth and understanding—the unlimited potential of the human mind—often unrealistically glorify science, as if this activity were radically different from everything else human beings do. Not only is this oversimplification faulty but it is also a disservice to science because many people become alienated from science as a result. In this chapter we will be interested in one of the most intriguing, and much debated, aspects of the scientific enterprise, the creation of ideas. As we will see, this is hardly a cold, logical process.

In many ways a scientist is no different from a priest, an artist, or a businessperson. We all face the same problems and uncertainties that require a creative response, even though they may take different forms. There are human factors behind every scientific discovery. Although some general science textbooks may touch on the influence of these factors, few consider the *epistemological significance*. That is, to what extent do these factors determine what is considered a fact, knowledge, and truth at any given time? In this chapter it may seem that we are supporting the relativism of Protagoras and the sophists: that scientific truth is a myth, that every belief has its hidden agenda being no more than an expression of human bias, and that every age has its own truths. However, our aim is not to support relativism. The philosophical and religious conceptions of a time may influence scientific interpretations, but ultimately, in a curious and unexpected way, they actually help scientific growth. Our point will be that the cultural background of a time is indispensable to scientific method because it supplies the pool of ideas without which we would be blind to the significance of observational facts.

World views are social constructions and they channel the search for facts. But facts are found and knowledge progresses, however fitfully.

STEPHEN JAY GOULD

---

[1] Thomas S. Kuhn, *The Copernican Revolution: Planetary Astronomy in the Development of Western Thought* (Cambridge, Mass.: Harvard University Press, 1957), pp. 76, 131. Copyright © 1957 by the President and Fellows of Harvard College; 1985 by Thomas S. Kuhn. Reprinted by permission.

Plato's Homework Problem:
The Problem of the Planets

It is ironic that religious fundamentalists feel science presupposes an antireligious philosophy. History shows the opposite to be true. Prior to Plato, Pythagoras realized that mathematics could be used to describe and understand the motions we experience of the world. This amazing correlation between mathematics and what happens in the world served as a basis for a religion for Pythagoras and his followers. The world was ordered, and mathematics was the way to read that order. Mystical powers and truths were revealed to those who could cipher and calculate. It was not long before a very durable notion emerged from this: Mathematical symmetries are the language of universal design and harmony; when one studies mathematics, one studies the mind of God. This belief, which we will call *Pythagoreanism,* dominated the minds of knowledge seekers in one way or another for many centuries and was a crucial factor in the discoveries that gave birth to our modern view of the universe. In attempting to understand the universe, astronomers argued not only about what the observational facts demonstrated but also whether the mathematical devices used in the proposed theories were sufficiently pleasing aesthetically. They were convinced that God would construct a harmonious and symmetrical universe, a simple universe absent of superfluous, ugly details. In the words of Rheticus, an early supporter of the sixteenth-century ideas of Copernicus, "Should we not attribute to God, the Creator of nature, that skill which we observe in common makers of clocks? For they carefully avoid inserting in the mechanism any superfluous wheel, or any whose function could be served better by another with a slight change of position."

Because of their faith in order, the ancient Greeks were intrigued by the motion of the planets. Their use of the word *planet,* Greek for "wanderer," implied that the motions of the planets were unordered and irrational. For centuries ancient astronomers from several cultures had noted that the stars, the Sun, and the Moon all move east to west uniformly during a day or night, and the Sun and Moon also move in a uniform eastward motion during the course of a year or month. In other words, in relation to the stars, both the Sun and Moon are not in exactly the same place at the same time the next night but at a slightly eastward position. Ancient astronomers also noted that five points of light, looking in some respects much like stars, also move in an overall eastward direction through the course of a year, but wander curiously from time to time in a westward direction. Thus, in the course of several months one of these points of light would move eastward, change directions and loop backward in a westward direction, and then switch directions again, looping back eastward (see Figure 5-1).

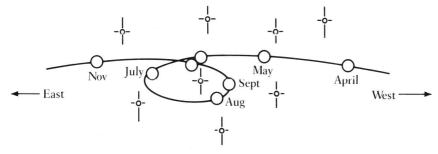

FIGURE 5-1

In plotting the course of the planets, ancient astronomers noticed that in addition to an overall eastward motion in the course of a year, planets also "wander" by changing directions and looping backwards in a westward direction. The number of these retrograde motions varies depending on the planet.

Although observations of this retrograde motion were recorded as early as 1900 B.C. by astronomers, prior to the Greeks there was little urgency to understand this motion. From a purely practical point of view—planting crops and navigating, for example—a precise knowledge of the positions of the stars, Sun, and Moon was sufficient. For the Greeks, however, the problem of the planets became more significant. Their view of mathematics and an ordered cosmos led them to believe that nature was again flaunting one of her many secrets.

Presupposing as they did a rational cosmos, this appearance of irrationality must be only an appearance. Thus, Plato, so we hear from legend, offered a perennial homework problem to the students of the Academy: Find a geometric scheme that would explain the apparent motion of the planets. Primitive man wanted predictive control over the heavens because the ability to predict the seasons was essential for survival. The philosophically minded Greeks were, of course, still interested in the practical application of this knowledge. Accurate navigation and a precise calendar were very important. But now more was at stake, one's world view. Because of Plato's great influence, several geometric models were soon constructed.

But before we look at these solutions, a thorough appreciation of the problem is essential. A successful geometric model had to give an account of numerous complex relative motions with precise locations and precise times for completing these motions. The Sun has three motions: its daily motion, east to west; its eastward motion in relation to the stars; and a yearly north-south motion, noticeable in the middle-northern latitudes as a change in seasons where the Sun is higher (more northerly) in the summer sky and lower (more southerly) in the winter sky. Also, any explanation must account for the Sun completing these motions and returning to any given position in just over 365 days. The Moon not only has an east-west nightly motion, a monthly eastward motion, and an even larger north-south motion than the Sun but phases as well. The appearance of the Moon changes perceptively as it moves, such that successive full moons will not occur in the same place.

Then there are the planets, only five of which are visible to the naked eye. When they are visible, they move nightly with the stars east to west. Over

The most incomprehensible thing about the world is that it is comprehensible.

ALBERT EINSTEIN

the course of a year, they lose ground to the stars in an eastwardly direction, as the Sun does. In addition, although the overall motion in the course of a year is eastward in relation to the stars, a noticeable retrograde westerly motion also occurs. Further compounding the problem of finding a simple geometric model to account for all this is the fact that the eastward motion of the planets occurs at different rates for each planet, and the number of retrogressions per planet year are different depending on the planet. For instance, Saturn moves the slowest in an eastward direction, almost keeping pace with the stars, and shows 28 retrogressions in the amount of time it takes to complete its eastward motion and return to its original location. Jupiter moves a little faster and shows 11 retrogressions, one approximately every 200 Earth days. On the other hand, Mercury moves much faster and shows 1 retrogression every 116 Earth days.

There are many complexities beyond this that need not detain us. It is clear, however, that only a strong faith in the order of the cosmos could sustain the students of Plato in tackling this problem. Many popular treatments of science prior to the seventeenth century imply that ancient civilizations failed to understand the real universe because they were dominated by a religious and philosophical dogmatism. On the contrary, the religious and philosophical ideas of the ancient Greeks encouraged precise observation of the eccentric motion of the planets and sustained the belief that the use of reason would eventually result in an explanation. The solution that dominated the minds of most astronomers up to the sixteenth century was a geocentric (Earth-centered) cosmology—a view we know to be false today. The Earth does move, and the Sun is the center of our solar system. An accurate understanding of the evidence available to ancient astronomers, however, shows that their acceptance of an immovable Earth was not unreasonable or unscientific. Their problem was to not only account for the observations of the motions of the planets but to also find ideas that made sense, given the state of knowledge at the time on other matters.

Science is facts. Just as houses are made of stones, so is science made of facts. But a pile of stones is not a house and a collection of facts is not necessarily science.

HENRI POINCARÉ

Solutions to the Problem of the Planets

By the second century B.C. several geometric models were offered as explanations for the problem of the planets. These were great intellectual achievements—the result of only the naked eye, human imagination, and intelligence used to solve Plato's homework problem. Countless hours worth of observational data of the movements of the Sun, Moon, planets, and stars had to be studied and checked. Earlier recorded planetary positions often were found to be inaccurate. Then through imagination one had to invert one's perspective and see how it would look to a god.

Note, however, that prior to Plato the intellectual background for creative thinking on this problem was rich with cosmological possibilities. The most accepted opinion, and the easiest to reconcile with common sense, was a two-sphere universe with the stationary Earth as the center sphere and the stars as the outer sphere revolving around the stationary Earth. But as early as the fifth century B.C. the Greek atomist Democritus proposed a universe of infinite space. In such a universe there would be no center, no unique position, and every astronomical body including the Earth would be in motion. There would also be an infinite number of suns and earths. As modern as this view may seem, there was little scientific reason for accepting it at the time, and it was based not on observation but on a "logical deduction" from the atomist's metaphysics, the belief that reality consisted of an infinite number of atoms moving in an infinite space. As with the origin of many ideas in science, it remained in the background waiting to be plucked if needed.

The followers of Pythagoras suggested a second cosmological possibility. Rather than placing the Earth at the center of all motion, the Earth moved around an immense central fire, known as the Altar of Zeus. This central fire, it was believed, could not be seen by people, because as the Earth moved it always kept populated areas in a direction away from the position of this central location. Finally, in the fourth century B.C. Heraclides suggested a third possibility. Rather than the Earth being absolutely stationary at a central location and the stars revolving around it, he proposed that the Earth rotated and the stars were stationary. Also, although the Sun revolved around the Earth, the planets Mercury and Venus revolved around the Sun rather than the Earth.

Thus by the time of Plato and the Academy, all three ideas usually associated with our modern view of the universe had been proposed: a universe in which the Earth is not unique, a revolving Earth, and a rotating Earth. All of these views, however, originated from metaphysical and cosmological concerns. Plato's interest in mathematics caused a concentration on the practical problem of predicting the locations of the planets and simply saving the phenomena. By modern standards this is a scientifically mature approach. Although Plato was not interested in observing the physical world for its own sake, the attention he drew to the problem of planetary motion caused later generations of astronomers to pay attention to the smallest observational details of each planet's motion.

Eudoxus (408–355 B.C.) offered one of the first solutions in response to Plato's challenge. He developed a system that consisted of a stationary Earth at the center and a series of homocentric interconnected spherical shells that carried the Moon, Sun, planets, and stars around the Earth in a complex way. All together he proposed 27 perfect circular motions: 1 for the fixed stars, 3 each for the Sun and the Moon, 4 each for the 5 visible planets. By having these spheres move at different rates and in opposite directions, the complex motions noted earlier could be explained and made mathematically

No one in his senses, or imbued with the slightest knowledge of physics, will ever think that the earth, heavy and unwieldy from its own weight and mass, staggers up and down around its own center and that of the sun; for at the slightest jar of the earth, we would see cities and fortresses, towns and mountains thrown down.

JEAN BODIN, Sixteenth-century political philosopher

predictable.[1] As the system was used to make observations, further spheres were added by the followers of Eudoxus to account for observed discrepancies. Callipus added one more for each body, and Aristotle, as part of a physics to account for a mechanics of the real motion of the heavenly bodies, added another 22 spheres for a total of 56 circular motions. Although complex, each circular motion conformed to the Pythagorean rule thought to be essential for a truly harmonious motion: Each motion was perfectly *circular* with the physical Earth at the exact center, and the motion of each heavenly body was perfectly *uniform*, moving at the same speed at all times around the central Earth.

The Eudoxian system did not, however, account for one important observation. As they retrogress, planets appear brighter as if they were closer to the Earth. This was not easily accounted for in the Eudoxian system, in which each planet is always the same distance from the Earth. Thus, Apollonius and Hipparchus developed a modification in the third century B.C. They proposed what came to be a very durable astronomical device, the *epicycle*. In this system, with the Earth in the center, a planet is carried in its eastward motion by a large circle, called a *deferent,* while at the same time the actual planet is on a smaller circle, the epicycle, revolving around a central point on the deferent. (See the Ptolemaic use of this device in Figure 5-3, p. 145.) By varying the relative sizes and speeds of epicycles and deferents, the complex motion of the planets could be accounted for, as well as the fact that planets appear brighter when they retrogress because they would be closer to the Earth when they move in a westward direction.

There was, however, an immediately recognized problem with this system—a problem, as we will see, that was of immense importance in the development of our modern view of the planets. Although the planets' motion was uniform relative to the central Earth, it was not in relation to the center of the epicycle, and there was no physical body at the center of the epicycle, only a mathematical point. The idea of a perfectly circular and uniform motion was inseparable from the idea of a mathematically harmonious cosmos. Although the epicycle-deferent system accounted for retrograde motion and other important observations, it was not mathematically elegant.

By the middle of the third century B.C., a third view had been proposed by Aristarchus of Samos, the "Copernicus of antiquity." Although his was a heliocentric (Sun-centered) system, little else is known about its details other than that the Sun was at the center of a greatly expanded sphere of the stars

---

[1] To get a feel for how this system made sense of the planetary observations, imagine being at the stationary center of a merry-go-round and watching a ticket-taker move around it in the opposite direction of the merry-go-round's motion. Imagine plotting the course of the ticket-taker against the background as he moves with the motion of the merry-go-round but against the motion of the objects on the merry-go-round.

and the Earth moved around the Sun in a perfect circle. We don't know whether this system could mathematically account for the observed motions of the planets better than or even as well as Earth-centered systems. Even assuming that it did account for planetary motions as well, almost all thoughtful astronomers of antiquity did not take this and other modern-sounding cosmologies seriously. As Thomas Kuhn has noted so thoroughly in *The Copernican Revolution:*

The reasons for the rejection were excellent. These alternative cosmologies violate the first and most fundamental suggestions provided by the senses about the structure of the universe. Furthermore, this violation of common sense is not compensated for by any increase in the effectiveness with which they account for the appearances. At best they are no more economical, fruitful, or precise than the two-sphere universe, and they are a great deal harder to believe. . . .

All of these alternative cosmologies take the motion of the earth as a premise, and all (except Heraclides' system) make the earth move as one of a number of heavenly bodies. But the first distinction suggested by the senses is that separating the earth and the heavens. The earth is not part of the heavens; it is the platform from which we view them. And the platform shares few or no apparent characteristics with the celestial bodies seen from it. The heavenly bodies seem bright points of light, the earth an immense nonluminous sphere of mud and rock. Little change is observed in the heavens: the stars are the same night after night. . . . In contrast the earth is the home of birth and change and destruction. . . . It seems absurd to make the earth like celestial bodies whose most prominent characteristic is that immutable regularity never to be achieved on the corruptible earth.

The idea that the earth moves seems initially equally absurd. Our *senses* tell us all we know of motion, and they indicate *no motion for the earth* [emphasis added]. Until it is reeducated, common sense tells us that, if the earth is in motion, then the air, clouds, birds, and other objects not attached to the earth must be left behind. A man jumping would descend to the earth far from the point where his leap began, for the earth would move beneath him while he was in the air. Rocks and trees, cows and men must be hurled from a rotating earth as a stone flies from a rotating sling. Since none of these effects is seen, the earth is at rest. . . .

The Greeks could only rely on observation and reason, and neither produced evidence for the earth's motion. Without the aid of telescopes or of elaborate mathematical arguments that have no apparent relation to astronomy, no effective evidence for a moving planetary earth can be produced. The observations available to the naked eye fit the two-sphere universe very well (remember the universe of the practical navigator and surveyor), and there is no more natural explanation of them. It is not hard to realize why the ancients believed in the two-sphere universe. The problem is to discover why the conception was given up.[1]

Thus, by the end of the third century B.C., astronomers had four mathematical models to choose from: the orthodox geocentric model of Eudoxus, the modified system of Apollonius and Hipparchus, the partial heliocentric system of Heraclides, and the complete heliocentric model of Aristarchus. All of these systems worked: The motions of the planets and their locations could be accounted for approximately. All had certain advantages, and all failed in some way or another. The Eudoxian system accorded well with common-sense observations of the Earth's apparent stationary position, used perfect circles, and explained planetary retrogression. But it failed to explain why planets appear brighter when they retrogress. The epicycle-deferent system of Apollonius and Hipparchus accounted for planets appearing brighter when they retrogress, but violated the Pythagorean notion that all bodies must move uniformly about a central point. The Heraclidean system also accounted for the increased brightness of a retrogressing planet, especially Venus and Mercury, but left unexplained how the Earth could rotate without flying apart. Also because the Sun in this system revolved around the Earth as Venus and Mercury revolved around the Sun, there was no explanation for how the Sun's orbit could pass through the orbits of Venus, Mercury, and the Moon without major physical problems. Finally, as Copernicus showed centuries later, the system of Aristarchus could also account for the increased brightness of retrogressing planets, and be made mathematically accurate, but no more mathematically accurate than the epicycle-deferent system. [See the later Tychonic version of the Heraclidean system, Figure 5-6 (p. 149), and the Copernican explanation of retrogression, Figure 5-4 (p. 147).] As Kuhn has pointed out, however, it leaves in its wake numerous unexplained physical problems.

---

[1] Thomas S. Kuhn, *The Copernican Revolution: Planetary Astronomy in the Development of Western Thought* (Cambridge, Mass.: Harvard University Press, 1957), pp. 42–44. Copyright © 1957 by the President and Fellows of Harvard College; 1985 by Thomas S. Kuhn. Reprinted by permission.

Instrumentalism, Realism, and Paradigms

Several systems to choose from, but which one is right? What is the real physical universe like? For a true Platonist it did not matter! The question was irrelevant. As we have seen, Plato developed a metaphysics and epistemology that essentially ruled out the possibility of ever answering this question. Plato recognized the limitations of empirical knowledge and, because of the influence of Protagoras, he concluded that a complete knowledge of the physical world was impossible. No matter how much evidence one has for a generalization about the physical world, that generalization may still be shown to be false some day. Moreover, false empirical generalizations can "work." Accurate predictions can be made that are observed in the world of our experience. Plato recognized the logical problem discussed in Chapter 2: True conclusions can be deduced from false premises.

A modern follower of Plato could add to this discussion by noting that given a finite set of observations, be it planetary positions or data from laboratory experiments, an infinite number of theories can, in principle, be constructed to account for these observations. The only practical limit is human creativity. Given new observations, only one theory may still work of the original infinite set. But with these new observations, a new infinite set of mathematical models will work, including the old one. This can continue, in principle, ad infinitum. Each time new observations are made, we could create a new set of conceivable systems.

A mathematical analogy illustrates the problem of separating the reasonable from the merely conceivable (see Figure 5-2). Suppose we want to find a mathematical equation that will allow us to draw a line on a graph that will pass through the points $(0,1)$, $(1,2)$, and $(2,3)$. The simple equation

$$(1) \quad y = x + 1$$

will work. When x is 0, y will equal 1; when x is 1, y will equal 2, and so on. But if we are creative enough we can conceive of another equation that will also work for these points. For instance, the equation

$$(2) \quad y = x^3 - 3x^2 + 3x + 1$$

will also allow us to draw a line on a graph that will pass through these points.

Having two lines that cover the same points is analogous to a common situation in the history of science, where a series of factual observations have been made and two completely different theories both account for the obser-

FIGURE 5-2

Analogous to our conceptual attempt to "cover" observational facts with theories, given any number of points on a Cartesian graph an infinite number of lines can be theoretically constructed to cover or connect the points. Given the points $(0,1)$, $(1,2)$, and $(2,3)$ both equations (1) and (2) connect the points. By adding another point, $(3,4)$, equation (1) will connect this point, but not equation (2). However, a new equation (3) can be created that also connects all the points. (Graph by James C. Reeder, Honolulu Community College)

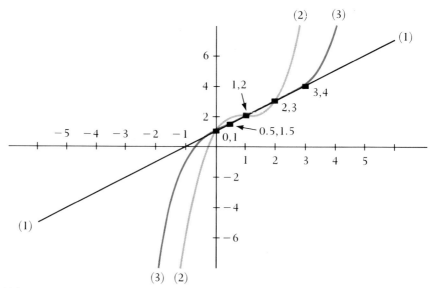

$(1)\ y = x + 1$

$(2)\ y = x^3 - 3x^2 + 3x + 1$

$(3)\ y = \frac{1}{72}\,(2x^5 - 13x^4 + 28x^3 - 23x^2 + 78x + 72)$

vations. A crucial experiment or observation usually follows, such that one of the theories is confirmed and the other is disconfirmed. At first, observations may be made that seem to favor one theory over the other, but the difference is not yet sufficient for the scientific community to be convinced that one theory is clearly superior. For instance, in our graph analogy suppose we want our line to also cover or pass through the point $(0.5, 1.5)$. Only the first equation allows us to draw a line that passes exactly through this point, but the second passes through the point $(0.5, 1.875)$, which from a practical point of view may be close enough to the other point. Ideally, eventually enough observations are made and agreement is reached. If we wanted our line to pass through $(3,4)$, then only the first equation will work. When x is 3, y is equal to 4 in the first equation, but y is equal to 10 in the second. In this way objective agreement is established that only the first equation covers or connects the facts. Unfortunately, the equation

$$(3)\quad y = \frac{1}{72}(2x^5 - 13x^4 + 28x^3 - 23x^2 + 78x + 72)$$

also will allow us to draw a line through the crucial points $(0,1)$, $(1,2)$, $(2,3)$, $(0.5, 1.5)$, and $(3,4)$. Given any set of points, in principle there will always be an infinite number of possible equations that will connect these points.

That a crucial experiment has confirmed one theory over another gives us no guarantee; a better theory may exist than the previous competitors to account for the very next observation.

How can we ever separate the reasonable from the merely conceivable, if infinite theories are always conceivable? The problem of the planets was similar to this mathematical situation. There were several competing theories, using different mathematical devices, that explained or covered the facts of planetary motion. All had strengths and weaknesses. Even if eventually one was found to be the best, how would it be known whether this was just temporary? Perhaps a better theory had not been thought of yet. Plato and his followers saw no way out of this problem. Thus, they believed that more than one model can work in describing planetary motions and that, in general, the best we can do is have models of the physical world that work.

Today this epistemological position is known as *instrumentalism* or sometimes as operationalism. This view states that scientific theories, especially those that involve abstract mathematical devices, are tools, instruments, or calculation devices and should *not* be interpreted as real. For instance, to plot the course of a projectile, we use a quadratic equation to predict its motion. We all know that when someone shoots a cannon, the ball goes up and curves back to Earth. The equation will describe this motion accurately. But the equation has two solutions. If taken literally, it also describes the motion going backward and curving through the solid Earth! Because we have never witnessed projectiles going backward through the solid Earth, the instrumentalist asserts that the equation is a device that enables us to predict where the cannonball will land, but that it should not be interpreted literally. We should not think of the mathematics as describing the actual motion of the cannonball.

In Chapter 8 we will cover the exciting and mysterious field of quantum physics. We will see that scientists use an abstract mathematical equation to describe the motion of an electron. If the equation is interpreted literally, the "real" electron would not be anything like a normal object. The equation describes the electron as a bizarre "smear" of energy that spreads from the small space of an atom to infinity! If this were true, it would mean that part of each of us is on Mars right now!

Because of the paradoxical nature of what the mathematics describes literally, most physicists choose not to deal with the problem of what an electron really is, opting for the instrumentalist position and dealing strictly with correlating the mathematics with the outcome of complex experimental arrangements. This implies, however, that not only is *physics* (from the Greek word meaning "the real") no longer in the business of telling humanity what kind of a world we are living in but also that we do not know exactly what an atom is, what it consists of, or whether it even exists.

Strange indeed. But not for Plato. Plato, so his modern followers claim, would have predicted this outcome: We will never be able to pin down an exact physical reality, because we are not dealing with an exact physical re-

If the purpose of scientific method is to select from among a multitude of hypotheses, and if the number of hypotheses grows faster than experimental method can handle, then it is clear that all hypotheses can never be tested. If all hypotheses can never be tested, then the results of any experiment are inconclusive and the entire scientific method falls short of its goal of establishing proven knowledge.

ROBERT M. PIRSIG, ZEN AND THE ART OF MOTORCYCLE MAINTENANCE

ality, but rather a shadowy, changing, incomplete, and imperfect appearance. Thus, any consistent follower of Plato in antiquity would conclude that there are many potential models for saving the phenomena of planetary motion, and even when one is found that works best, its mathematics should not be interpreted as describing the real motion of the planets. Plato and his followers concluded that the physical world was an illusion, that another dimension of ideas existed that we could participate with via mathematics and abstract thought. Once we discover the harmonious truths of this other dimension, we can then use these truths as practical tools in this changing realm of confusing appearances. But they will never work perfectly because we live in a shadowy world of imperfection.

For Aristotle, initially a student of Plato, Plato went too far in separating the formal truths of mathematics completely from the physical world. As we saw in Chapter 4, Aristotle taught that mathematical formulas were the formal relationships existing between physical bodies and they could not be separated from the physical world. In addition, Aristotle was a *realist*. An epistemological realist believes that scientific theories and the mathematical devices that work best, that accurately predict observational results, are not mere tools but can be said to characterize the way the world is. For Aristotle then, at most one of the systems describing planetary motion could be correct.

Aristotle was aware of the system of Heraclides and the ideas of Pythagoras and Democritus. Aristotle chose the geocentric view of Eudoxus as the real one for three reasons. First, the Eudoxian system was at the height of its popularity. Second, the observational evidence available supported it. Third, and most importantly, Aristotle developed a theory of motion and a physics that necessitated, he thought, an Earth-centered universe. According to Aristotle, the natural motion of a weighted physical body would be toward the center of a circle, so only a deviation from this natural motion would need to be explained further by reference to a force. Rocks naturally fall toward the center of the Earth, and throwing a heavy object into the air requires a force. If the Earth moved, thought Aristotle, then there would be no way to account for these simple facts.

Another important consideration for Aristotle and all astronomers of antiquity was Pythagoreanism. It was well accepted that whatever scheme was put forth, the best scheme must not only account for the facts but must also fulfill the Pythagorean requirement that the speed of each planet be *uniform* and move in a *perfect circle*. The real universe must follow the most harmonious mathematical path; the most aesthetically pleasing shape must be the true path. This is an example of what philosopher of science Thomas Kuhn has called a *paradigm*. Every historical period has its world view, its set of background beliefs that govern a great deal of intellectual behavior. Scientists often argue over which view best fits the facts and which view best fits the accepted world view of the time. During Aristotle's time, the Eudoxian model best fit the paradigm. It consisted entirely of perfect circles and uniform motions.

When intersected by a plane, the sphere displays in this section the circle, the genuine image of the created mind, placed in command of the body which it is appointed to rule; and this circle is to the sphere as the human mind is to the Mind Divine.

JOHANNES KEPLER

Ptolemy:
The Completion of the Geocentric Universe

As so often happens, an idea that is "fruitful," to use Kuhn's term, in focusing our attention on new phenomena is also fruitful in leading to its own destruction. Aristotle's views proved to be very persuasive. In spite of not being able to account for the apparent increase in brightness of retrogressing planets, the combined Aristotelian-Eudoxian system accounted for so many other things, it served as the basic guide for astronomy for several centuries. But by the first century many observational inconsistencies with the orthodox Eudoxian model were known. Hence, Ptolemy (A.D. 85–165), living in Alexandria, a city still heavily influenced by Greek thought, developed a modified geocentric system that maintained the focus of Aristotelian physics and governed astronomical thinking for the next 1,400 years.

This system was essentially a greatly extended version of the epicycle-deferent system of Apollonius and Hipparchus. With the Earth the approximate center of all celestial motion, the planets revolved around the Earth in their westward and eastward motions and also on an epicycle around an invisible point. Now, however, to account for observational discrepancies, there were not only epicycles upon epicycles but also two other geometric devices, known as *eccentrics* and *equants*. By using the eccentric, Ptolemy had the Earth displaced slightly from the exact central point of a circle. The central point was in turn placed on a circular orbit that revolved around the Earth. Thus, a celestial body revolved on an epicycle around a central point on another circle, which revolved around a central point, which in turn revolved around the Earth! Using the device of the equant, the motion of the celestial body was not uniform relative to the central Earth, but rather to a displaced point. See Figure 5-3.

Altogether Ptolemy used over 80 circles and various combinations of epicycle-deferents, eccentrics, and equants in representing the motions of the Sun, Moon, stars, and 5 known planets. The details need not detain us. What is important is that this convoluted geometric scheme was much more successful than all previous systems in predicting the motions of the planets. In fact, this system can still be used today with only a single degree of error for many calculations. Although the various devices used were not original with Ptolemy, as Kuhn has noted,

Ideas not only offer direction for viewing the world, but veil the world as well.

Ptolemy's contribution is the outstanding one, and this entire technique of resolving the problem of the planets is appropriately known by his name, because it was Ptolemy who first put together a particular set of compounded circles to account . . . for the observed quantitative regularities and

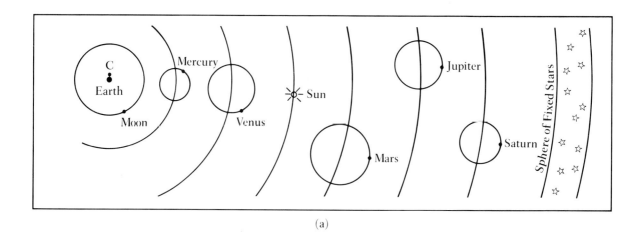

(a)

FIGURE 5-3

The Ptolemaic universe (a) with a central Earth and an *epicycle-deferent* system for each planet. To account for as many observations as possible (b), the Earth is displaced from a central point C. This device, known as an *eccentric,* has a planet on an epicycle revolving around a central point on a deferent, which in turn revolves around C, which in turn revolves around the Earth. Also, the planet revolves uniformly in relation to an *equant* point, rather than in relation to the Earth. Further complicated variations of this scheme could include epicycles on epicycles and eccentrics on eccentrics.

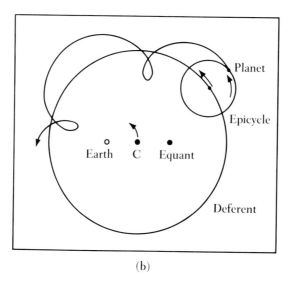

(b)

irregularities in the apparent motions of all the seven planets (Sun, Moon, and five planets). His *Algamest,* the book that epitomizes the greatest achievement of ancient astronomy, was the first systematic mathematical treatise to give a *complete, detailed,* and *quantitative* account of all the celestial motions. . . . For its subtlety, flexibility, complexity, and power the epicycle-deferent technique . . . has no parallel in the history of science until quite recent times. In its most developed form the system of compounded circles was an astounding achievement.[1]

But was it real? Did the planets truly move this way? Although the Ptolemaic system used only circles there was a troubling violation of the uniform motion paradigm. By using the equant, not every circular motion would be uniform relative to its center. Furthermore, Aristotle had proposed that as it circled the Earth, the mechanism of support for each planet was a crystalline sphere. How could the planet circle a point on this physical sphere without crashing into it? Aristotle had developed the best physics and Ptolemy had developed the best astronomy, but they were not completely compatible. So the majority of astronomers and natural philosophers did what most scientists would do today: They accepted what made sense (a central Earth) as real and used what did not make sense (the physical epicycle) as a "calculation device," as a tool for making predictions.

This viewpoint dominated much of the Middle Ages. The Catholic Church, the most powerful political and intellectual force of the time, adopted the Earth-centered model not only because of the authority of Aristotle, common sense, and scripture—where else would God put his special creatures but in the center of things?—but also because it worked. The puzzling mathematical aspects were relegated to instruments of prediction.

Copernicus:
The New Heliocentric Universe

By the sixteenth century, the general paradigm that guided the educated person can be described as follows. Man, as God's special creature, was the center of the physical universe in several ways. The Earth was the physical center of a mathematically planned universe, and man was given the precious gift of being able to read this mathematical harmony. Man could know of God's work through faith and through reason. Unlike the attitude during the Dark Ages, there now existed within the church-dominated intellectual circle of opinion an unbounded faith in the power of human reason to solve the problems of the natural world. By this time the Church had an established tradition of supporting scientific research.

---

[1] Thomas S. Kuhn, *The Copernican Revolution; Planetary Astronomy in the Development of Western Thought* (Cambridge, Mass.: Harvard University Press, 1957), pp. 71–72. Copyright © 1957 by the President and Fellows of Harvard College; 1985 by Thomas S. Kuhn. Used by permission.

Historically, we call this time the Renaissance. There was a great intellectual excitement over the rediscovery or "rebirth" of ancient thought. Of particular interest was a renewed focus on Platonism and Pythagoreanism. Nowhere was this excitement more evident than in the life and work of Nicolaus Copernicus (1473–1543). According to Copernicus, following the ancient masters, an adequate scientific astronomy must satisfy two conditions: It must "save the phenomena"—account for the observed motions of all celestial bodies—and it must not contradict the Pythagorean axioms that the motions of the celestial bodies were circular and uniform. On both accounts, according to Copernicus, Ptolemaic astronomy had failed.

In the many centuries that separated Ptolemy and Copernicus, the Ptolemaic system had undergone many modifications. As more and more observational discrepancies were discovered, more and more epicycles, eccentrics, and equants were added to save the appearances. Thus, for Copernicus, under the influence of a renewed interest in the importance of mathematical harmony and economy, the Ptolemaic system could not possibly be true. The future of mankind and the scientific revolution that bears his name depended *not* upon any new great observational discovery but on the fact that to Copernicus, the Ptolemaic system was *not pretty*. It was not aesthetically pleasing. Surely God could do better than this.

But why choose the Sun as the central figure in a new universe? Why did Copernicus spend hour after hour, day after day, over the course of a lifetime developing the right mathematics to work with a heliocentric system? Why not a new mathematics for a geocentric system? Why not a new mathematics for a modified system like that of Heraclides? As is so often the case in science, Copernicus focused his attention on one possibility, not because of a great observational discovery, but because of a philosophy. The facts are important, but they often come after a commitment has already been made to an idea.

A later version of Platonism, referred to historically as Neoplatonism, combined elements of Christianity and Platonism and made popular the belief that a vital figure such as God, although eternal and nonmaterial, would have a "materialized copy" of Itself. Just as God was a creative force of immense potency responsible for sustaining all life, so the Sun, responsible for light, warmth, and fertility, could be the only appropriate material manifestation of God. According to Copernicus, highly influenced by this Neoplatonic belief, "in this most beautiful temple" of a universe, there is no better place but the center to place this "luminary . . . from which He can illuminate the whole at once." In other words, Neoplatonism demanded that man be replaced with God as the central concern.

Thus, with the Sun as the center of attention and a rigorous fulfillment of the Pythagorean requirements of perfectly circular and uniform motion, Copernicus devised a model that reduced the number of circles needed to account for planetary appearances from well over 80 to 48, a brilliant achievement in simplification. From Copernicus's point of view we either attribute three basic motions to the Earth—a yearly revolution around the Sun, a daily rotation on its axis, and a small gyration of the Earth's axis—or five times that many by having each planet possess three basic motions. God would have surely chosen the simpler plan, just as a watchmaker chooses the simplest design.

Simplification, however, is a relative notion. Each system, the Ptolemaic and the Copernican, relied on numerous minor circular motions. To maintain the Pythagorean presuppositions and account for the most recent planetary observations, the complete Copernican system not only required 48 circles but also some of the same questionable mathematical devices of the Ptolemaic system, such as deferents, epicycles, and eccentrics. In fact, strictly speaking, the Copernican system was not even Sun centered! To account for the observation of the Sun, three circles were needed to describe the motion of the Earth: one for the Earth, one for a central point of the Earth's orbit, and one for another central point for the circle of the central point of the Earth's orbit. The central point for the circle of the Earth's central point in turn revolved around the Sun. A similar complex relationship of circles was needed to account for the observations of the planets, even though retrogression was accounted for in a more straightforward way (see Figure 5-4).

What was gained from all this? No equants violating Pythagoreanism, fewer circles, hence astronomical computations were easier, but—and from a purely empirical point of view this is most important—predictions of planetary locations were no more accurate in the Copernican system than the Ptolemaic. Both possessed an error of approximately 1 percent. The revolutionary system of Copernicus did not account for the facts any better than the centuries-old Ptolemaic system did.

But for a mathematician, the Copernican system was more "elegant" than the Ptolemaic system. In the geocentric system, each planet had an east-to-west motion, a motion in the opposite direction (west-to-east), and an epicycle motion opposite the east-to-west motion. In the Copernican system, the Earth and all the planets moved around the Sun in the same direction. There was little disagreement on this point: Astronomical calculations were easier using the Copernican system. But was it real?

Seek simplicity and distrust it.

ALFRED NORTH WHITEHEAD

FIGURE 5-4

The Copernican system was not as simple as is often advertised. To account for the observations of the Sun (a), the Earth revolved around a central point C_e, which in turn revolved around a point CC_e, which in turn revolved around the Sun. In (b) Mars is on an epicycle revolving on a deferent. The deferent's center, C_m, keeps a fixed geometric relation to the point C_e, the central point of the Earth's orbit, which moves as in (a). The Copernican explanation of retrograde motion is shown in (c). The relative motions of the Earth and Mars around the Sun cause the appearance of retrograde motion against the backdrop of the stars.

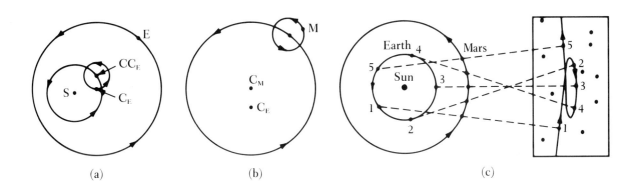

(a) (b) (c)

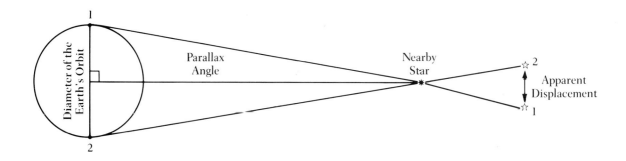

FIGURE 5-5

If the Earth revolves around the Sun, then its position at a six-month interval, its movement from (1) to (2), should cause an apparent displacement, or parallactic motion, in a star, just as holding a finger in front of your nose and then alternating closing one eye then the other will produce an apparent motion of your finger. If after six months no motion is observed, two explanations are possible: either the Earth does not revolve around the Sun or the stars are so far away in relation to the diameter of the Earth's orbit that the apparent motion cannot be detected by the naked eye.

Did the Earth's revolution around the Sun really involve a circular motion around two invisible points, one for the circle of the Earth and another for the center of the center of the circle of the Earth? If the Earth rotated, would this not create an incredible constant east wind? When stones are thrown up into the air, they are not blown away. If the Earth rotated, would it not fly into pieces from the tremendous centrifugal forces generated? Imagine riding on a merry-go-round at 1,000 miles per hour and not being strapped down. Finally, if the Earth revolved around the Sun and the stars were fixed, in the course of six months the distance displaced by this motion should be so large that the apparent positions of the fixed stars should change. That is, the large displacement space created by the Earth's revolution should create an angle in relation to any star. We would see the star move, which astronomers call a parallax, or parallactic motion (see Figure 5-5). The best observers of the time could detect no such parallactic motion. The serious astronomer of the sixteenth century had difficulty reconciling these "facts" with a moving Earth. The most intelligent thing to do was to accept the new Copernican system as a calculation device but not something that was literally real.

Today, with the exception of the mistaken circular revolutions around invisible points, we know how it is possible for the Earth with its gravitational force to rotate and not fly apart or have a constant unbearable east wind. We also know that there is a parallactic motion for the stars, detectable with powerful telescopes, and that it is impossible to see this motion with the naked eye because the stars are so far away. To the credit of those who objected to the Copernican system, they considered the possibility that no parallactic motion was observed because of the stars' distance. One of the best astronomical observers of the time, Tycho Brahe, even computed a hypothetical distance. After carefully searching for a change in the position of various stars at six-month intervals and finding none, he concluded that if the Earth really moved, the sphere of the stars would need to be 700 times farther from the planet Saturn than Saturn was from the Sun. In the sixteenth century this

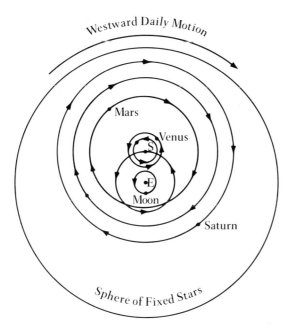

Westward Daily Motion

Mars

Venus

S

E

Moon

Saturn

Sphere of Fixed Stars

FIGURE 5-6

A third alternative between the geocentric and heliocentric system is the mixed system of Tycho Brahe. The Sun revolves around the Earth, but the planets revolve around the Sun.

made no sense at all. In a harmonious, elegant universe without any superfluous gadgetry, what could be God's purpose for all this "wasted space?"[1]

Tycho Brahe (1546–1601) is of particular interest here. As one of the best scientists of his time, he knew that an objective appraisal of the Copernican and Ptolemaic systems left little doubt that the Copernican system was better mathematically. However, an objective appraisal of the physical and factual problems of the Copernican system left little doubt that it could not be physically real. Thus, we find Tycho reviving and extending the Heraclidean system (see Figure 5-6). All of the planets revolved around the Sun except the Earth. The Earth was stationary at the center of the universe, and the Sun, the Moon, and the stars revolved around it. It was another marvelous intellectual achievement. Although it too required epicycles, both major and minor, eccentrics, and equants, many of the Copernican simplifications were maintained, at the same time avoiding the physical problems of a moving Earth. Tycho's system was mathematically equivalent to the Copernican system, but did not violate Scripture and common sense. It was not, however, consistent with Neoplatonism. The Sun was not given a mystical, central significance.

Contemporary empiricists, had they lived in the sixteenth century, would have been first to scoff out of court the new philosophy of the universe.

E. A. BURTT,
THE METAPHYSICAL FOUNDATIONS OF NATURAL SCIENCE

[1] Tycho also pointed out that if the stars were indeed this far away, then some of the brighter ones would need to be about as large as our entire solar system. In the sixteenth century this was absurd.

Kepler, Galileo, and the Church

By the time of Kepler's work in the early 1600s, three systems existed for astronomers to improve.[1] Relatively speaking, all "worked"; all three could account for the observed motions of the celestial bodies within the same margin of error. It was difficult, however, to give a realistic interpretation to any of these systems. Thus, the original reaction of the Catholic Church, which had long supported astronomical investigations, was to interpret the new Copernican system instrumentally. There was no problem in having another mathematical device with which to map the motions of God's universe. In fact, in Copernicus's book, *On the Revolutions of the Celestial Orbs,* a preface written by Osiander stated that the book's contents should be interpreted as only a mathematical device.[2]

However, some philosophers, primarily for philosophical and religious reasons, argued that the heliocentric system was real. If the stars are at rest relative to the Earth, as they are in the Copernican system, then a much larger universe, perhaps even an infinite one, is conceivable. If the stars move as in the Ptolemaic and Tychonic systems, then they must all be the same distance from the Earth; if they are at rest relative to the Earth and the Earth moves instead, then all of the stars do not need to be the same distance from the Earth. Thus, both Nicholas de Cusa (1401–1464), before Copernicus, and Giordano Bruno (1548–1600), after Copernicus, revived the work of Democritus, arguing that the Sun was only one of an infinite number of stars. Why? Because only an infinite sphere would be consistent with the greatness of God. Both also argued that some of these other stars would have planets and would be populated. Although Galileo's use of the telescope, revealing stars where none were seen before, later helped solidify this view, the arguments of de Cusa and Bruno were primarily philosophical and deductive. An infinite universe is the only one consistent with the infinite perfection of God, therefore a heliocentric system must be true. It must be real.

Bruno also argued that consistent with this scheme would be a new relationship among God's creatures. God granted each creature its own inner source of power, and these powers were more or less equal, leaving no justification for domination and servitude. This was the final straw, so to speak, for the Church. These men not only were transforming the Earth and the

> *How odd it is that anyone should not see that all observation must be for or against some view if it is to be of any service!*
>
> CHARLES DARWIN

---

[1] Actually there were four, if we count the system of William Gilbert, court physician to Queen Elizabeth. Gilbert's system was like Tycho's, except that the Earth rotated on its axis daily.

[2] The book was published in 1543, the same year Copernicus died. There is little doubt among scholars that Copernicus did not approve of Osiander's preface because there is every reason to believe that Copernicus thought of his system as real.

entire cozy, homely system of the planets into an insignificant speck but they also were questioning the hierarchical structure the Church depended on for its authority. The concept of hierarchy had become inextricably bound with the geocentric cosmology. The Aristotelian-Ptolemaic universe was a homely up and down structure, where "up" meant an ascension to greater perfection and greater control as well. God and heaven existed outside the celestial sphere of the stars, and there was a gradation of existence and control from this perfect existence to the central imperfect Earth. God delegated power to various angelic beings who controlled the movements of the planets and observed and guided various earthly events. Similarly, plants and animals served humans who in turn served God through the ecclesiastical hierarchy of the Church.

The Church was forced to choose. And thus occurred one of the greatest political blunders ever recorded. The Catholic Church, which had supported scientific research as a method for studying the wonders of God's creation, could have accepted this new system as evidence of God's infinite greatness. But its leaders chose the safe path of tradition, and the rest is history. Copernicus's book was edited to make sure the instrumentalist interpretation predominated, Bruno was convicted of heresy and burned at the stake (for his religious view on equality), and Galileo was persecuted for advocating the heliocentric system as real. Copernicus, Bruno, and Galileo were all religious men; they thought of themselves as attempting to read the mind of God. Their difference with the Church was not so much a battle between science and religion, as it is so often portrayed, but part of a larger battle over different conceptions of epistemology, God, and world view.

Note that Psalms 93 and 104 read, respectively, "You have made the world firm, unshakable. . . . You fixed the earth on its foundations, unshakable for ever and ever." Are these clear statements that the Earth is the center of planetary motion? Nicolas Oresme, a devout Parisian defender of the faith, argued that these passages are not meant to be taken literally, no more than those that describe God as angry or pacified, and Galileo, borrowing the thought from Cardinal Baronias, said, "The Bible teaches how to go to heaven, not how the heavens go."

The mathematical force of the Copernican system had already gone too far for the Church's decision to make a difference. By the beginning of the seventeenth century and the work of Johannes Kepler (1571–1630), astronomers had been using this system for decades as an easier guide in calculating the motions of the planets and devising accurate calendars. Many, however, still thought of the system as a mathematical device. Thus, of all the roles played in the so-called Copernican revolution, we must attribute the greatest to Kepler. He is by far the most interesting and revolutionary character. As a Protestant and religious individualist, he was not swayed by the attitude of the Church or the views of other Protestants, such as the Lutherans, who were actually the first to attack the Copernican system as heresy. For the most part, he lived in his own world of an ardent mystical Neoplatonic faith. Because of the problems and inaccuracies of the Ptolemaic, Copernican, and

I have declared infinite worlds to exist beside this our earth. It would not be worthy of God to manifest Himself in less than an infinite universe.

GIORDANO BRUNO

I do not feel obliged to believe that the same God who has endowed us with sense, reason, and intellect has intended us to forgo their use.

GALILEO

Tychonic systems, Kepler was convinced that no one had yet succeeded in reading the harmonies of the world. He desired passionately to be the first to read the mind of God.

It was impossible, however, for Kepler to be completely objective. Tycho had put together the best data ever assembled on the motions of the planets. To match theory with observation, system after system of circles had to be tried. Day after day, year after year, the full creative powers of the human mind needed to be brought to bear on assimilating Tycho's data. But as with our mathematical analogy discussed earlier, the choices were infinite. There was no time in the course of one life to develop all the possibilities. Kepler had to choose. He had to be convinced ahead of time that the Sun was the center of planetary motion.

Popularizations of science during this period often attempt to portray scientific discoveries such as Kepler's as an objective, logical progression from what was happening at the time. In 1609 Galileo Galilei opened another layer of the majesty of the universe by using the telescope for the first time. New stars were discovered where none were seen before, and the idea of an infinite universe seemed more plausible. Galileo saw pits and craters on the Moon and spots on the Sun; the traditional idea that the Earth was unique as a center of change and decay became harder to believe. The newly discovered four largest moons of Jupiter appeared to move around the planet just as the Moon moved around the Earth in Copernican astronomy, providing also a visible minimodel of the heliocentric system. Observations of Venus clearly showed a cycle of phases (including a full phase) like that of the Moon, consistent with that predicted by Copernicus. Viewed from a modern perspective these facts appear overwhelming, and it is difficult to understand how any but the most dogmatic religious fanatic could believe anything but the Copernican system after 1610 when Galileo announced his discoveries. But it is always easier to see the folly of past generations rather than our own.

Kepler did not know these facts until after he made his major discovery, and even if he had known them earlier they would have had little effect on him. Kepler too knew of the telescope—he even designed one, but did not bother to construct it. As a Neoplatonist, what mattered most to him were not such qualitative physical features as new moons, new stars, or phases of a planet. What mattered were the precise quantitative data of Tycho and finding the elegant mathematical relationships that would instantly relate every possible observation. Furthermore, as Kuhn has pointed out, the qualitative observations of Galileo "contributed primarily to a mopping up operation":

The evidence for Copernicanism provided by Galileo's telescope is forceful, but it is also strange. None of the observations discussed above, except perhaps the last [the phases of Venus], provides direct evidence for the main tenets of Coper-

Galileo was a scrambling social climber. . . . Fame . . . brought power of a kind, perhaps the power to persuade the entire Catholic hierarchy to adopt the Copernican system. At least Galileo was egotistical enough to expect that it would. In [his] rush to assert a claim of priority he was sometimes more aggressive than might seem prudent.

OWEN GINGERICH, "THE GALILEO AFFAIR"

nicus's theory—the central position of the sun or the motion of the planets about it. Either the Ptolemaic or the Tychonic universe contains enough space for the newly discovered stars; either can be modified to allow for imperfections in the heavens and for satellites attached to celestial bodies; the Tychonic system, at least, provides as good an explanation as the Copernican for the observed phases of and distance to Venus. Therefore, the telescope did not prove the validity of Copernicus's conceptual scheme. But it did provide an immensely effective weapon for the battle. It was not proof, but it was propaganda.[1]

Thus we find Kepler a convinced Copernican, narrowing his efforts to finding the right mathematics, even though he knew that the quantitative evidence did not yet support it, and even though he was highly critical of the messiness of the numerous circles created by Copernicus. Kepler himself then tried numerous arrangements using circles. Finally, he became convinced of a truly revolutionary idea: Perfect circles would not work; God must have used a different mathematical shape. This was scientific heresy.

All the great minds of astronomy since the time of Pythagoras had agreed that the shape of the planetary orbits must be a perfect circle, because geometrically, only the circle allowed for an elegant, symmetric motion. Other mathematical relationships might exist to account for various features of the planets, such as relative distances and sizes—in fact, Kepler had suggested that there was a special relationship between the five perfect geometric solids and the six known planets—but only the circle, it was thought, could be the actual path taken by a planet's motion. Even the great Galileo ignored Kepler's solution and held to the end that only the circle provided the mathematical property capable of explaining the mechanical motion of a revolving planet. That Kepler was not interested in observing his God's universe through the telescope and Galileo could not break out of the ancient Pythagorean paradigm is one of the greatest ironies in the history of science.

In studying Mars, the most eccentric wanderer of all the planets, Kepler tried for the first time various ovals. Still finding discrepancies between his models and Tycho's data, he then noticed the discrepancies had a pattern, a familiar pattern that could be covered by using an *ellipse*. This required one more heretical step: To account for the data, the planet's path around the Sun must be an ellipse and it must move around the Sun at a *variable speed*.

[The Sun] . . . which alone we should judge to be worthy of the most high God, if He should be pleased with a material domicile, and choose a place in which to dwell with the blessed angels.

JOHANNES KEPLER

[1] Thomas S. Kuhn, *The Copernican Revolution: Planetary Astronomy in the Development of Western Thought* (Cambridge, Mass.: Harvard University Press, 1957), p. 224. Copyright © 1957 by the President and Fellows of Harvard College; 1985 by Thomas S. Kuhn. Reprinted by permission.

This idea worked like nothing before. The 80 or so circles of Ptolemy, the 48 of Copernicus, and all of the epicycles, deferents, eccentrics, and equants vanished in favor of 7 slightly squashed circles with the planets moving around the Sun and the Moon around the Earth at variable speeds. Kepler had solved the problem of the planets. He had, so he thought, finally read the mind of God. He had discovered the true harmony of the motions of the worlds (see Figure 5-7).

An Epistemological Reflection:
Is There a Logic of Discovery?

If science reveals much contingency to life, why should we be surprised to find much contingency in the application of scientific method?

The twists and turns leading up to Kepler's discovery raise significant epistemological issues. The ellipse was not a mathematical object created by Kepler. It was discovered by the ancient Greeks. Often in the history of science, a mathematical relationship, object, or device is discovered with no apparent application to the physical world, and then years, decades—in this case, centuries—later it is used to solve a crucial scientific problem. A century before Einstein discovered his theory of relativity, a bizarre geometry was discovered by Georg Reimann. Because it contradicted the geometry of our three-dimensional common sense, no one had the slightest idea how it could be used. Later, Einstein, in what is known as the General Theory of relativity, showed that this geometry worked in understanding the large-scale gravitational relationships discovered by twentieth-century astronomy. Other bizarre imaginary and irrational numbers, such as the square root of -1 have been indispensable in tracking the behavior of the atom. Why does mathematics work so well? Is there an evolutionary resonance between the mind of human beings and the universe? Does the fact that our brains are made of atoms obeying certain mathematical laws drive us in some sense to these discoveries?

For Kepler there was no mystery in how he was capable of solving the problem of the planets. There was nothing special in his mastery of this problem. God had created human beings with the gift of reading the mathematical harmonies in His mind. Because this gift was part of human nature, it was only a matter of time for someone to discover God's plan. We will see in Chapter 9 that some contemporary scientists wish to replace this *creationary resonance* between the mind of human beings and God with an *evolutionary resonance* between the brain of human beings and the laws of nature. Advocates of this view argue that, given the infinite experiments of evolution, it is inevitable that creatures will evolve capable of reading the laws of nature. Similarly, commentators on Kepler's discovery of the ellipse believe that his discovery was essentially inevitable. If Kepler had not been successful in dis-

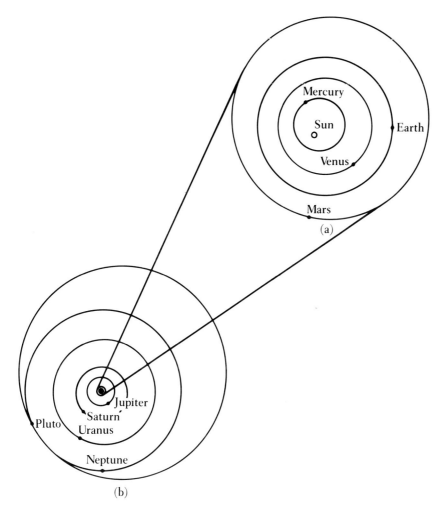

covering that the ellipse worked, then someone else surely would have. As proof of this, they will point to simultaneous discoveries by independent thinkers, as in the case of the discovery of natural selection by Darwin and Wallace, and the mathematical calculus by Newton and Leibniz that proved indispensable for the theory of gravitation. Thus, it is argued, if the time is right and the right accumulation of facts has occurred, scientific progress is inevitable, new discoveries will be made, false ideas will fall by the wayside, and the reasonable will be separated from the conceivable.

Very much related to this notion are modern discussions on and efforts in creating computers with artificial intelligence. If discoveries are the result of logical and inevitable processes, can computers be programmed to create scientific hypotheses that work? Can the hardware and software be developed such that given Tycho's data, the solution of the ellipse is provided by a computer as the best solution? Is creativity actually a logical process in disguise?

Most modern scientists find it easier to think that Kepler's mystical Neoplatonism had little to do with his discoveries, that it was just some cultural baggage that came along for the inevitable revolutionary ride. There is another possibility, however, a more unsettling and humbling one. Perhaps, as Plato and Protagoras have argued, there are many technically capable paths by which the universe can be modeled. Perhaps, if Kepler had not been an ardent Neoplatonist and had not applied the ellipse to the problem of the planets, we might be using some other system today.

Although an extreme version of this possibility implies relativism, what is suggested here is that Kepler's solution is indeed the best one and that without his Neoplatonism and many other fortuitous circumstances surrounding his life, this "best solution" may have never occurred or, at the least, been delayed many years. After all, the Ptolemaic system lasted for over 1,400 years. It is possible that the Copernican system may have won the day, but for many years, perhaps centuries, would have been nothing more than another system of perfect circles modified again and again to match new data. The way to the truth may have many paths, some short, some long, and we could still be on the longer path adhered to by Galileo, who was committed to the traditional paradigm of perfect circles and uniform motion for the planetary paths. His telescopic observations did not imply the ellipse directly. (Galileo also rejected Kepler's proposal of an "attractive force" between the Sun and the planets, a forerunner to Newton's theory of gravity.)

This suggestion is analogous to what we have learned from the theory of natural selection. Just as innumerable coincidental events had to be just right to produce *Homo sapiens,* so too numerous events must be just right to produce the creative environment for a great scientific discovery. In addition, just as the necessary events in evolution are unrelated in any rational or purposeful sense—a change in the weather, cosmic rays from a distant supernova, chance mutations—so the events surrounding a great discovery have no logical connection either with themselves or the discovery. Kepler's Neoplatonism may not have been logically necessary for his discovery of the ellipse, but historically it may have been absolutely necessary.

Does this undermine the image of science as a self-corrective enterprise? Not necessarily, but it does imply that, as with evolution, the self-correctiveness is a messy one. In the course of time we may have found ideas that are better, but this may not have been inevitable, and there is no guarantee that the future holds inevitable scientific progress. We are confronted once again with the ultimate epistemological question: How do we separate the reasonable ideas from the merely conceivable ideas? If it is granted that we are able to eventually discover the best ideas, how do we do this when there is an infinite number of conceivable ideas? Is it just luck?

Let us look at one final example before closing this chapter. In announcing his discovery, Kepler described three laws of planetary motion. First, each

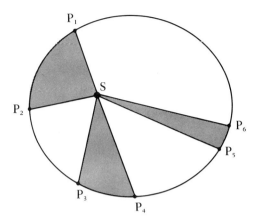

FIGURE 5-8

Kepler's second law. A planet sweeps out equal areas in equal times. Although a planet will move at a slower rate from P_5 to P_6 than it does from P_1 to P_2, it will take the same amount of time to cover these distances, and the areas swept out by connecting these points with the Sun (the shaded areas) are the same.

planet revolves around the Sun in the path of an ellipse with the Sun not in the center but at one focus. Second, each planet proceeds at a variable speed such that if an imaginary line is connected between the planet and the Sun, the planet sweeps through equal areas during equal times (Figure 5-8). Third, there is a mathematical relationship between the distances the planets are from the Sun and the time they take to revolve around the Sun, such that if one takes the sidereal period (the time measured for one complete revolution relative to the fixed stars) of one planet and divides this value by the value of the sidereal period of another planet and squares the result, this value will be precisely equal to the cube of the value that results from dividing the average distance of one planet from the Sun by the average distance of another planet from the Sun.[1]

The first two laws solidified the workability of the Copernican revolution and solved the problem of the planets. But what good was the third law? Who cared if the cube of a planet's average distance was proportional to the square of its orbital period? It had no practical value. At this time it did not suggest any new observational prediction. As a Neoplatonist, however, Kepler found it fascinating. The fact that such a simple mathematical equation could be so sweeping in its power, that it could so elegantly "save the phenomena," that it could cover all of the observations made over the centuries of planetary positions, that it could summarize so inclusively books of recorded data on planetary positions, even correcting some inaccurate citations, proved to Kepler beyond any doubt that the heliocentric system was real. That such a simple

---

[1] A more elegant, but specialized, way of stating this would be: The ratio of the squares of the orbital periods is equal to the ratio of the cubes of the average distances from the Sun. Better still would be $(t1/t2)^2 = (d1/d2)^3$, and this is why mathematicians prefer mathematics to talking.

equation worked was much more of an explanation than any observation Galileo could make with the telescope. There was no logical relationship, however, between this law and the acceptance of Copernicanism. It was not needed to prove the heliocentric system. The observational data, the ellipse, and the notion of variable speed were sufficient. But although of no initial practical value to the astronomers of Kepler's time, the third law, as we will see in the next chapter, was of immense importance later in the development of Newton's theory of gravity.

Was this another lucky result of Kepler's Neoplatonism? Or was this an inevitable discovery, the result of a resonance between the brain of human beings and the laws of nature, plus the cumulative work of centuries of inquiry? Although these ruminations on science and creativity are of great importance, we cannot follow them further at this point. We will address these important questions in Chapter 9, where we will consider issues related to intelligence, the possibility of extraterrestrial life, and communicating with extraterrestrial life.

Beliefs and Objectivity:
Does the World Kick Back?

For now it is time to ask ourselves what we have learned from all this. Was the degree of cultural influence on scientific events leading to the Copernican revolution unique? Hardly. Every age has its paradigm, its mental set, a way of organizing the apparent chaotic, unrelated events of life into a meaningful whole consistent with the ultimate concerns of the time. Unlike many textbook presentations of scientific method, we see that the intrusion of philosophical considerations into the creation of theories and the interpretation of facts is not only inescapable but also ultimately beneficial. From a modern point of view, the astronomers of antiquity may have had many "silly" beliefs. But these beliefs were crucial in providing direction to their inquiries and fuel for the creativity mill necessary for the production of ideas on how the universe works. (Our foolish beliefs are hard to see because we have no distance from them.)

Facts have little meaning without ideas to interpret them. Without ideas, the world of appearances is full of chaotic, disconnected motions. Without their faith in order and mathematics as the tool to read the motions of the cosmos, the ancient Greeks would not have thought that the problem of the planets was important or capable of a solution. Without believing in a solution, they would not have spent so much time observing the planets in such detail. Without being an ardent Pythagorean and Neoplatonist, Copernicus would

The great problem: How can we admit that our knowledge is a human—all too human—affair, without at the same time implying that it is all individual whim and arbitrariness?

KARL POPPER,
CONJECTURES AND REFUTATIONS

The cosmos has a way of intruding upon our most cherished beliefs.

not have objected to the mathematical ugliness of the Ptolemaic system, nor would he and Kepler have concentrated on the Sun as the central figure in a new solution. Without his concern for mathematical harmony and the importance of the facts matching an elegant mathematics, Kepler would not have stressed the importance of Tycho's accurate data.

Ideas are precious; it matters not so much where they come from, only that they work. As we have seen, many of the ideas we consider modern were not proposed on the basis of the facts, but were deductions from philosophies. Democritus first proposed an infinitely large universe because it was logically consistent with a metaphysics of atomism. Pythagoras suggested that the Earth was a sphere because the circle was the most perfect mathematical object. De Cusa and Bruno argued that a heliocentric system must be real, because of that idea's consistency with a larger universe and a greater God.

It is impossible for scientists to operate with a machinelike objectivity. Often they must be convinced ahead of time that an idea is true, and often the crucial facts supporting an idea come after a commitment, not before. Copernicus, de Cusa, Bruno, Kepler, and Galileo were all convinced that the heliocentric solution was real before there was factual proof. Their commitment produced the mental set and the direction that made the discovery of the facts possible. As Joseph Weizenbaum has stated so well in *Computer Power and Human Reason*:

The rightful distinction of science lies in the fact that in spite of and because of our biases, because science makes us confront the world, the method provides us with a great opportunity to transcend our biases.

> The man in the street surely believes . . . scientific facts to be as well-established, as well-proven, as his own existence. His certitude is an illusion. Nor is the scientist himself immune to the same illusion. . . . He must . . . suspend disbelief in order to do or think anything at all. He is rather like a theatergoer, who, in order to participate in and understand what is happening on the stage, must for a time pretend to himself that he is witnessing real events. The scientist must believe his working hypothesis, together with its vast underlying structure of theories and assumptions, even if only for the sake of argument. Often the "argument" extends over his entire lifetime. Gradually, he becomes what he at first merely pretended to be: a true believer.[1]

Lest the cynic, however, think that all this supports relativism, consider that the history of the Copernican revolution also shows that regardless of what one believes, "reality kicks back." Each idea caused the astronomers of

---

[1] Joseph Weizenbaum, *Computer Power and Human Reason: From Judgment to Calculation* (San Francisco: W. H. Freeman, 1976), p. 15.

this time to pay more attention to the facts and the discrepancies. Kepler was a committed Copernican, and because of this, not in spite of this, he could not accept any solution with a central Sun that did not match the facts. Although discrepancies can be ignored or rationalized for a time, the cosmos has a way of intruding upon our most cherished beliefs. Our ideas about the world may be crucial in our observation of the world, in making sense of anything, but they must eventually yield to the facts.

In our romance with the universe, as in any love affair, an interaction between two personalities is needed. Although we may attempt to impose as many human features as possible, the personality of this mysterious place cannot be ignored for long. So like romantic gestures we throw our ideas for acceptance toward this great being, and most of the time they are rejected mercilessly. But our love is fanatical and we keep trying nevertheless.

Concept Summary

In Chapters 2 and 4 we discussed the point that science attempts to walk a very difficult epistemological path between relativism and absolutism: Its method attempts to achieve reasonable beliefs about the world without being absolutely assured that its beliefs are true. In this chapter our cultural roots are further explored, and the role of cultural background is presented from the point of view that it is ultimately an aid to the goal of scientific method, rather than a validation of relativism. We see science as both a meandering parade of ideas and an expansion of our understanding of the universe and our place in it.

The ancient Greek faith in order, and the philosophies of Platonism and Pythagoreanism, produced a mental set early in the history of Western culture that drew attention to the need for explanation of the motion of the planets. Although many systems of explanation, both geocentric and heliocentric, were proposed, Ptolemy's geocentric system was considered the most reasonable until the sixteenth century. Persuasive theoretical and observational reasons existed for the acceptance of Ptolemy's system. Hence, a complex mix of philosophical, religious, cultural, and observational reasons was needed for its replacement by the Copernican and Keplerian systems.

By the time of Copernicus several problems with the Ptolemaic system were well known. Although the system was functional, it was not observationally accurate and the mathematics was not sufficiently consistent with what astronomers of this time were convinced to be a harmoniously constructed universe. As a Neoplatonist, Copernicus was convinced that the Sun must replace the Earth as the center of the universe. The Copernican system was simpler, but it still maintained the Pythagorean principles of perfectly circular and uniform motion and the mathematical devices of deferents, epicycles, and eccentrics. Although the system was more pleasing mathematically, it was no more accurate observationally than Ptolemy's system. It was also accepted and explored further by important Renaissance figures like Galileo and Kepler in spite of problems of realistic interpretation. (Could the Earth really revolve around a central point that revolved around another central point that in turn revolved around the Sun?)

Although empirical evidence existed for the new model, like many turning points in the history of science, the acceptance of this new model was based upon a mix of scientific and extrascientific ideas. It is not an exaggeration to say that the staying power of this new system was based as much on religious and philosophical concerns as it was on its pure scientific merit. For instance, Tycho Brahe, the acknowledged best observationalist of his time, was committed to a non-Copernican system, whereas Kepler, interested in the facts of planetary motion primarily because the discrepancies proved to him that no one had yet read the mind of God, became a committed Copernican. Like Copernicus, Kepler was convinced that the Sun must be given a central significance. Others were also committed Copernicans because it was thought that this new system was consistent with the greatness and infinite perfection of God.

The story of Kepler and the Copernican revolution shows that although those who have a passion to know must confront and be intimate with the world, there is an ironic, serendipitous, messy humanness to this relationship. The world views intermingled in Kepler's culture forced him to confront the world from a point of view. It was impossible, and would have been counterproductive to even try, for him to be totally objective; Kepler was faced with too many choices to spend a lifetime objectively exploring each one. He had to choose an idea to test, or like Sartre's self-taught man discussed in Chapter 2, he would have wallowed in a sea of infinite, aimless facts.

Was the discovery of the ellipse inevitable? Is scientific progress inevitable? If so much of science depends on having the right idea at the right time, and because the origin of ideas can depend on so many contingent factors—such as subjective hidden agendas and biases, philosophical and religious dogmas, cultural traditions—and if ideas not only offer direction for viewing the world, but veil the world as well, what guarantee is there that science will always find the right idea at the right time? If science reveals much contingency to life, why should we be surprised to find much contingency in the application of scientific method?

Although there is disagreement today as to what extent science is objective, perhaps the answer that deserves serious thought is that the overwhelming presence of the world makes the veil of our ideas, no matter how strongly held, inescapably translucent and penetrable. In other words, the rightful distinction of science lies in the fact that in spite of and because of our biases, because science makes us confront the world, the method provides us with a great opportunity to transcend our biases.

Suggested Readings

A History of the Sciences, by Stephen F. Mason (New York: Collier Books, 1970).

A good general overview of the history and development of science. Covers ancient developments of Babylonia and Egypt to American and Soviet science in the 1950s. Contains several chapters on the roles of Greek philosophy, Copernicus, Kepler, Galileo, and the Church in the Copernican revolution.

Watchers of the Stars: The Scientific Revolution, by Patrick Moore (New York: Putnam, 1974).

Intended for a wide audience; contains a history of astronomy through the Copernican revolution. Also includes illustrations and is rich in the personal details of the major players.

The Structure of Scientific Revolutions, by Thomas S. Kuhn (Chicago: University of Chicago Press, 1962 and 1970).

A must-read book for anyone interested in the history and philosophy of science. Although controversial, this book has played a major role in exploding the myth that scientific method is a cold, logical affair that objectively stalks the truth oblivious of the culture and philosophical environment of a time. The 1970 edition is recommended because it contains responses to criticism and to misunderstanding in a postscript.

The Copernican Revolution: Planetary Astronomy in the Development of Western Thought, by Thomas S. Kuhn (Cambridge, Mass.: Harvard University Press, 1957).

Another essential book by Kuhn. Although thematic—it approaches the topic from the point of view that scientific revolutions require a complete shift in the mental sets of scientists—this book is still simply the best available in terms of a methodical, comprehensive discussion of every important component of the Copernican revolution. Contains excellent illustrations and a scholarly bibliography.

Also see "The Galileo Affair," by Owen Gingerich, *Scientific American* 247, no. 2 (August 1982): 132–143. An enlightening discussion of what was really at stake between Galileo and the Church.

A History of Astronomy from Thales to Kepler, by John Louis Emil Dreyer, 2nd ed. (New York: Dover, 1953).

A reprinting of a seasoned, conventional introduction to the history of astronomy. Rich in scholarly detail and corrective of traditional misconceptions. Because this book approaches the history of our evolving view of the solar system from the more traditional approach—science as a steady accumulation of objective observations and truths—it is instructive to compare this book's chapters on the Copernican revolution with Kuhn's treatment.

The Nature of Scientific Discovery: A Symposium Commemorating the 500th Anniversary of the Birth of Nicolaus Copernicus, by Owen Gingerich, the National Academy of Sciences, and the Copernicus Society of America (Washington, D.C.: Smithsonian Institution Press, 1975).

Articles from the fifth international symposium of the Smithsonian Institution on the work of Copernicus and issues related to the Copernican revolution. A scholarly celebration with contributions from major scientific figures and international historians of science attempting to add to our understanding of this important time of our cultural roots and the implications for the future of science. Particularly relevant to the discussion on the possible lack of inevitability of scientific discovery, and the many contingencies related to producing the environment for a great discovery, are the articles by the Harvard astrophysicist and historian of science Owen Gingerich.

CHAPTER 6

Cultural Roots:
3. Science as a Religion, the World as a Country

The World is my Country, Science my Religion.

CHRISTIAN HUYGENS

Science without religion is lame, religion without science is blind.

ALBERT EINSTEIN, *OUT OF MY LATER YEARS*

God [before Copernicanism] . . . was the ultimate object of purpose. In the Newtonian world . . . all this further teleology is unceremoniously dropped. . . . We are to become devotees of mathematical science; God, now the chief mechanic of the universe, has become the cosmic conservative. His aim is to maintain the status quo. The day of novelty is all in the past; there is no further advance in time. Periodic reformation when necessary by the addition of the indicated masses at the points of space required, but no new creative activity—to this routine of temporal housekeeping is the Deity at present confined.

E. A. BURTT, *THE METAPHYSICAL FOUNDATIONS OF NATURAL SCIENCE*

Voyagers

J ust as ancient mariners were eventually able to use Eratosthenes' new model of the Earth to circumnavigate the globe, so the unexpected discovery of elliptical planetary orbits by Kepler, later followed by Newton's theory of gravitation, has enabled modern humankind to explore our solar system. We have traveled far both mentally and physically since the seventeenth century. At this moment somewhere several billion miles from our Earth, *Voyager 1,* a small, 1,819-pound robot spacecraft launched in the fall of 1977, is exiting our solar system and entering the vast sea of our galaxy.

With the *Mariner, Pioneer, Viking,* and *Voyager* flights we have learned some astounding things about our solar system. The Greek faith in mathematics and the application of this faith by Kepler and Newton has allowed us to transport our awareness billions of miles and know things about places that we as individuals will undoubtedly never visit.

On Mercury we have found rocky cliffs over a mile high and almost 1,000 miles long. On Venus we have discovered an atmosphere 100 times more dense than that of Earth and an average surface temperature of 900° F. We have begun to radio-map the continents of this hellish world and have seen lightning in its sulfuric acid clouds, which speed around the planet at 225 miles per hour. With the *Viking* spacecraft we have tested for life on Mars and were startled by the puzzling results; we have seen a volcano there, larger than many of Earth's volcanoes put together, and a grand canyon that could stretch from New York to Los Angeles; we have seen tantalizing geological features resembling shorelines, gorges, riverbeds, and islands suggesting that seas and rivers once existed on this planet.

Of the four small points of light first seen by Galileo revolving around the planet Jupiter, we now have close-up pictures. As *Voyager* hastened through the Jupiter system at over 45,000 miles per hour, active volcanoes were photographed on Io with plumes over 100 miles high, and the other three Galilean moons, Europa, Ganymede, and Callisto were seen to be large ice balls that could possibly harbor enormous underground oceans. On Saturn we discovered a planet with winds of over 1,000 miles per hour and thousands of more rings than were thought to have existed based on observations from Earth. Of Titan, Saturn's largest moon, we were both disappointed and intrigued by its thick methane atmosphere. Visions of methane rain and snow, of methane rivers and lakes can now be imagined with a little scientific respectability.

Finally, with *Voyager 2* we now have close-up pictures of Uranus, a strange, oddly tipped planet whose pole facing the Sun is cooler than the pole away from the Sun, and the chaotic geological features of Miranda, perhaps the most bizarre object yet seen in our solar system. It was a great tribute to the intellectual giants who gave us the faith in an ordered universe and the mathematical tools to explore it that *Voyager* arrived at its predetermined point,

50,000 miles above the cloud tops of this mysterious planet, 1,842,610,000 miles from its initial starting point, on course and a minute ahead of schedule!

We all wish to make some sort of mark in life, ideally a lasting one, one that our children's children could contemplate. For some the path to a piece of immortality is making money or writing a book or music. For most of us our mark in life will be recalled only in family photographs. Gutzon Borglum, the main artist of the granite sculptures of the four U.S. presidents on Mount Rushmore, made his mark in life. Since granite erodes about one inch for every 1,500 years, this monument to democracy should endure for thousands of years. Eventually though, this shrine to our need for permanence and the perseverance of the human spirit will be gone. The *Apollo* astronauts who walked on the Moon will have a special place in our history books, but they left something there that will surely outlast most books and perhaps the entire human race. Because the Moon lacks an atmosphere and an active weathering environment, their footprints are likely to persist for over 1 million years.

But it is on the *Voyager* flights that the largest piece of immortality is engraved for all of us. On these little modern cosmic mariners is a copper phonograph record sprayed with gold. It should be playable for 1 billion years. There may never be any players, but just in case, it comes complete with playing instructions. The instructions are in a scientific code that scientists hope portrays a universal language, a language understandable to any form of life that may have evolved consciousness and developed a technology.

The record consists of a 120-minute message of greetings from people in many languages, a 12-minute montage of sounds of Earth, such as rain, wind, a baby's first cries, the sound of someone chopping wood, the screech of bus brakes, a whale song, and even a vital-sign recording of the human body and brain. There is also 90 minutes of music—from Bach's *Brandenburg Concerto,* Beethoven's *String Quartet in B flat,* and the Chinese *Flowing Streams* to Chuck Berry's *Johnny B. Goode* (see box, pages 168–169).

Voyager has a long way to go. It will be tens of thousands of years before it reaches the vicinity of a close star and 100 million years before it circumnavigates our galaxy. Never to touch the limitless intergalactic spaces and enter a single one of the billions of other galaxies, even in our own galaxy it is a message in a bottle with very little hope of ever finding a reader. *Voyager* is not likely to pass near even one of the billions of stars of our galaxy, and we know not whether there are any other readers besides ourselves in the first place. But it is our way of crying out, of proclaiming to the cosmos that we are here, that we exist, that we are unique but are lonely and would like to have some company in this grand adventure called life. As Ann Druyan, a member of the team that assembled the message, stated, "The *Voyager* record is earth's cultural audition for the universe and in another sense, our application for citizenship in that vastness."

Voyager 1 was actually launched two weeks after *Voyager 2.* The initial mission of both spacecraft was to explore only the planets Jupiter and Saturn. Because of different trajectories, calculated on the basis of the discoveries by

Kepler and Newton, *Voyager 1* arrived at Jupiter first, 625 million miles from the Sun. After taking thousands of pictures of Jupiter and its moons, it moved on to Saturn, over a billion miles and almost twice the distance from Earth as Jupiter, and then out toward the depths of our galaxy. *Voyager 2* took a different route, a slightly longer one, but one calculated to pass not only Jupiter and Saturn, taking thousands of pictures that *Voyager 1* missed, but also, if funds could be found, to continue on to Uranus, and finally Neptune 2,797,000,000 miles from the Sun. Funds were found, and the technical feat to reprogram *Voyager 2* to follow a path to Uranus has been likened to sinking a 1,500-mile golf putt: a marvelous achievement, but impossible without the right conceptual tools.

Yet at our present level of technology it takes so long to travel these relatively short distances. To those without vision, these explorations seem like an enormous waste of time and resources. We have dealt with this problem before, however. Below is a chart representing the relative times and distances traveled by the explorers of the sixteenth and seventeenth centuries and our robot space explorers of today.

Nature and nature's laws lay hid in night: God said, let Newton be! and all was light.

ALEXANDER POPE

Sixteenth and Seventeenth Centuries	Today	Time
Spain to Azores	Earth to Moon	A few days
Europe to America	Earth to Venus, Mars	A few months
Europe to China	Earth to Jupiter	1 to 2 years

At present many known obstacles to human space travel make this analogy questionable. The most important one is that the human body has not evolved to experience long periods of zero gravity. But apparently insurmountable obstacles often become self-fulfilling prophecies when there is a lack of will. In the sixteenth and seventeenth centuries, countries that lacked the will to overcome the obstacles of long sea voyages soon found themselves to be spiritually stagnant and second-rate world powers. (When explorers of the sixteenth and seventeenth centuries said good-bye to their families, they would casually remark that they would return in five years.)

We have already discussed the practical relevance of space exploration (see Chapter 1). In this chapter we must see how the Copernican revolution was completed by Newton's theory of universal gravitation and investigate the world view that made possible the mathematical discoveries upon which the success of the *Voyager* flights are based. We must also prepare ourselves properly for the shock of twentieth-century physics. For as we will see shortly, the greatest irony of the success of the *Voyager* flights is that just as the ancient navigators of the Earth's oceans used the mistaken Aristotelian-Ptolemaic conception of the planets as a navigation guide to travel successfully from port to port, so we believe today that the Newtonian system of universal

PICTURES (in sequence)

calibration circle	Guatemalan man
solar location map	Balinese dancer
mathematical definitions	Andean girls
physical unit definitions	Thai craftsman
solar sys. parameters (2)	elephant
the sun	Turkish man with beard and glasses
solar spectrum	old man with dog and flowers
Mercury	mountain climber
Mars	Cathy Rigby
Jupiter	Olympic sprinters
Earth	schoolroom
Egypt, Red Sea, Sinai Pen., Nile (from orbit)	children with globe
chemical definitions	cotton harvest
DNA structure	grape picker
DNA structure magnified	supermarket
cells and cell division	diver with fish
anatomy (8)	fishing boat, nets
human sex organs (drawing)	cooking fish
conception diagram	Chinese dinner
conception photo	licking, eating, drinking
fertilized ovum	Great Wall of China
fetus diagram	African house construction
fetus	Amish construction scene
diag. of male and female	African house
birth	New England house
nursing mother	modern house (Cloudcroft)
father and daughter (Malaysia)	house interior with artist and fire
group of children	Taj Mahal
diagram of family ages	English city (Oxford)
family portrait	Boston
continental drift diagram	UN building (day)
structure of earth	UN building (night)
Heron Island (Australia)	Sydney Opera House
seashore	artisan with drill
Snake River, Grand Tetons	factory interior
sand dunes	museum
Monument Valley	X-ray of hand
leaf	woman with microscope
fallen leaves	Pakistan street scene
sequoia	India rush-hour traffic
snowflake	modern highway (Ithaca)
tree with daffodils	Golden Gate Bridge
flying insect, flowers	train
vertebrate evolution diag.	airplane in flight
seashell (Xancidae)	airport (Toronto)
dolphins	Antarctic expedition
school of fish	radio telescope (Westerbork)
tree toad	radio telescope (Arecibo)
crocodile	book page (Newton's *System of the World*)
eagle	astronaut in space
S. African waterhold	Titan Centaur launch
Jane Goodall, chimps	sunset with birds
sketch of bushmen	string quartet
bushmen hunters	violin with score

Pictures, greetings, and sounds on the *Voyager* record.

GREETINGS IN MANY TONGUES (alphabetically)

Akkadian
Amoy (Min dial.)
Arabic
Aramaic
Armenian
Bengali
Burmese
Cantonese
Czech
Dutch
English
French
German
Greek

Gujorati (India)
Hebrew
Hindi
Hittite
Hungarian
Ila (Zambia)
Indonesian
Italian
Japanese
Kannada (India)
Kechua (Peru)
Korean
Latin
Luganda (Uganda)

Mandarin
Marathi (India)
Nepali
Nguni (SE Africa)
Nyanja (Malawi)
Oriya (India)
Persian
Polish
Portuguese
Punjabi
Rajasthani
Roumanian
Russian
Serbian

Sinhalese (Sri Lanka)
Sotho (Lesotho)
Spanish
Sumerian
Swedish
Telugu (India)
Thai
Turkish
Ukranian
Urdu
Vietnamese
Welsh
Wu (Shanghai dial.)

SOUNDS OF EARTH (in sequence)

whales
planets (audio analog
   of orbital velocity)
volcanoes
mud pots
rain
surf
cricket, frogs
birds

hyena
elephant
chimpanzee
wild dog
footsteps and
   heartbeats
laughter
fire
tools
dogs (domestic)

herding sheep
blacksmith shop
sawing
tractor
riveter
Morse code
ships
horse and cart
horse and carriage
train whistle

tractor
truck
auto gears
Saturn 5 rocket liftoff
kiss
baby
life signs:
   EEG, EKG
pulsar

MUSIC (in sequence)

Bach: Brandenberg Concerto # 2, 1st m.
Java: court gamelan—"Kinds of Flowers"
Senegal: percussion
Zaire: Pygmy girls' initiation song
Australia: horn and totem song
Mexico: mariachi—"El Cascabel"
Chuck Berry: "Johnny B. Goode"
New Guinea: men's house
Japan: shakuhachi (flute)—"Depicting the Cranes
   in Their Nest"
Bach: Partita #3 for violin
Mozart: "Queen of the Night" (from "The Magic
   Flute")
Georgia (USSR): folk chorus—"Chakrulo"
Peru: pan pipes

Louis Armstrong: "Melancholy Blues"
Azerbaijan: two flutes
Stravinsky: "Rite of Spring," conclusion
Bach: Prelude and Fugue # 1 in C Major
Beethoven: Symphony #5, 1st m.
Bulgaria: shepherdess song—"Iziel Delyo hajdutin"
Navajo: night chant
English 15th cent.: "The Fairie Round"
Melanesia: pan pipes
Peru: woman's wedding song
China: ch'in (zither)—"Flowing Streams"
India: raga—"Jaat Kahan Ho"
Blind Willie Johnson: "Dark Was the Night"
Beethoven: String Quartet #13, "Cavatina"

gravitation, although convenient for navigating between the planets, is in a sense "false."[1]

The era between the Copernican proposal of a heliocentric cosmology and the completion of that cosmology with Newton's theory of universal gravitation, known as the Renaissance and Age of Enlightenment, was a time of exciting intellectual "newness," geographical and astronomical exploration, and philosophical consolidation and attempted reconciliation between science and religion. It was a time when tiny enlightened Holland became a world power and produced such free thinkers as the astronomer Christian Huygens, the first microbiologist, Anton van Leeuwenhoek, and the philosopher Benedict de Spinoza. It was a time when imagining what was possible was being pushed to new frontiers. Huygens introduced a new theory of light, improved the astronomical telescope, and discovered the rings of Saturn. Van Leeuwenhoek used the developing knowledge of light and lenses to construct the first microscopes, observing blood cells, protozoa, bacteria, and spermatozoa and adding the microcosm to the rapidly expanding horizons of exploration. And Spinoza, who made a simple living grinding and polishing lenses for the new microscopes and telescopes, astounded the world with a new conception of God. It was also the time of the French philosopher René Descartes, who was busy creating a new mathematics and a new world view that attempted to reconcile his Catholic faith and the new world of science, and the German philosopher Gottfried Leibniz, who boldly echoed the confidence of his contemporaries by proclaiming that this new world revealed by these physical and intellectual explorers was "the best of all possible worlds."

Newton and a Greater God

Every solution to a puzzle creates in its wake new problems. Kepler solved the astronomical riddle of planetary motion, but in the process generated far-reaching physical and philosophical questions. The Aristotelian-Ptolemaic universe was a cozy place imbued with direction and purpose literally from top to bottom. Based on what philosophers call *teleological causation,* in which a future goal causes actions in the present, God's existence made for a goal-directed universe, and everything that happened, including planetary motion, resulted from a striving to fulfill a natural purpose ultimately reducible to the mind of God. Because humans are familiar with having future goals that cause actions in the present, and because much of our lives can be understood

---

[1] To say that a theory is false assumes that one is a realist. Even if one is an instrumentalist, it must at least be said that Newton's theory of gravitation has a limited realm of applicability, that it cannot accurately represent all of the facts of observation.

from this point of view, why not understand God and the universe this way as well? (This is also a strange kind of causation. The cause is always in the future of an event happening now.)

In this teleological universe there were *natural motions, natural places* or spaces, and a natural *hierarchy,* an up and down, with certain places and motions closer to the perfection of God. Thus, the natural place for heavy things was "down" toward a natural stationary geometric center, and the natural place for light things was "up" because such things were thought to be closer to a spiritual substance. The higher one went, the less material things became until the celestial sphere of the stars was reached, the demarcation between God's creation and a spiritual realm.

In this cosmology God is separate from His creation, or transcendent, to use philosophical terminology. Most important, however, is that in spite of His otherness, God is in this universe in an intimate way. The "thoughts" of God are in the universe in a sense. His "purpose" is the continual, sustaining cause of all motion. Remove God and purpose from this universe and everything would fall apart in chaos.

There was no need to explain planetary motion further in this cosmology. Circular motion around a geometric center was considered a natural motion; it was an explanation in itself. As the most perfect motion from the geometric point of view of God, nothing further needed to be said. Planets move the way they do because they are striving to fulfill the goal of perfect motion. For this reason, having circular paths for the planets was very important for Copernicus. He attempted to preserve the notion of a circular motion as a natural motion, but as we have seen, his system required circular motion around invisible points that in turn revolved around other central points, and few of his contemporaries failed to see major problems with such a scheme. With Kepler's conception of an ellipse it was no longer possible to maintain that the motion of the planets was a natural motion requiring no explanation. Now a physical mechanism that moved the planets and kept them in their orbits was needed, and the lack of such a mechanism could no longer be swept under an old cosmological rug.

A related problem concerned the moving Earth. If the Earth is spinning on its axis and is not the geometric center of the universe, what kept objects from flying away when they are thrown into the air? The genius of Isaac Newton (1641–1727) and his theory of universal gravitation unified celestial and terrestrial events into one comprehensive explanation. However, this scientific breakthrough needed a conceptual guide. In this new universe places called "up" and "down" no longer existed, and a new cosmology was needed for the people of this time to understand all the ramifications.

In Chapter 5 we saw how Copernicanism had the unexpected consequence of conveying a renewed interest in the doctrines of the ancient atomists. Also, Bruno had argued that an infinite universe, populated with an infinite number of suns, was more consistent with a greater God. With no center and no up or down, hierarchy and special places vanished. This was a radical idea to the people of the seventeenth century.

[The] new philosophy casts all in doubt. . . . The Sun is lost, and the earth. . . . 'Tis all in pieces, and all coherence gone.

JOHN DONNE, *ANATOMY OF THE WORLD*

Although Galileo's use of the telescope had revealed stars where only empty space was seen before, it was not until the nineteenth century that observations with the telescope were refined to the point of being able to demonstrate that different stars were different distances away. For Copernicus, Kepler, and Galileo the heliocentric universe was a finite universe with the Sun in the center and the sphere of the stars the end of the physical universe. Although Galileo had argued that biblical Scripture should be interpreted in the light of scientific discoveries, rather than the other way around, and had stressed the importance of observation, quantitative measurement, and experiment in the pursuit of understanding God's creation, rather than relying on traditional, authoritative ideas, traces of ancient explanations persisted nevertheless in his thought as well as that of Kepler. Galileo could not shake the hold of the idea of circular motion, and Kepler, because of his Neoplatonism, was not comfortable with the idea of an infinite universe in which the importance of the Sun would be nullified. Thus, further conceptual help was needed, and the notion of a greater God contributed greatly to the preservation and discussion of radically new ideas necessary to complete the Copernican revolution.

Having an infinite universe with no special places suggested a liberating implication in terms of understanding the simple motion of a physical body through space. Specifically, it led to the idea of inertial motion: a physical body at rest would remain at rest and one moving in space would continue to move in a straight line until a "force" acted upon these bodies to alter their motion.[1] If there was no special place, then no explanation was needed for a body doing what it was already doing. Only an alteration of motion or "acceleration" needed explanation.

Descartes:
On the Shoulders of Giants

If I have seen farther than others it is by standing on the shoulders of giants.

ISAAC NEWTON

The French philosopher René Descartes (1596–1650) was the first scientist-philosopher to take these strains of thought and organize them into a comprehensive, influential world view. For Descartes, the originator of analytic geometry and what is referred to today as Cartesian coordinates, the universe

---

[1] But how do we know that a body is at rest or moving "in a straight line"? At rest in relation to what? Straight in relation to what? See the following discussion of absolute space.

was a huge mechanical machine full of material atoms, or "corpuscles," and all change could be explained by the free movement of these particles of matter or collisions between them. Of particular importance for the later development of Newtonianism and the theory of gravity was Descartes's clear statement of the law of inertia. A physical body that was at rest would remain at rest, and one that was moving in a straight line would maintain a constant speed in the same direction until deflected by a "force" caused by a collision with another physical body. (Today, this is known as Newton's first law of motion.) Thus, motion and change of motion were explained not by reference to natural places as in the Aristotelian cosmology but by forces caused by collisions between material things. The goal of natural science was now clear: to find the mathematical laws that govern the motions, interactions, and combinations of material particles.

With the concept of inertia clearly stated, because the planets move in elliptical orbits and not in straight lines, a force in the direction of the Sun needed to be explained to account for the curved motion of the planets. Thus, we find a direct conceptual connection among the rejection of a teleological explanation for planetary motion, a renewal of atomism, the idea of a greater God and an infinite universe, and the ultimate solution of Newton, a quantitative representation of the force of gravity. Descartes attempted to solve this problem himself with a "vortex" theory of colliding particles, whose collisions supplied the force that pushed the planets toward the Sun. His theory failed, and we need not be concerned with the details here except to note that Descartes's force is caused by the direct contact of colliding physical particles of matter. Descartes's philosophy and attempt at a solution were very influential in the later seventeenth century.

Eventually, in contemplating Kepler's third law, Newton deduced mathematically that if the squares of the speeds of the planets were related to the cube of their distances, then the force of attraction between the planets and the Sun must decrease with an inverse proportion to the square of the distance of any given planet. It was a mathematical achievement unparalleled in the history of thought, and over a decade later Newton completed this insight by showing that if this inverse square law was generalized, it explained Kepler's first and second laws as well. If this force was universal throughout the solar system, then the planets must follow the path of an ellipse and move just as Kepler said they did, at variable speeds, covering equal areas at equal times in relation to the Sun.

The idea of an attracting or emanating force of some sort was not new. What was revolutionary was a precise mathematical treatment of this force that explained why Kepler's laws worked in accounting for the motions of the planets and why physical projectiles fall back to Earth. It was exactly what these inheritors of Copernicanism, who denied a special difference between the Earth and the celestial motions of the planets, and the mathematically based philosophies of the past demanded—a simple mathematical law that unified celestial and terrestrial motions.

As surely as the person born with six fingers or the calf with two heads, Isaac Newton was a mutant. . . . His intellect was too profound, his capacity for rage too great, his desire for seclusion from the outside world too obsessive, his passion for original thought and scholarship too exclusive. He was the incarnation of the abstracted thinking machine that some scientists predict humankind may become through the natural process of evolution.

GALE E. CHRISTIANSON,
IN THE PRESENCE OF THE CREATOR: ISAAC NEWTON AND HIS TIMES

Gravity, Action at a Distance, Absolute Space and Time

But what was gravity? Unlike Descartes's theory that involved contact forces, the gravity of Newton mysteriously "acted at a distance." How could the Moon and the Earth attract each other when there was no physical connection or contact? It was either an "occult" quality or a special inherent force possessed by all physical objects. What was gained scientifically by replacing mysterious special places with a mysterious special force?

Newton tried very hard to persuade his contemporaries that nothing was gained and attempted to maintain a positivist, or instrumentalist, interpretation of the concept of gravity. It is not the job of scientists, he argued, to assert "hypotheses" about the ultimate nature of things. The job of a scientist is to capture what we see in a precise mathematical web of relations; it is to explain the "how" of experience, not the "why": how one set of observations at a particular time will result in a new set of observations at a future time. Thus, Newton is credited with being the first authentic scientific *positivist*. In other words, he recognized the necessity of banning all metaphysical speculation from science and proceeding only with a strict empirical methodology. Only that which can be verified by reference to our experience of the world is admissible into science. Gravity cannot be seen or touched; it is just a name we give to the results of what physical bodies do. It is an instrument that practical scientists use to "save the phenomena." What is important is what physical bodies do. Only the physical observable results matter; scientists need not speculate about what gravity "really is."

By now, however, it should be clear to any reader of this book that eliminating all philosophical assumptions from science is neither possible nor desirable. This completion of the Copernican revolution was as much a change in the European conception of God as it was a change in a view of the physical world. Rather than a hierarchical-teleological man/universe/God relationship, we now find the universe thought of as a perfect machine with God the creator of the machine and less involved in its day-to-day activities. Mathematics is still the basis for the machine's operation, and human beings still possess the special gift to read the basic floor plan for the universe, but explanations no longer need to be made in terms of some purpose. (Descartes specifically stated that it is not possible to know God's purposes.) Now, simple mathematical relationships between particles or aggregates of moving matter are sufficient. Everything knowable is assumed to be reducible to mathematical-mechanical relationships of points of moving physical matter. Although a very different conceptual scheme than that of the Middle Ages, it is a metaphysics nonetheless. In this light, we find Newton, in spite of his professed positivism, actively contributing to the metaphysical foundations of the science that bears his name and speculating again and again on the ultimate "why" questions relating man, God, and the universe.

For instance, the Cartesian-Newtonian emphasis upon the universe being a perfect machine raised a major question about the role of God. In the Aristotelian universe of the Middle Ages, God was present in the universe in an intimate way: The thoughts of His purpose caused all natural and sustained motion. In the new mechanical universe was any room left for God? Was anything left for Him to do? On these questions we find Newton commenting on as many theological as physical issues. As we will see, God is "scientifically required" to complete the Copernican revolution.

A complete discussion of the theological underpinnings of Newtonian science would require a more detailed discussion than is feasible here. Instead we will confine our discussion to a few examples that are especially relevant to Chapter 7 and Einstein. A central controversy for Newton involved notions of *space* and *time*. In Newtonian physics, in order to mathematically capture an object's motion it is necessary to give a spatial location to that object in three-dimensional Cartesian coordinates and a time, so that, using a mathematical equation of transformation, one can identify a future or past position of that object. In this way we can know the path of cannonballs, the location of comets past and future, the occurrences of eclipses past and future, and where *Voyager* will be in the year 2000.

In practice, however, the spatial and temporal positions of these bodies are always relative. For instance, if we could place a special satellite half way between the Earth and the Moon in such a way that the satellite maintained an equal distance between the Moon and the Earth as the Moon revolves around the Earth, how would we measure the satellite's place? Suppose we were able to impart to the satellite one half the orbital speed of the Moon, such that the satellite required the same amount of time to revolve around the Earth as the Moon. Because the Earth spins on its axis once every 24 hours, from Earth the satellite would appear to move as the Moon does. However, because the Moon rotates on its axis in about the same amount of time as it revolves around the Earth (27⅓ days), the Moon always shows the same face to the Earth, such that from the Moon the satellite would appear to be at rest or stationary. From one point of view the satellite is moving in space and time, and from another point of view it is at rest in time. Which is the truth? Is there a real motion, what physicists refer to as an "absolute" motion?

From a practical point of view this is not a severe problem, because an observer on the Moon and one on the Earth would be able to communicate with each other about the motion of the satellite. In spite of their different perspectives they would be able to know that their observations are symmetrical. By knowing the speed of the Moon around the Earth, and using what is known as the Galilean transformation, they would be able to mathematically deduce the way the other was seeing the satellite's motion.

For another example, consider two observers, one on a train moving at 60 miles per hour and another stationary with respect to the ground. If the observer on the train walks in the direction of motion of the train at 3 miles per hour, the observer on the ground would measure his speed at 63 miles per hour. From one point of view they would disagree on the motion of the

Absolute, true, and mathematical time, of itself, and from its own nature, flows equably without relation to anything external, and by another name is called duration: relative, apparent, and common time, is some sensible and external . . . measure of duration by means of motion, which is commonly used instead of true time; such as an hour, a day, a month, a year.

ISAAC NEWTON

observer on the train. The observer on the train experiences his speed at only 3 miles per hour. Again though, by knowing the speed of the train, and using the Galilean transformation, which involves only simple arithmetic, the observers could deduce the perspective of the other and know that their observations are symmetrical. In this way they could measure an event and know that it was the same event.

Intuitively though common sense seems to want something more. Is there a real motion in these examples? Is there a vantage point where it can be stated categorically what the satellite's motion is? On January 24, 1986, *Voyager* made its closest encounter with Uranus. At the time it was traveling at a speed of 49,000 miles per hour in relation to the Sun. In relation to the Earth, it was traveling at over 80,000 miles per hour. We know this by using the simple addition and subtraction of the Galilean transformation and knowing the relative speeds and positions of the Earth and Uranus as they revolve around the Sun. But is there a real speed for *Voyager,* some spatial vantage point that would allow us to know its true speed?

Another puzzle for Newtonians was the measurement of time. January 24, 1986, is a relative measurement of time. On Earth we would measure the duration of our special satellite's motion relative to the Earth's rotation, approximately 24 hours, or a complete day. But suppose we observed the satellite from the planet Mercury with a powerful telescope. The days on Mercury are very long, requiring 176 Earth days for one Mercury day. Thus for someone accustomed to Mercury time the satellite would require only about ⅙ of a day to revolve around the Earth. This could get confusing. An event that appears to take an entire day on Earth, due to the Earth's rotation, could take only a small portion of a day on Mercury. What is the real time? We could, of course, synchronize two watches on Earth, and then fly one of them to Mercury, so that any observer on Mercury would have the benefit of real Earth time. But then how would we know that the two clocks stay synchronized? An observer on Earth and another on Mercury would need to send messages back and forth to verify that the watches were still synchronized. This presupposes that we can easily account for the speed and time that it takes the message to travel to Mercury. How would we measure this time? What would be our reference point?

Our common sense usually enters here and says this is silly. Time is time, isn't it? Barring any malfunction in the watches, time marches on regardless of what human beings are doing with their watches or where they are. Time is objective, isn't it? As with the intuitive demand for an absolute space, our common sense demands that there is a real time, a standard time, what physicists refer to as an "absolute" time. All other times are simply practical reference points, needed for measurements, but not the real time. Would it make any sense to say that time can slow down? Could time speed up in one part of the universe and slow down in another? Just as Tycho Brahe realized the implications of a moving Earth and found the whole concept absurd because it implied a large amount of wasted space, so most Newtonians realized the implications of the possibility of relative space and time, but rejected these preposterous violations of common sense.

At stake was a world view that preserved common sense and the notion of a God-initiated universe of absolute space with an infinite amount of room for material things to move around with a machinelike precision. Started by God, and ticking away at precise mathematical intervals ever since, time flowed in this universe from past, present, to the future. Each point in time constituted a definite allocated tick of a universal clock-time. The problem for Newton and his contemporaries was that experience did not support absolute time. Every measurement of experienced space and time was and is a relative measurement, necessarily relative to some reference point. The proposal of an absolute space and time was thus a metaphysical assumption that went beyond experience. What we experience is an Earth-centered time and an Earth-centered space, time and space measurable from the point of view of a stable Earth. In Newton's day, even though scientists were well aware that the Earth moved, there were no experiences or measurements from a perspective beyond the Earth, so it was easy to believe that there was but one time and one space. As we will soon see, Newton and most of his contemporaries were fooled once again by the appearance of the Earth as a stable, central platform. (As we will see in the next chapter, Einstein recognized that physicists, in asserting an absolute space and time, were making an assumption that had no empirical justification.)

It was important, however, for Newton to preserve the notions of an absolute space and time not only for the world view and common sense but also to give God something to do in this machinelike universe! If the universe was a perfect physical self-sustaining machine, what role did God play? God must become the divine reference point, the absolute vantage point we seek to keep time and space stable and avoid paradox.

Both Galileo and Descartes had already addressed this problem. For Galileo, although God ceased to be the immediate cause of all motion, He was at least retained as the first cause, the creator of all the atoms, the supreme mechanical inventor. But compared to God's previous intimate role, this was a minor involvement, almost an embarrassing heretical afterthought. What was to be done with all the important nonmaterial concerns of the human race? What were to become of values, of moral laws, of all the spiritual considerations of humankind's ultimate concern that the Church had always affirmed a direct understanding?

Metaphysically, many of the immediate successors of the new science adopted an extreme materialism. Human thoughts became what philosophers refer to as "epiphenomenal events." Thoughts and ideas were demoted to *secondary*, almost illusory, results of the *primary* process of the motions of atoms. Just as steam is a result of, and could not exist without, boiling water, just as heat is not primarily real, but the secondary result of moving and colliding particles of matter—so thoughts, ideas, feelings, and values were relegated to "phantasms." What started as a heuristic or methodological device—nar-

I do not know what I may appear to the world; but to myself I seem to have been only like a boy playing on the seashore, and diverting myself in now and then finding a smoother pebble or a prettier shell than ordinary, whilst the great ocean of truth lay all undiscovered before me.

ISAAC NEWTON

rowing our focus of attention to specific physical items for mathematical manipulation—had now become a metaphysics. Here is Galileo's own description of the new situation:

I feel myself impelled by the necessity, as soon as I conceive a piece of matter or corporal substance, of conceiving that *in its own nature* it is bounded and figured in such and such a figure, that in relation to others it is large or small, that it is in this or that place, in this or that time, that it is in motion or remains at rest . . . but that it must be white or red, bitter or sweet, sounding or mute, of a pleasant or unpleasant odor, I do not perceive my mind forced to acknowledge it necessarily accompanied by such conditions. . . . Hence I think that these tastes, odors, colors, etc., on the side of the object in which they seem to exist, are nothing else but mere names, but hold their residence solely in the sensitive body; *so that if the animal were removed, every such quality would be abolished and annihilated.* [Emphasis added.][1]

Thus, if human beings are removed from the universe, all the secondary qualities of color, emotion, and thought disappear. But this notion then applies to all thoughts—including moral values and the thoughts of scientists! By extending Copernicanism to the point of removing the human viewpoint from fundamental importance, by acknowledging the human point of view as only a perspective, it was a short logical step to throw the baby out with the bath water. Might not science just be a perspective?

Descartes's Compromise:
Mind-Matter Dualism

Descartes was very much aware of the connections among the persecution of Galileo, the war that had erupted between the domains of physical science and religious orthodoxy, and the philosophical problems of the new science. As a Catholic who would not think of denying an ultimate religious inter-

---

[1] *Opere Complete di Galileo Galilei,* quoted in W. T. Jones, *Hobbes to Hume: A History of Western Philosophy,* 2nd ed. (New York: Harcourt Brace Jovanovich, 1969), pp. 115–116.

pretation of the universe and as a philosopher-scientist who made significant contributions to the new revolution in scientific thought, Descartes reveals in his philosophy the tension of an age in transition.

For Descartes the only solution to this tension is a philosophical compromise. There is indeed no room for God, or at least very little, in this machinelike material world. The greatness of God, however, can be preserved because besides this material world, which God created and set in motion, there is another world, another dimension of existence parallel to the material world of extended three-dimensional space and mathematical time: the world of unextended thinking spirits. The material universe and a spiritual universe are parallel to, but independent of, each other. The material universe started by God does what it does regardless of what human beings wish or think. It is just there. On the other hand, there is an inner realm, a realm of thinking, of ideas, of feelings, of willing, of imagining, a separate spiritual sanctuary where the ultimate concerns of God and humankind communicate and play out the traditional religious drama of moral striving and salvation. This *dualistic* metaphysics has room for both the theologian and the scientist: The former is in charge of the spiritual universe, of the questions of ultimate meaning and right and wrong, while the latter is in charge of the physical universe and questions of the nature of things, the motions of the planets, and the structure of the material universe.

Thus, with Descartes the first systematic attempt is made to reconcile the various strains of thought emerging in the seventeenth century as a consequence of the success of Kepler and Galileo. Descartes was an atomist who made room for a spiritual God, a scientist-mathematician who recognized the importance of the empirical element in the new science, but ultimately a Renaissance Neoplatonist who relied on the apparent certainty of mathematics to preserve a sense of certainty in our knowledge. As a Renaissance man he was also thoroughly Greek in his outlook: Happiness is a marvelous game of fulfilling our potential; the human species has an unlimited ability to discover and explore. Descartes was also a Christian, who could not think of denying God's intimate role in the affairs of human beings.

Of all the things that may be said of Christianity, we cannot deny that it was the first truly democratic philosophy. The ancient Greeks had taught that happiness was the result of developing our potential to know, but they did not think of this being equal in all individuals. (In Athens, for instance, although there were 400,000 people, only 40,000, or 10 percent, were considered citizens.) Descartes was one of the first Europeans to advocate the concept of universal education: Just as all members of the human species are equal in the eyes of God, so the potential to know and wonder are equal as well—and the world will not be a better place until all are allowed to participate in developing this potential.

Try to understand what Descartes was attempting to preserve with this metaphysical dualism. Earlier, the British philosopher Thomas Hobbes had looked at the success of Kepler and Galileo and boldly claimed that the *primary qualities* of physical atoms were the only reality. The thoughts of people and

Whoever has undergone the intense experience of successful advances made in [science], is moved by profound reverence for the rationality made manifest in existence.

ALBERT EINSTEIN

all *secondary qualities* are "phantasms" caused by the motion of atoms. From this an epistemological paradox followed: If our thoughts are secondary fabrications, then how do we *know* that reality is only atoms and empty space? If the thoughts of scientists were nothing more than the motions of atoms, then the thought that the world is made of atoms has the same status, and we cannot say we know that atoms are real! The world view implied by the new science cannot have the validity and priority that the success of the mathematical-experimental method conveyed. Science becomes just a perspective, a "recommendation" for Hobbes. It can only be recommended that if we adopt this point of view we will be more successful. Thus, Hobbes adopted an instrumentalist interpretation of scientific theories: We cannot say that the new science is true, only that it works.[1] Could perspectives other than that of the new science work? Yes, said Hobbes. We cannot be sure that the universe has been constructed on the basis of the mathematics that we "make up." That our mathematics works is probably a "coincidence." Hobbes saw that this implied relativism, and the only way he saw out of the problem was to suggest that every society have a supreme ruler who would command a particular perspective.

Descartes perceived the epistemological, religious, and moral problems inherent in this view. If it were true, then not only all religious sentiment and all human value but also our scientific knowledge of the motions of atoms must be illusions. If all the thoughts of thinking beings were the secondary results of moving matter only, then all the thoughts of scientists would have the same status as that of "color." If our experience of color is not real in the sense of accurately reflecting what objects are "in themselves," then how do we know that the mathematical conclusions of science are real?

According to Descartes, with a spiritual universe (or a universe of "mind") parallel to a physical universe, the problems of finding a place for religion and value, as well as establishing a secure foundation for scientific knowledge, are solved. The reason mathematical ideas found in human minds work so well in mapping the physical world of atoms is that there is a preestablished parallelism and harmony between the structure of the physical universe and the human mind. Mathematical ideas are "innate," planted by God in our minds, as it were, in the beginning. How do we know this? If God is good, thought Descartes, he would not deceive us! Although not a solution that will win many converts in the twenty-first century, it was a philosophy that was very influential at the time.

---

[1] Notice how every time there is a problem of interpretation or a paradox resulting from a successful scientific application, scientists and philosophers will adopt an instrumentalist interpretation. If an epicycle does not make sense, it must be a calculation device; if gravity is mysterious, it must be just a name to summarize observations. In Chapter 8 we will see this happening again as scientists argue today about how to interpret the mathematical equations dealing with the atom.

Descartes, however, had been vague about the extent to which God is still involved in the material universe. In spite of the Cartesian compromise providing religion with a universe of its own, it was perhaps too early and too dangerous to completely separate God from the day-to-day motions of the material universe. Thus, we find Descartes speaking of God maintaining the vast machine by His "general concourse," as if the laws of nature and all physical motion would cease if God were to take His divine awareness from them. Later this view was criticized as being actually demeaning to God. Could not God create a perfect machine in the beginning that required no divine maintenance? For instance, the German philosopher Leibniz, who simultaneously with Newton developed a mathematical calculus, claimed this view of divine tinkering was absurd:

According to their doctrine, God Almighty wants to wind up his watch from time to time, otherwise it would cease to move. He had not, it seems, sufficient foresight to make it a perpetual motion. Nay, the machine of God's making is so imperfect according to these gentlemen, that he is obliged to clean it now and then by an extraordinary concourse, and even to mend it as a clockmaker mends his work. . . . According to my opinion, the same force and vigor remains always in the world, and only passes from one part of matter to another, agreeably to the laws of nature and the beautiful preestablished order. . . . Whoever thinks otherwise, must . . . have a very mean notion of the wisdom and power of God.[1]

Newton and the Divine Sensorium

Newton vacillated between these two positions. When his empiricism was strongest he treated the universe as a neutral, self-sustaining machine that does not require religious references or philosophical hypotheses to be dealt with successfully. But when certain puzzles presented themselves from the

Come celebrate with me in song the name Of Newton, to the Muses dear, for he Unlocked the hidden treasures of Truth. . . . Nearer the gods no mortal may approach.

EDMOND HALLEY, "ODE TO NEWTON"

[1] As quoted in Edwin Arthur Burtt, *The Metaphysical Foundations of Modern Physical Science,* rev. ed. (Garden City, N.Y.: Doubleday, 1954), pp. 292–293, from David Brewster, *Memoirs of the Life, Writings, and Discoveries of Sir Isaac Newton,* 2 vols. (Edinburgh, 1855), II, p. 285.

implications of universal gravitation, God entered the picture to make things clear and preserve the perfect order. For instance, if gravity is universal, then, according to Newton, unless God placed the stars at an immense distance from each other and intervened to keep them at such a distance, the universe would eventually collapse at a common center. (This answered Tycho's wasted space problem.) Also, because the gravitational pull of each planet and satellite body cause various irregular motions in each other, and because comets move in all kinds of eccentric orbits, the ordered motion of the solar system would eventually become quite disordered if God and His "reforming" influence did not intervene.

It is with the notions of an absolute space and time, however, that God's greatest role was reserved. To avoid paradox, the new science of mechanics seemed to require the notions of absolute space and time. The very notions of velocity, of acceleration, of motion and rest seemed to lose their meaning, if space and time were not stable. Unfortunately, as we have seen, it was impossible to justify these notions empirically. Thus, Newton sought a theological rescue, appealing to the notions of God's *omniscience* and *transcendence*. Few people in Newton's day questioned the all-knowing capability of God. According to Newton, unless there was a divine reference point for the infinite theater of God's creation, unless there was a "divine sensorium" for the divine consciousness, God would get "confused" and be incapable of knowing everything at once! It would be impossible for any form of consciousness to keep track of the enormous number of motions in the universe, if there was not a divine reference point, a central place from which to view everything.

God, of course, could never be confused, so there must be an absolute space, a central reference point, perhaps unknowable to the limited minds of scientists, but "out there," nevertheless, from which all motions could be measured. From such a reference point the motion of our hypothetical satellite could be measured absolutely; it would have only one real motion. A similar line of reasoning applied to time. In order for God to keep track of everything, the universe must be started by God at time zero and be clicking away at uniform, mathematical intervals ever since. From time zero until now there has been a precise series of universal slices of "nows" throughout the universe. At each "now" every event occurring at that time has the same time and is known by God based upon the ability of this great intellect to know all the past events and past "nows." Just as the stable platform of Earth allowed a projection of a universal space and time, so the way human beings think allowed Newton to state how God must think.

A half century later, the implications of this conception of space and time were immortalized in a statement by the Newtonian the Marquis Pierre-Simon de Laplace (1749–1827).

An intelligence knowing, at any given instant of time, all forces acting in nature, as well as the momentary positions of all things of which the universe consists, would be able to

Newton's belief that God was on his side never led him into that fatal . . . illusion that God would do the work. . . . His mind functioned like a powerful achromatic lens, collecting disparately tinctured rays of thought and bringing them to a single unblemished focus.

GALE E. CHRISTIANSON,
IN THE PRESENCE OF THE CREATOR: ISAAC NEWTON AND HIS TIMES

comprehend the motions of the largest bodies of the world and those of the smallest atoms in one single formula, provided it were sufficiently powerful to subject all data to analysis; to it nothing would be uncertain, both future and past would be present before its eyes.

Although an agnostic,[1] Laplace underscores how important it was for a Newtonian to be part of a stable, deterministic universe in which the relative perspectives of space and time, necessary for practical observations, do not influence what is real. With a stable space and time, to a superintelligence such as God the entire universe can be represented by one unimaginably complex mathematical formula. The formula would describe every motion, past, present, and future. Every particle that makes up my body, your body, the smallest piece of a star in the Andromeda galaxy—the motions of these particles would be known instantaneously to this superintelligence, how each interacted in the past, and how each will interact in the future. Every thought caused by the motions of particles in my brain, every action I will ever undertake, would already be known to this superintelligence the moment the paths of the first motions of the universe were plotted.

It all made sense again. What before was a universe of purpose with special places of "up" and "down" is now a perfectly crafted mechanical place where every location is its own special place and every time is its own special time. To complete the world view, absolute space and time replaced "up" and "down."

Epistemological Implications of Newtonianism

The most important epistemological aspect of Newtonianism, and the world view that completes the Copernican revolution, is that *what is real does not depend on us*. For our purposes this notion is the essence of Newtonianism. God does not change the universe by knowing it, and we do not change the path of a planet by knowing it. For many this notion was simply another extension of Copernicanism—an epistemological removal of humankind from importance.

Like looking through a pane of glass, the material world is independent of us, and our thoughts and values do not affect this reality.

---

[1] Laplace purportedly once replied to Napoleon when asked why he did not mention God in his book about the Universe, "I have no need for this hypothesis."

This conception of the relationship between the minds of human beings and reality should not be difficult to understand. To a large extent our everyday thinking is governed by the Newtonian paradigm. When I observe a tree or any object of the outside physical world, I do not think that my observation or thought has any effect on the real tree. My observation is passive, and if my senses are operating properly I assume that I will see the real tree. My thoughts of the tree are powerless. No matter how hard I try, I cannot "wish" the tree to be an apple tree if it is a lemon tree. My point of view does not affect where it is or how long it has existed. Similarly, depending upon my location, I may measure a spatial location and time differently from another observer, but I assume that this difference is only due to the angle of my point of view. It is not real. There is only one space, one time, and one universe.

Thus, Newtonianism consists of epistemological as well as metaphysical assumptions. In this world view the observations that make up our knowledge of the world are assumed to be like someone watching the world behind a thick pane of glass. We are objective, passive observers, and our measurements do not influence the reality of what we are measuring. It is a testimony to the success of Newtonian physics and its close alliance with common sense that few people ever think of such obvious thoughts as philosophical assumptions. Although there were philosophers of Newton's time who recognized these assumptions, it was not until the twentieth century that science had uncovered enough problems to bring these assumptions into clear focus for a thinker such as Einstein.

Science as a Religion

That deeply emotional conviction of the presence of a superior reasoning power, which is revealed in the incomprehensible universe, forms my idea of God.

ALBERT EINSTEIN

As with Kepler and his Neoplatonism, it is easy for the modern scientific positivist, living at a more secular time, to claim that all this theological and metaphysical silliness was unnecessary for the tremendous success of Newtonian physics. We do not find God as a variable in any of the mathematical equations used to plot the trajectories of the *Voyager* flights. Unfortunately, a controlled scientific experiment cannot be undertaken to see if the success of Newtonianism and the completion of the Copernican revolution could have taken place without such cultural baggage. It is true that Newton, unlike many before, strongly emphasized the importance of just observing the world and finding the correct mathematical laws to relate our observations, avoiding speculations about what invisible realities might be causing these observations. But this is easy to do, if you already believe that the universe is a perfect machine whose surface features follow a precise mathematical plan. Just as the facts of planetary motion were important to Kepler, because only in knowing the facts could he read the mind of God, so the observational facts and a little mathematics were all Newton needed to read the machine's operation.

The notions of an absolute space and time, of a mechanical universe of atomic particles in motion, of a preestablished parallelism between the mind of man and the mathematical laws of the material universe, of a valid, but deemphasized, mental realm for the secondary qualities of human nature—these were the basic and generally accepted elements of a new world view. It was a world view constructed to answer the problems generated by the Copernican revolution: the problems of value, knowledge, humankind's place, and God's role. This strong emphasis upon the external physical world as primary also conveyed a greater dignity and value to it. This new dignity was simply a continuation of the Copernican principle—human nature and its place and idiosyncrasies are less special and important. The greatness of God demands more attention paid to His entire creation, not just to one insignificant speck of substance occupied by only one conscious creature of perhaps many.

Thus, the astronomer Huygens continued, for very similar reasons, the speculations of Bruno. In this new universe there must be an infinite number of suns and an infinite number of planets, otherwise God would have "wasted" a sun. And on each planet there must be creatures of some sort, otherwise God would have wasted a planet. Following the Copernican principle, because there is nothing special about us, these creatures would most likely be very different from us. They would not likely have the same physical features, and of course their cultures would be very different, just as the New World explorers were finding Indian cultures that differed from that of seventeenth-century Europeans. They would, however, according to Huygens have writing and geometry, otherwise they would not be able to appreciate God's great creation!

If the greatness of God's creation demands more attention, then it is a small step for another idea to emerge, an idea, not at all orthodox for its time, but significant nonetheless, that when one studies the physical universe, one is studying not just a dry ugly material substance but also something of great value, perhaps God Himself. The philosopher Spinoza had shown that a logical analysis of the traditional notion of God as transcendent, as a completely different substance from the material universe, was "demeaning" to God and contradictory. For instance, if God created an imperfect universe, then this is paradoxical, because God has created an imperfect thing. If God created a perfect universe, then this is also paradoxical, because there is then something perfect that is not God. Likewise, if the material world was completely *determined* by mathematical laws and God was aware ahead of time of the results of all motions, how is it possible that human beings are "free" to act out the great drama of salvation and be "responsible" for the result? If God has created me, and knows ahead of time exactly how I will behave, how am I responsible for any mistakes I make?

A person has the right to worship God according to his or her own metaphor.

KENNETH BURKE

In essence, Spinoza's solution involved the claim that God's nature is not transcendent but *immanent*. In other words, God is not some silly separate humanlike fatherly figure who sits on a gold throne in some heavenly dimension. God is a mystical, but rational, force that is "in" and "is" the universe itself. God is everything. The physical world is a "mode" of God. For Spinoza the language that best embodies this force, that captures its meaning for our limited human understanding, is the very same mathematical laws discovered by science. Thus, when one studies the cosmos mathematically, one studies God Himself, an object of "intellectual love." Religion is finally reconciled with science by making science a religion.

For Christian Huygens, an intellectual colleague of Spinoza, who looked out onto the beautiful mystery of space from cosmopolitan Holland, discovered the majestic rings of Saturn and other marvels, and felt the awe of being part of something profoundly meaningful and rational, but infinitely elusive, science was his religion and the entire universe his country. And when Albert Einstein, several centuries later, was asked whether he believed in God and replied that he believed in "Spinoza's God," we can understand how, for him, although religion without science is blind, science without religion is lame.

Concept Summary

The completion of the Copernican revolution with Newton's theory of universal gravitation not only made possible magnificent discoveries, but also produced a world view or cosmology midway between the Aristotelian-Ptolemaic cosmology and that produced by twentieth-century physics.

The Aristotelian-Ptolemaic universe was a purposeful, goal-directed universe. Events such as planetary motion were understood in terms of a striving to fulfill a natural purpose. Thus, this universe was conceived of in terms of natural motions, natural places, and a natural hierarchy. God played an important "intimate" role in this universe; His thoughts were the continual, sustaining cause of all motion. For instance, because from God's perspective circular motion is the most perfect motion, it is a natural motion, and the circular motion of a planet needs no further explanation.

Shifting the center of motion from the Earth to the Sun, and making planets revolve in elliptical rather than circular orbits, produced shocking implications for understanding humankind's place in the universe and God's relationship to His creation. New scientific, philosophical, and religious concepts were needed to place these implications in an understandable context. If no special motions existed, then a theory was needed to explain why the planets were attracted to the Sun and did not fly off into space; if the universe was to be understood as a mathematically plotted physical machine, then a new understanding of thought and values was needed; if God was not "in" the universe in the same intimate way, a new conception of His special role was needed. With help from the revival of atomism and the conception of a greater God, work by Descartes, Galileo, and especially Newton completed the Copernican revolution.

As the scientific details (inertial motion, the mathematization of nature, and universal gravitation) were worked out, a new epistemological, metaphysical, and religious world view was conceived. The Newtonian-Copernican cosmology that replaced the Aristotelian-Ptolemaic was a God-initiated, machinelike universe full of material things in motion; all motions could be plotted mathematically in terms of precise and (at least from God's point of view) unique (absolute) times and locations. This cosmology was coupled with a new perspective on the role of scientific explanation. The success of the concept of universal gravitation raised the question of the nature of gravity. Gravity seemed to "act at a distance" with no physical contact between gravitating bodies. Newton's response was to assert that it was not the business of scientists to worry about the ultimate nature of things. Rather, the job of a natural scientist was to be concerned only with what can be observed and to mathematically map empirical relationships. A natural scientist need not waste time on metaphysical hypotheses.

Nevertheless, Newton contributed extensively to the metaphysical foundations of this new science. That the universe is assumed to be a machine is a metaphysical postulate, one in fact which justifies a positivistic attention to the empirical surface details of the machine. If it is believed that the universe is a perfect deterministic, clocklike machine, then there is no need to worry about its inner mechanism; the surface behavior of the machine is all that is needed for practical interaction with the machine. Also, because any practical measurement of space and time, an essential part of the new physics, required a relative measurement of space and time, a metaphysical postulate was needed to preserve the objectivity of space and time. If a moving object did not have a place and a time to be in, paradox and confusion seemed to result. According to Newton, the infinite theater of God's creation required an absolute, celestial perspective, otherwise it would be impossible to keep track of all the motions in the universe. Common sense, the new science, and religious considerations all required that at least God would have this perspective.

Another problem concerned the role of thought and ideas in this new universe. If reality could now be completely reduced to the primary motions of material objects, then thoughts, ideas, feelings, and values became secondary, miragelike qualities. In addition to heightening the confusion of the role of God and religion, this conclusion seemed to undercut the validity of science as well: The thoughts of scientists would have the same subjective status of all secondary qualities. Science would be just a subjective perspective.

A famous metaphysical and epistemological solution to this problem was Descartes's mind-body dualism. A world of thought and spirit exists independent of, but parallel to, the material world of the natural scientist. This parallel mental realm preserved not only traditional concerns of ultimate meaning, but also scientific certainty and objectivity, because a harmonious correspondence must exist between the God-inaugurated, mathematical thoughts of scientists and the motions of the physical world.

Ultimately, however, the success of Newtonianism implied a stronger emphasis on the importance of the external material world. The material world is independent of us, and our thoughts and values do not affect this reality. At best we are good spectators when our thoughts and values stay consistent with this now dignified, glorified creation of God.

Suggested Readings

Seven Ideas That Shook the Universe, by Nathan Spielberg and Bryon D. Anderson (New York: Wiley, 1987).

For the general reader a revised edition of an earlier college text concentrating primarily on the history of Western physics from the Copernican-Newtonian paradigm to the present quark theories. Two chapters give a very readable summary of the transformation from the Aristotelian-Ptolemaic view of the universe to the Copernican-Newtonian view.

In the Presence of the Creator: Isaac Newton and His Times, by Gale E. Christianson (New York: Free Press, 1984).

Although this book focuses on biography rather than science or philosophy, it is a scholarly, entertaining, and at times dramatic compilation of just about everything you would ever want to know about Isaac Newton and England in the seventeenth century. For a more demanding version of Newtonian biography, see Richard Westfall's *Never at Rest: A Biography of Isaac Newton* (New York: Cambridge University Press, 1980).

Murmurs of Earth: The Voyager Interstellar Record, by Carl Sagan (New York: Random House, 1978).

A series of essays about the implications and complex making of the *Voyager* record. The effort, pictures, and greetings are quite stirring and are perhaps best summed up by Jimmy Carter's greeting: "This is a present from a small distant world, a token of our sounds, our science, our images, our music, our thoughts, and our feelings. We are attempting to survive our time so we may live into yours. We hope someday, having solved the problems we face, to join a community of galactic civilizations. This record represents our hope and our determination, and our goodwill in a vast and awesome universe."

A History of Philosophy, vol. 4, *Descartes to Leibniz,* and vol. 5, *Modern Philosophy: The British Philosophers,* by Frederick Charles Copleston, new rev. ed. (Garden City, N.Y.: Image Books, 1962).

Part of an eight-volume history of Western philosophy from the ancient Greeks to the twentieth century. Although there are many good introductions to the history of Western philosophy, this time-honored work is still one of the best in terms of completeness and objectivity. For a somewhat more readable account, see *A History of Western Philosophy,* vol. 3, *Hobbes to Hume,* by William Thomas Jones, 2nd. ed. (New York: Harcourt Brace Jovanovich, 1969–75).

The Newtonian Revolution: With Illustrations of the Transformation of Scientific Ideas, by Bernard I. Cohen (New York: Cambridge University Press, 1980).

Although this book has a controversial theme (it excuses Newton's theological concerns and the influence that had on the science of the time, while it magnifies the importance of the Newtonian positivistic "style"), it is nevertheless an excellent source for a detailed analysis of Newton's scientific thought. Written by a principal Newtonian scholar, the book focuses more on the transitional details and sequential influences of one scientist on another rather than on the macroscopic features of world view change.

The Metaphysical Foundations of Modern Physical Science, by Edwin Arthur Burtt, rev. ed. (Garden City, N.Y.: Doubleday, 1954).

An essential book covering the philosophical transformation from the Aristotelian-Ptolemaic conception of the universe to the Newtonian-Copernican. Covers Newton's theism and the philosophical roles of Hobbes, Descartes, Galileo, and other major figures in creating a new foundation for science. Particularly relevant is the excellent discussion of the epistemological problems created by the new world view.

CHAPTER 7

Our Time:
1. Understanding the Theory of Relativity

The dogma [absolute space and time] was so ingrained that the thought of questioning it never arose. It was Einstein's great genius to question the self-evident. The theory of relativity is not basically difficult. It may seem hard to come to grips with it and understand it, but the reason is neither mathematical nor conceptual. It is simply that the theory of relativity apparently violates common sense; that is, it overthrows the dogma. In actual fact, it does not violate common sense in everyday life at everyday speeds. . . . But it poses that threat, and since everyone tends to be basically conservative (that is, fears the unknown), it is much easier to say 'this is too hard for me' than to come to grips with it.

CLAUDE KACSER

In principle one could reach the year 2000 in a few hours.

PAUL DAVIES

Modern scientific epistemology . . . justifies discoveries of such far-reaching consequences as would, in former times, have been merely empty speculation, fantasies without empirical foundation.

HANS REICHENBACH

Slow Time and Fast Time

When you're sitting in a room listening to a very boring speech or lecture, it seems to take forever, and time "drags"; when the lecture is so interesting that it all seems to take place in a matter of moments, time "flies." But we all "know" that this is only a subjective experience, that the real time measured by an objective instrument such as a clock clicks along at the same rate whether or not we are interested in what is happening.

But do we really know this? Our knowledge of the objective world must be pieced together on the basis of our subjective experiences. For all we know at the moment we become engrossed in some interesting activity, at that precise moment everything in the universe "speeds" up, including the clock! Or perhaps at the precise instant that we become bored, the universe decides to "slow" everything down, including the clock! So when we look at the clock and see that an hour has passed, we conclude that an objective hour has passed, but how do we really know if it was a fast hour or a slow hour? If everything in the universe could speed up or slow down, then the clock would either speed up or slow down as well, and there would be no way to tell.

Preposterous you say. All that is needed is for two people to be in the room, one bored and the other not, and then have both see that based upon the objective clock time, one hour has passed. Case closed. As Socrates and Plato noted long ago, it is impossible for a single individual to be objective. The experiences of others and communication with others about what they experience are necessary conditions for understanding the world. Part of what we mean by "objective" knowledge is a public knowledge, a knowledge established by a community of observers. Also implied is a definite state of existence independent of our observations. In the case of time, it is what it is independent of our emotional states.

But wait. Suppose someday we finally discover another intelligent form of life separated from us by a great distance. Suppose that just by chance one day their entire planet is bored and everyone on our planet is not. Where will we now look to find an objective clock? How would we know that the universe does not slow down for them and speed up for us? We could synchronize two clocks on Earth and then transport one to this distant planet. How would we know that the two clocks stay synchronized? We would assume that if nothing is physically wrong with the clocks, they would stay synchronized. But unless we have a way of directly comparing the time these clocks are measuring, we would not know whether they are still synchronized.

This example is far fetched, but its possibility demonstrates that unless a universal standard of time measurement exists, we cannot say we know that time flows on uniformly as our common sense dictates and does not speed up

If you do not ask me what is time, I know it; when you ask me, I cannot tell it.

ST. AUGUSTINE

or slow down. Instead we must admit that we are *assuming* that time behaves throughout the universe the way we experience it here on Earth. We are assuming that on some other planet "now" is the same as "now" here on Earth, that at any given moment there is a *slice of simultaneity* throughout the universe.

This assumption is, of course, rather safe and reasonable for most practical purposes on Earth. But is it true? The history of science has shown us repeatedly that we should be very careful in projecting those views of reality that are practical, and work, as real. Most of what human beings do on this Earth can be accomplished by assuming the same set of beliefs accepted in the Middle Ages. We see the Sun moving everyday and we do not feel the Earth moving. We have learned, however, that our experience on this planet encompasses but a small portion of all that exists and that the universe is not required to conform to our view of things.

Strange indeed. The fact that we have no way of knowing if the universe is slowing down or speeding up at any given moment, that we must "assume" that it is not, is a paradox. It is the kind of thinking that the average person would not take seriously. It is similar to a perennial philosophical problem: that logically no one has ever been able to prove the existence of the external world! With problems like this, small wonder that few people major in philosophy at universities. History shows us, however, that great thinkers have always taken such paradoxes seriously, seeing them as nature's way of waking us up to some possible secret and revealing the fallibility of our "normal" thinking. In this chapter we will see that Albert Einstein's realization that Newtonian scientists assumed, like the rest of us, that time stays normal stimulated a great discovery.

It never occurs to me that if I leave my home at 8:00 and arrive at my office at 8:30 that I am assuming *that it is not then 9:00 at home.*

David Hume:
The Problem of Causality

In the eighteenth century the Scottish philosopher David Hume (1711–1776) discovered another paradox, a paradox that clashed with the astounding success of the new science of his time and its rudimentary technological applications. We saw in Chapter 6 that by the beginning of the eighteenth century Newton had discovered the laws of gravity, demonstrating and reinforcing the elliptical orbits of Kepler. Scientists were using these same laws and the machine paradigm to further our understanding of the physical world and begin the industrial revolution. The mechanical world view of Newtonianism would soon produce machines of mass production, power looms, factories, and the steam engine for ships and trains. In the light of such practical success, few would question this world view.

Hume discovered, however, that the "concepts" that served as the foundation for the success of Newtonian science were a philosophical puzzle. Take the fundamental notion of *causality*. We assume that the job of science is to successfully understand the real world, in part, by revealing the causes of things. Newtonian physics apparently did this job brilliantly. Gravity was the cause of both terrestrial and celestial motion. Into this arena of scientific certainty Hume stepped and arrogantly questioned this supposed knowledge with impeccable logic. According to Hume, we cannot know the cause of anything. For instance, if I throw a rock at a window and then watch the window break, there is no "logical" way to prove that the impact of the rock "caused" the window to break. It might be a coincidence. Night always follows day, but surely we do not want to assert that day causes night. We cannot see causation nor can we logically deduce causation, no matter how many rocks we throw at windows. All we can honestly say is that when we do such things again and again, there is a "constant conjunction" of events—throwing the rock, breaking the window.

Another paradox: The science of the time worked and implied that we knew not only the causes of the motions of the planets but also many important events relevant to daily life, and it held out the promise of a unified knowledge of almost everything important to us. As with causation, gravity could not be seen, and its action at a distance was not understood. Nevertheless, scientists were confident that a final understanding of the universe was at hand. The great French mathematician Jean Le Rond d'Alembert claimed, "The true system of the world has been recognized, developed and perfected." The influential physicist, physician, and mathematician Hermann von Helmholtz declared at an important scientific meeting that the job of science was almost over, that "its vocation will be ended as soon as the reduction of natural phenomena to simple forces is complete and the proof given that this is the only reduction of which phenomena are capable." For Helmholtz, the future of science amounted to no more than a "mopping-up" operation; Newton had already made all the important discoveries. All that remained were a little logic and experimental tinkering.

With Hume, philosophy was embarrassing the whole show. It could not be proved that the fundamental concepts of this understanding represented reality! In terms of logic, all the useful discoveries could be nothing more than a massive amount of coincidences. (Hume's work showed that the problem of induction could be applied to every Newtonian concept.) Why not? It had happened before. For thousands of years human beings had watched the "coincidence" of the Sun moving from east to west everyday. Belief in a stationary Earth worked when applied to navigation and astronomy, and still works, but was not, and is not, true. Such logical ruminations convinced Hume, perhaps not unlike Protagoras, that he should retire from philosophy, spending his

As the Sophists had challenged the ancient Greeks centuries before, Hume was challenging all Newtonians to demonstrate how this new science provided a durable knowledge, to show how the "truths" of this new science would not be overthrown by future experience.

Physics is finished, young man. It's a dead-end street.

Advice from Max Planck's teacher

time on more important things like politics and a successful career. But for a philosophically minded German physicist, Hume's work was the challenge of a lifetime.

Immanuel Kant:
Turning Subjectivity into Objectivity

Concepts without percepts are empty. Percepts without concepts are blind.

IMMANUEL KANT, *THE CRITIQUE OF PURE REASON*

Immanuel Kant (1724–1804) will never be noted for having been an exciting person. He never traveled more than a few miles from his home during his entire life, and he followed such a rigid routine that his neighbors could set their clocks based upon his daily habits. He chose instead the life of the mind and traveled far in the depths of inner space. His major book, *The Critique of Pure Reason,* will probably always be read by those who take philosophy, science, and the life of the mind seriously.

Kant was shaken by Hume's critique of Newtonian concepts. Hume awoke Kant from a "dogmatic slumber," to use Kant's words. Kant taught physics and philosophy at the University of Königsberg and had contributed to the theoretical applications of Newton's work by proposing a hypothesis of the origin of the solar system, known as the "nebular hypothesis." He also noted that the tides would, over a long period of time, slow the Earth's rotation, and proposed that earthquakes were caused by shifting and faulting of large sections of rock. But he quickly saw that Hume's work questioned the very foundation of not only Newtonian physics but all of science as well and any attempt to know the causes of things.

Kant wrote his book primarily for scientists and scientifically minded philosophers to allay their fears that any attempt at a unified body of knowledge was a hopeless dream. There was a solution to the problem, according to Kant, but the solution would have to be a radical one. Just as the Sun-centered system of Copernicus was a radical revision of our normal thinking about our place in the universe, a switch from being central to being simply just another planet, so the epistemological solution to Hume's problem would also be an "inversion" of our thinking about reason and knowledge.

According to Kant, Hume was right to an extent: There would be no way to logically deduce an objective reality of causal connections independent of our thinking about reality. We can, however, deduce from human experience a kind of "metaknowledge." We can deduce that causality and other fundamental scientific concepts, such as three-dimensional space, a universal time, and the basic principles of mathematics, will always be the framework for which all human observations of the world will take place. These concepts will always be with us and will never be overthrown by any future experience of reality because these notions are an inherent part of the human mind. They are the "filters" through which we will always view the world. No matter where we travel in this great universe, space, time, causality, and mathematics will be the same, not necessarily because the universe itself is always this way but because wherever we go, every interpretation, every observation made by human beings will always conform to our filters, the "categories of understanding" as Kant put it. The objective universe might conform exactly to the way we filter it, but there would be no way to know this. When we observe reality, we must do so through our filters, and there is no way of stepping outside of our minds and filters and seeing how reality is when we are not looking at it.

At first this idea was very difficult for the people of Kant's time to understand. In fact, Kant wrote two more books attempting to communicate the same basic theme to scientists. To the modern person his concept should be a little easier to understand. What he is saying is similar, but not identical, to the following well-known scientific examples of human perception.

When I view my room right now and listen to the sounds around me of birds singing and children playing, I assume this is a completely objective reality: that the colors I see of the objects around me and the sounds coming through my window are "there" just the way I perceive them. The "blue" of my coffee cup appears to be really there, on my coffee cup. Scientists, however, tell us that the "blue" is not really "out there." What is really there is a physical substance that absorbs certain wavelengths of light and reflects others. My coffee cup reflects a particular wavelength of light, which in turn strikes the rods and cones, the light-sensitive receptors in my eye, sending an electrical impulse to my brain, where the "blue" of my coffee cup appears. It is not really out there. It is an interpretation due to the neurophysiology of a normal human being. A creature with a different neurophysiology would not view reality the same way. Similarly, the sounds I hear are the result of a physical medium (the air) that vibrates as children play and birds sing. These vibrations are carried by the medium to my ear where three little bones are subtly rattled by the vibrations and eventually another electrical impulse is carried to my brain where I finally "hear." Without the physical medium of the air, I would not hear anything. In a vacuum there is no sound. Without light, and the different possible wavelengths of light, there would be no color. Without the right physiology there would also be no color and no sound. The colors of the objects in my room and the sounds coming in from my window are "interpretations," appearances that serve us well for practical purposes, but interpretations nonetheless due to our neurophysiological filters.

What Kant is saying about causality, space, time, and the practical application of mathematics is similar but more fundamental. Hume had shocked Newtonians by suggesting that the basic principles of science were secondary rather than primary—that causality was not a designation of a primary quality of objects but rather the subjective "result" of human perception with the

We ourselves introduce that order and regularity in the appearance which we entitle "nature." We could never find them in appearances had we not ourselves, by the nature of our own mind, originally set them there.

IMMANUEL KANT, *THE CRITIQUE OF PURE REASON*

same epistemological status as color. Kant responded by reasserting the primary status of the scientific concepts of causality, space, and time, but he inverted the notion of primary. The primary concepts are not claimed to designate the way objects are in themselves—we cannot know this—nor are they the subjective results of human perception. Rather, *they are primary in the sense that they make perception possible;* they are concepts we "impose" on the world; they are the categories by which we organize the world. The perceptual features produced by these essential concepts will always be, according to Kant, a necessary part of a "phenomenal realm" of appearance. These concepts are built-in filters by which the human mind understands the world. They cannot be overthrown someday by future experience, because they make all human experience possible.

Logically, we cannot know if the world itself matches these fundamental concepts; it might. Rather, we can only be sure that reality will always appear this way for us. It is our view of things. It is as if we were all born with rose-colored contact lenses as a natural part of our visual physiology. Suppose that without the lenses we would not be able to perceive anything meaningful; we would be effectively blind. With the lenses we see an organized, but always rose-colored, world. For all we know the real world is rose colored also, but there will never be any way to tell. In this way Kant thought he had turned subjectivity into objectivity. We do not need to worry about proving on the basis of our experience that causality is a valid concept or that it will not be overthrown someday; we know that it will always be a valid concept and never be overthrown, because it is one of the fundamental concepts that makes any experience of the world possible. Causality does not depend on experience; it is not an "a posteriori" concept, to use Kant's terminology. It is "a priori," independent of experience. In other words, we do not derive the concept from the world of experience; we bring it to the world, just as we would not derive our experience of rose-coloredness from the world, but impose it upon the world.

Kant then tried to show that the basic laws of Newtonian physics can be deduced from these basic "categories of understanding." Thus, Kant could now say in answer to Hume, we may not know whether the Newtonian world is indeed the real world, but we can say that it is, and always will be, the real world for us—the way it appears to us. Every observation, every new discovery scientists make in the future, will simply fill in details around a general framework, a mopping-up operation as Helmholtz later suggested, just more facts consistent with the Newtonian world view. No observation or discovery will be inconsistent with the Newtonian filters, because they are a priori. According to Kant, the answer to Hume's skepticism is that we are all born Newtonians, just as if we were all born with rose-colored contact lenses. We will discover new facts to fill in our concepts, but our concepts will never change. Without our concepts we could not discover new facts. Concepts without the perception of new facts are empty, said Kant, but without concepts we are blind.

Things which we see are not by themselves what we see. . . . It remains completely unknown to us what the objects may be by themselves and apart from the receptivity of our senses. We know nothing but our manner of perceiving them.

IMMANUEL KANT, *THE CRITIQUE OF PURE REASON*

Concepts and Other Worlds

Kant was wrong, but he was wrong in a very important way. Both Newton and Kant prepared the way for Einstein. It is intellectually fashionable today to highlight the more dogmatic aspects of Kant's work as another example of armchair philosophy slowing down scientific progress. After all, if Kant were correct, this would imply an end to science. If no future experience could refute the basic principles of Newtonianism, if the principles were irrefutable and a priori, then there would be no future fundamental discoveries. Kant, in a sophisticated way, surrendered to the temptation discussed in Chapter 2: He was so worried about proving his beliefs true that he made them irrefutable so no future experience could overthrow them. His philosophy reflected the confidence scientists of his time had in Newtonian science. He was not alone in believing that the laws of nature as outlined by Newton would never be overthrown.

To concentrate, however, only on the provincialism of Kant's reaction to Hume's skepticism misses the revolutionary implications of Kant's epistemological inversion and how it prepared the way for much of modern thinking. In drawing attention to the importance of concepts as the organizing force behind perception, Kant awoke other philosophers from their own dogmatic slumbers. The human mind is not a blank tablet upon which truth imprints itself. It is an active instrument in sorting out the infinite amount of data that floods our perception and threatens to overwhelm our attempts at organizing the world into a meaningful perspective. Without thoughts, a mass of empirical data is nothing at all. (Remember Sartre's self-taught man in Chapter 2.)

Kant's philosophy also prepared the way to think about the possibility that there are many "real" events, other worlds, happening all the time right in front of us, so to speak, even though we are incapable of directly experiencing them. It would soon be commonplace for scientists to believe that there are many realities beyond the perceptual window allowed by Newtonian conceptual filters. Kant was right about our normal perceptual window. It is Newtonian. Every observation scientists make, whether it be in an Earth laboratory or of deep space, will be framed within a normal three-dimensional window. Although Kant himself believed this would never be possible, it would soon be commonplace to believe that there could be indirect methods of deducing other realities, that observations made in our normal mode could indicate or point to a world totally different from what we normally observe. Scientists would soon be routinely setting up laboratory conditions to reveal the heretofore unimaginable and invisible realities of electromagnetic radiation, of electricity, of X rays, of AM and FM transmissions.

Ironically, in his attempt to establish Newtonianism as a priori, Kant prepared the way for us to think the unthinkable: that we could conceptualize,

The great thing [about Kant's philosophy] was to form the idea that this one thing—mind or world—may well be capable of other forms of appearance that we cannot grasp and that do not imply the notions of space and time. This means an imposing liberation from our inveterate prejudice.

ERWIN SCHRÖDINGER, *MIND AND MATTER*

understand, and even have knowledge of new unimaginable realities, that our common notion of three-dimensional space, our normal experience of the unidirectional flow of time, and our thoughts about causality could be but a human point of view.

Einstein and a Philosophical Discovery

Einstein has not . . . given the lie to Kant's deep thoughts on the idealization of space and time; he has, on the contrary, made a large step towards its accomplishment.

ERWIN SCHRÖDINGER, *MIND AND MATTER*

Let us return now to our discussion about time and clocks. Recall that we must *assume* that a clock measures time the same regardless of our perspective of it. Suppose we synchronize two clocks and separate them. Suppose one clock stays on Earth and another is taken to Mercury. How do we know they stay synchronized? Again following Newton, we assume that time flows on objectively, everywhere being the same, and that assuming that the clocks are both working correctly, they will objectively measure this flow the same way. If the clock on Earth says two o'clock, I assume that this is also the time on Mercury according to the other clock. If two full hours have passed on Earth since I last looked at the clock, I assume that exactly two full hours have passed on Mercury as well.

One of the great tasks of philosophy is to reveal the assumptions we make when we assert something to be true. Often this is very difficult precisely because it is so easy. Our assumptions are usually so obvious, so fundamentally embedded in our outlook, that we cannot recognize that we are making them. Philosophical analysis is a vital part of scientific method. As we saw in Chapter 2, many ideas are involved in deducing possible results that can then be tested by experiment. Behind every experiment is a hypothesis set, consisting of the main hypothesis, many minor hypotheses, and assumptions. This set, along with the conditions of the experiment, serve as premises for inferring what should happen when nature is subjected to our probing. Logically, if the result of an experiment is negative, if what we expect to happen does not happen, then this proves only that at least *one* of the ideas of the hypothesis set must be false. Thus, it is important to identify as many as possible of the ideas that make up the premises for the predicted result.

Einstein recognized that to assume that two separated clocks stayed the same involved a philosophical bias. Why should they stay the same? If we are to be empirically honest and subject all our assumptions to tests, then we need some way to measure what time the clocks record. Because there is no big cosmic clock in the center of the universe for all to see simultaneously, the only way we can measure the time of our clocks and see if they stay synchronized is by directly comparing them, and at great distances this necessarily involves the speed of light. To be sure that a clock on Mercury is still keeping the same time as one on Earth, we must communicate with an observer on Mercury by sending an electromagnetic signal traveling at the speed of light. Also, because both planets are moving in relation to each other and because the speed of light is finite, we must take into account the speed of light and what effect, if any, the relative motion of the planets might have on this speed.

At the turn of the century when Einstein as a young man was thinking about such things, the speed of light had also become a paradox. It was known that the speed of light had a finite, but very great, velocity (186,000 miles per second), and it was assumed that the measurement of this velocity, like any other velocity, should change depending upon how fast the source of the light is moving and in what direction. For instance, if we are on an object moving at a speed of 66,600 miles per hour, the speed of our Earth around the Sun, and we send a signal to Mercury in the direction of the motion of the Earth around the Sun, then should not the speed of the electromagnetic signal be the regular speed of light plus the speed of our motion? If a train is moving at 60 miles per hour and a person on the train is walking at 3 miles per hour in the same direction as the train is moving, then that person's accumulated speed, as measured by a stationary observer, is 63 miles per hour. The two velocities are added together. Should not light be the same?

By the turn of the century experiments showed that light did not behave the way our common sense says it should. No matter what the speed or direction an object moved, a beam of light from that object was always the same—186,000 miles per second. Whether a beam of light was traveling in the direction of the Earth's motion, at right angles to its motion, or in the opposite direction of its motion, the speed of the light beam was the same.

The implications of this result were not easy to digest. If we could travel on a special vehicle, say at only 100 miles per second less than the speed of light, and we turned on a flashlight and pointed it in the direction of our motion, the flashlight beam would move away from us at the regular speed of light. If we increased our speed to 50 miles per second less than the speed of light, we would not gain on the beam of light. Finally, if our vehicle could reach the speed of light, the speed of our flashlight beam would still be the normal speed of light. In the case of light the normal addition of velocities does not work; one plus one is one.

Einstein boldly accepted these paradoxes as axioms: Time must be "tested," and the speed of light is the same regardless of the speed of its source. Einstein then recognized that for science to establish universal laws of nature, laws that remain the same regardless of one's point of view, then a price had to be paid. We must accept the fact that when we test time, it will speed up or slow down relative to moving frames of reference. We must accept as commonplace that because Mercury and the Earth are moving in relation to

It requires a very unusual mind to undertake the analysis of the obvious.

ALFRED NORTH WHITEHEAD

each other, and any electromagnetic communication device will transmit a signal at the speed of light unaffected by the relative motion of the planets, the clocks on these planets will not be synchronized when they are compared.

Cosmic Trains

I can now rejoice even in the falsification of a cherished theory, because even this is a scientific success.

JOHN ECCLES

As Claude Kacser points out at the opening of this chapter, it is easy to believe such thinking is too hard. Let's use the same example Einstein used, which is easier to visualize than the Earth-Mercury example. In Figure 7-1 imagine a special train moving toward point A and away from point B. On the train is a person we will designate as X. As the train passes by, imagine a person Y midway between points A and B. Suppose as Y watches the train pass, at exactly the moment X is opposite Y, two bolts of lightning strike the ground from Y's point of view simultaneously at points A and B. How would X view the two bolts of lightning? Let's imagine that the train is moving very fast, at about three-fifths the speed of light. Since X is moving toward point A at such a great speed, the light from A will be received significantly before the light from point B, which will have to catch up to the swiftly moving train. Thus, whereas from Y's point of view the two events were simultaneous (happening at the same time), from X's point of view they were not simultaneous at all—the bolt of lightning struck A before B. Who is right?

It is tempting to respond immediately that Y is right because X is moving. It is comfortable to think that Y's reference frame is the right place from which to view the "actual truth of the matter." Because X is moving so fast, it is easy to believe that his experience is an illusion due to his motion. If X got in the "right place," if he slowed down, then both observers would see the same thing, the bolts of lightning striking the ground at the same time. But wait. Y also is moving. Y is, in fact, moving many different ways, depending upon which reference frame is adopted. If Y is close to the equator of the Earth, he is moving at about 1,000 miles per hour. From the point of view of the Sun, Y is moving at approximately 66,600 miles per hour. And from the point of view of the center of our galaxy, he is moving at a speed of over 500,000 miles per hour. Where is the right place? Why can't X assume that Y is the one who is moving?

A Newtonian might object that X could use the Galilean transformation to detect his motion, and on the basis of this he could calculate the simultaneity of the lightning flashes. If the addition of velocities assumed in the Galilean transformation is universally valid, then X should be able to measure the velocity of light coming from A as being the normal speed of light plus another three-fifths the speed of normal light and the speed from B as only two-fifths the speed of normal light. By using these values and measuring the time he

200   CHAPTER 7

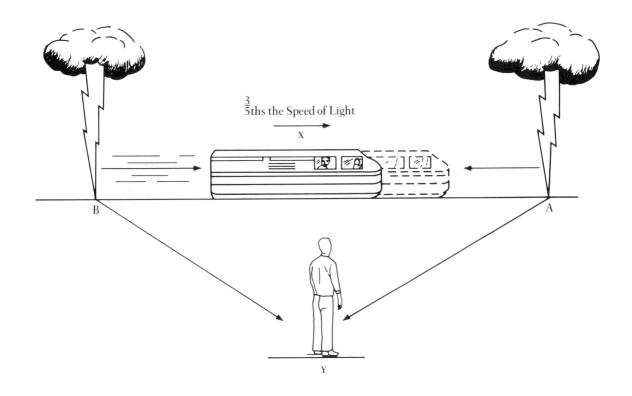

$\frac{3}{5}$ths the Speed of Light

receives each flash, he could calculate the "real" time of the original flashes, which would then agree with Y. Unfortunately, in the case of light, nature does not cooperate with the Galilean transformation and the simple addition of velocities. If X had the proper equipment to measure the speed of the incoming light signals from A and B, he would find that the speeds of each beam are the same, the normal speed of light. Similarly, if Y had the proper equipment, he would also find the beams coming from A and B to have the same speed. Thus, both observers are entitled to adopt the perspective that they are at rest and the other is moving.

Einstein's solution states that we must obey the facts. The speed of light, as a law of nature, is the same everywhere for every observer, and this is true no matter how each observer is moving relative to another. Furthermore, unless we are willing to make an unwarranted metaphysical assumption, time must be tested; that there is a "right" place where time is absolute is just an assumption for which there is no evidence.

Time is very much related to our relative place in space. Our time measurements (clocks, calendars) are actually spatially "local" things. On Earth when we measure one hour, we are really measuring a portion of our space, a portion of the rotation of our Earth (approximately 15°). On Mercury this convention would be inconvenient, because that planet rotates once every 59

FIGURE 7-1

In Einstein's special train example, the light from A will arrive at X before that from B. Hence X will observe the lightning at A as happening before that at B. Y, however, will observe the bolts of lightning to be simultaneous. An example of how observations from reference frames moving at great speeds relative to each other reveal a different timing of events.

Earth days and revolves around the Sun in 88 Earth days. The combination of these rotation and revolution periods makes one Mercury day equal to two of its years![1]

Einstein's theory of relativity teaches us that when we are fairly close together and moving together—when, as on Earth, the speed of our relative motions is very small in comparison to the speed of light—then time will behave itself. In our train example, if the train is moving at a normal train speed, then X also measures the bolts of lightning as simultaneous. (Also, because the relative speeds of Earth and Mercury are very small compared to the speed of light, we would require very precise atomic clocks to show that our clocks are not synchronized.) But when astronomical objects are widely separated and move at great speeds relative to each other, time does not obey our Earthly standards. Our assumption of an absolute time, the intuitive feeling that time clicks along at a steady rate throughout the universe, that "now" on Earth is the same "now" for all locations, is due to the fact that normally we do not move at such great speeds relative to the speed of light. These relative measurements of time show up only when relative speeds are attained appreciably close to the speed of light. This is the essence of Einstein's great discovery: partly a philosophical discovery (time must be tested) and partly an empirical discovery (the speed of light is the same in all reference frames). For the most part, the rest was logic and mathematical deduction.

It is important to understand that the relativity of time measurements is necessary to preserve the laws of nature. Although two observers will measure the temporal occurrence of events differently, from their respective reference frames they will not notice anything unusual. The theory of relativity does *not* prove that everything is relative. Both observers in our train example will find that the laws of nature apply normally, regardless of what they might think or wish to be true. If observer X conducted experiments in a laboratory on the train, he would obtain the same results as Y would in a laboratory located in his reference frame.

According to Einstein, as he relates in his *Autobiographical Notes*:

After ten years of reflection such a principle resulted from a paradox upon which I had already hit at the age of sixteen: If I pursue a beam of light with a velocity *c* (velocity of light in a vacuum), I should observe such a beam of light as a spa-

*Imagine someone living on Mercury. The time from noon to midnight for this person is one Mercury year. One complete Mercury day, from noon to noon, takes two Mercury years. . . . Suppose we choose to land at a location on Mercury where the sun is about to rise. We will be treated to one of the most interesting sights in the solar system. . . . The sun rises for a while and then stops and sets again!
. . . After perihelion passage, the sun rises again and this time slowly travels toward the meridian.*

CHARLES E. LONG,
*DISCOVERING THE
UNIVERSE*

---

[1] After one rotation, Mercury will be ⅔ of the way around its orbit, but will no longer have the same side of the planet facing the Sun. It requires 176 Earth days to go from one noon until the next noon (one Mercury day).

tially oscillatory electromagnetic field at rest. However, there seems to be no such thing . . . on the basis of experience. . . . From the very beginning it appeared to me intuitively clear that, judged from the standpoint of such an observer, everything would have to happen according to the same laws as for an observer who, relative to the earth, was at rest. For how, otherwise, should the first observer know, that is, be able to determine, that he is in a state of fast uniform motion?[1]

Einstein concludes this passage by acknowledging his debt to the skepticism of Hume, referring to Hume's insight that the major assumptions of Newtonian physics had no empirical foundation.

Time Dilation

Einstein then realized that if the laws of nature are the same from every standpoint, then for physicists to be able to continue to do physics in a universe of relative moving objects, they will need to mathematically transform how time will be viewed from different reference frames. Einstein used a mathematical equation called the Lorentz-Fitzgerald transformation. It worked perfectly. The equation is a relatively simple algebraic relationship:

$$T = T' \sqrt{1 - (v/c)^2}$$

The idea that time can vary from place to place is a difficult one, but it is the idea Einstein used, and it is correct—believe it or not.

RICHARD FEYNMAN

In this equation, T is the time of an event in a reference frame moving at a velocity v in relation to an observer whose time is measured by the variable T'. The c is a constant, the speed of light. Let's see how it works.

Suppose in our train example that X and Y both possessed clocks that some time ago were synchronized when the train was stationary. Y then moved at a normal speed, very slowly compared to the speed of light, to a point between A and B. Suppose at a prearranged time X departs and that for the past 15 minutes by Y's reckoning the train has left the initial starting point and been

---

[1] Albert Einstein, *Autobiographical Notes*, trans. and ed. Paul Arthur Schilpp (La Salle, Ill.: Open Court, 1979), pp. 49–51.

moving to the point where Y is at an average speed of three-fifths the speed of light.[1] What time will the clock of X read according to the Lorentz-Fitzgerald transformation? If we plug in the data,

$$T' = 15 \text{ and } v/c = \frac{3}{5}, \text{ so } T = 15 \sqrt{1 - \left(\frac{3}{5}\right)^2}$$

When X passes Y, X's clock will show that the train has been moving for only 12 minutes! From X's point of view, time has *slowed down* relative to Y. It would not just seem to slow down; real physical measurable effects would be seen when X and Y compare their clocks. If X and Y both lit cigars at the prearranged time, X would find that his cigar has burned less than Y's as they pass each other, even if a smoking machine was used.

X will experience nothing unusual. The laws of nature are the same, including that for burning cigars. For X everything will appear normal including the movement of his clock. At no point will he see the clock suddenly slow down dramatically. It will appear normal. Likewise, Y will not suddenly see the last few minutes of his experienced 15 minutes fly by like the clocks in bad aspirin commercials. It will also appear normal. Nor will either notice anything strange about the rate their cigars burn. Time slows down in reference frame X only in relation to reference frame Y. Within their respective reference frames everything is normal.

This slowing down of time of a reference frame relative to another reference frame is called *time dilation*. As unbelievable as it may seem,[2] it is one of the most accepted scientific facts of our time. It has been tested in numerous ways. Very precise atomic clocks have been synchronized and then compared again after one was flown around the world in a speeding jet. The clock on the speeding jet slowed down in relation to the clock that stayed on the ground. A similar test was conducted using one of the U.S. space shuttles with the same result. Scientists now apply time dilation routinely in sophisticated laboratory situations. In the billion-dollar particle accelerator laboratories all over the world, physicists can keep special particles of matter "alive" far longer than would normally be expected because of the time dilation effects that result by accelerating particles close to the speed of light. In this way special forms of energy can actually be stored for use in crucial experiments.

[1] Y would need to be 100,440,000 miles away, and it would have taken him a little over 191 years to travel to this point at 60 miles per hour!

[2] A more pronounced effect would be recorded if the train was far enough away and had been moving for five hours by Y's reckoning. X's clock would then register only four hours since he left at the prearranged time.

As far as most physicists are concerned, time dilation is a fact of nature. Rather than the large fishbowl of time implied by common sense and Newtonian science, Einstein has revealed to us a "chunky" universe of relative reference frames and times. There is no universal "now"; there are only "nows" relative to different reference frames. For the modern physicist, this cosmology is commonplace.

Trains, of course, do not move at 111,600 miles per second (three-fifths the speed of light), and hence, it is easy to see why time dilation effects were not noticed by common sense. Theoretically spaceships could. What kind of astronautical scenarios are then possible? Suppose we know twins 20 years of age, one an astronaut who will take a space voyage that will take 20 years by Earth time. Suppose that the astronaut twin averages in his rocketship a speed of three-fifths the speed of light. How old will each twin be when they meet again 20 Earth years from now? Using the time dilation equation we have

If my home and office were separated by many light-years, my home (depending on the distance and the speed of my travel) could be many thousands of years in the future when I arrive at work.

$$T = 20 \sqrt{1 - \left(\frac{3}{5}\right)^2} + 20,$$

which would equal 36. The twin who stays on Earth will be, of course, 40 years of age. And yet his brother will now be 36!

What applies to time will also apply to space. Just as the twins would discover that there is not one time that marches along in a strict Newtonian fashion, so they would also discover that there is not one large fishbowl of space where the distance between two points is the same for every observer. Just as time slows down, so space contracts. Because of the speed he is traveling, between the points of his departure and arrival the astronaut twin will measure his voyage as 14 trillion miles shorter than his twin would with instruments on Earth.

This example is not science fiction. We believe, based on the best laboratory data available and other corroborating evidence, that this effect on an astronaut would indeed happen. If it is so hard to believe, it is because we have difficulty realizing that the things we take for granted on Earth do not necessarily apply throughout the cosmos. With Einstein, the cosmos is now our laboratory, and we must adjust to the conditions of this new laboratory.

Stranger still, consider this scenario. The star Vega, a star very much like our Sun and a possible candidate for a planetary system, is approximately 32 light-years away. Suppose in the near future a 30-year-old mother, who has a 5-year-old son, went on a space voyage to explore this star, averaging the colossal speed of 99.5 percent the speed of light. At such a speed it would take her about 64⅓ years Earth time to make the trip. When she returned, her son, remaining on Earth, would be into his 69th year. Both would be in for a great shock. The 69-year-old son would embrace his long-awaited 36½-year-old mother! Their personal histories would have seemed to be normal in all other respects, but it would now be clear that traveling at great speed slows our histories down from one point of view and allows us to speed to the future from another. If such a voyage were taken, the mother astronaut, if she left in the year 1990, could get to the year 2054 in a little under 6½ years.

An event requiring only 6½ years for the mother would require over 64 years for the son. In our train example, an event that happened before another event for one observer (lightning striking point A before point B for X), happened at the same time for another observer (Y). It would also be possible then for another observer moving in the opposite direction of reference frame X at a great speed to record the lightning striking at point B first. Thus, one person's past could be another's future. Would it then be possible for the mother to return at an age before her son was born?

Not according to Einstein's theory, not if the speed of light is a law of nature. Because the speed of light is an absolute that cannot be exceeded, causal connections, such as mothers' causing the birth of babies, are preserved in their normal sequences. According to Einstein's theory, the measurement of "before" and "after" may involve a wide latitude, but the order of events will not be changed. The time between the mother's "before" and "after" of her space voyage is much shorter than that experienced by her son, but both would experience her leaving before she came back. If and only if the speed of light can be exceeded will the sequencing of causal events be changed, and it is a basic consequence of Einstein's theory that the speed of light cannot be exceeded. According to his theory, it would take an infinite amount of energy to accelerate any object (even an electron) up to the speed of light and thus require more than an infinite amount of energy to exceed the speed of light.[1]

Traveling at great speed slows our histories down from one point of view, and allows us to speed to the future from another.

Epistemological Implications

The pane of glass that separated the observer from reality in Newtonian physics has been shattered. To some extent what is real does depend on us.

The epistemological implications of relativity theory are very significant. The role of the observer is much different from that of Newtonian science. In Newtonian science the variety of perspectives of human observation, and the observer himself, could be ignored, excused as irrelevant to our descriptions

[1] Note, however, that the speed of light would only need to be exceeded by .004358 of a percent for the mother, if she left on the day of her thirtieth birthday, to return on the beginning of the second day after her twenty-fourth birthday, at least one year before her son was born and a few months before his conception! What would happen if she were then involved in a fatal car accident?

of the real world. The different results of observation due to different reference frames were considered to be simply practical inconveniences that could be reconciled by Galilean transformations. But in relativity theory, the observer is intimately involved in scientific measurement, and what is measured can be different depending on one's reference frame. Our knowledge of the world must unfold from empirical measurement of it. In the destruction of the notions of absolute space and time, Einstein showed that an honest empiricism must involve the observer, that *to some extent what is real does depend on us.* According to the physicist Paul Davies:

> The essential element injected into physics by the theory of relativity is subjectivity. Fundamental things like duration, length, past, present and future can no longer be regarded as a dependable framework within which to live our lives. Instead they are flexible, elastic qualities, and their values depend on precisely who is measuring them. In this sense the observer is beginning to play a rather central role in the nature of the world. It has become meaningless to ask whose clock is "really" right, or what is the "real" distance between two places, or what is happening on Mars "now." There is no "real" duration, extension or common present.[1]

However, Einstein did not think he had proved that each observer is involved in creating reality. Einstein did not doubt the existence of an independent physical reality, or whether there must be some absolutes. In fact, the intent of his theory was to preserve absolutes, the laws of nature. Einstein thought the secret structure of nature resembled the internal mechanism of a special, mysterious, cosmic clock. We have no direct access to the internal workings of this cosmic clock. We are forever limited to seeing the outside motions of the hands and can only submit hypotheses about how the internal mechanism produces the movements of the hands. But limited as we are, we can judge which hypotheses are better on the basis of which ones predict best the motions that we observe.

Einstein was not an epistemological relativist, and he believed as passionately as Socrates and Plato that some ideas are better than others. For Einstein, knowledge of how the clock works is possible, even though we will never see the mechanism directly or know with logical certainty that our best hypothesis is correct. Einstein was a realist: If our experiences of the world

The universe plays fair. Its tricks may operate by principles of incredible subtlety, and we may never discover all of them, but it keeps performing its illusions over and over again, always by the same method.

MARTIN GARDNER

---

[1] Paul Davies, *Other Worlds: A Portrait of Nature in Rebellion—Space, Superspace and the Quantum Universe* (New York: Simon and Schuster, 1980), p. 42.

overwhelmingly support a particular hypothesis, no matter how bold and radical the ideas contained in this theory, we are entitled to believe that these ideas refer to real things. Like the man who found his way out of Plato's cave, the going may not always be easy. New ideas often encounter great resistance, engender fear, and lead to ridicule, but according to Einstein, there is a world out there that "beckons like a liberation," and we must "believe in the possibility of knowing the things-in-themselves," no matter how strange the result. As Socrates stated many centuries earlier, we will be "better, braver, and less idle" if we do.

Einstein's basic insights concerning space and time served as only the first premises in the development of the many marvels of twentieth-century physics. From these insights, known as the Special Theory, Einstein later showed in his General Theory that if his ideas on space and time were true, many hard-to-believe things must be true of our cosmic laboratory.

For instance, after Einstein, physicists understand what Newton could not understand: how gravity can act on bodies great distances apart. To do so, however, we must make use of a new idea—"warped space." Large massive bodies like the Earth and Sun create multidimensional indentations in the fabric of space, much like a bowling ball in the middle of a trampoline. Because of the curvature of this space, gravity is now like the force on a ball that is constrained in its motion to roll toward the center of the indention. With warped space there is a new geometry, where the shortest distance between two points is a curved line rather than the intuitive straight line. We must also believe in strange places such as that of a "singularity," a place with no spatial dimension from which space and time can both emerge (as in the case of the Big Bang) and disappear (as in the case of black holes). Here is Paul Davies's description of a black hole.

A black hole . . . represents a rapid route to eternity. In this extreme case, not only would a rocket-bound twin reach the future quicker, he could reach the *end of time* in the twinkling of an eye! At the instant he enters the hole, all of eternity will have passed outside according to his relative determination of "now." Once inside the hole . . . he will be imprisoned in a timewarp, unable to return to the outside universe again, because the outside universe will have happened. He will be, literally, beyond the end of time as far as the rest of the universe is concerned. To emerge from the hole, he would have to come out before he went in. This is absurd and shows there is no escape. The inexorable grip of the hole's gravity drags the hapless astronaut towards the singularity where, a microsecond later, he reaches the edge of time, and obliteration; the singularity marks the end of a

We cannot predict what comes out of a singularity. . . . It is a disaster for science.

STEPHEN HAWKING

one-way journey to "nowhere" and "nowhen." It is a non-place where the physical universe ceases.[1]

The human mind balks at such thoughts. How can space be warped? Isn't empty space just space? How can an empty something be warped? How can the shortest distance between two points be a curved line? How can a point exist that is not *in* space, and how can space be *inside* the point? Our minds rebel against such ideas as illogical and impossible.

A Neo-Kantian perspective explains why. Our normal perspective of the world is Newtonian: there appears immediately to us only one time and a three-dimensional space. For the most part experience takes place "within" this normal perspective. But Einstein discovered that Kant was wrong. Indirectly we can break out of our filters by radicalizing some of the conditions of our normal perspective. We can move at very great speeds, for instance, and experience results that show how limited our normal perspective is.

Davies's description of a black hole is not just something physicists have made up to stay competitive with modern fashions. Black holes are a mathematical deduction from the assumptions of Einstein's theory of gravity; scientists believe they are the result of the deaths of very massive stars. We know such stars exist and have now confirmed the existence of black holes from the gravitational effects on other stars and the radiation they leak. Cataloging the locations of black holes has become almost routine.

Although Einstein did not believe that his theory proved that human observers create reality, he did show that the observer begins to play a crucial role in what is real and that Copernicanism had gone too far, or in a sense not far enough. The completion of the Copernican revolution in Newtonianism fulfilled the dream of a unified science; the laws that governed terrestrial motion were the same that governed celestial motion. This enabled other ideas to get in through the backdoor: Our common-sense notions of space and time, which worked so well on Earth, were assumed to be true for the entire universe. This assumption was so pervasive that it was not recognized as an assumption, especially given the success of Newtonian physics. In revealing this assumption as a metaphysical postulate and not an empirical fact, *Einstein showed that what was masquerading as a removal of humankind and subjectivism from science was actually a projection of a subjective human point of view, which is close enough to the truth at a certain level to enable us to fail to recognize it as a human point of view.*

Think about this last statement carefully. Think about it for a long time. It is crucial for understanding the paradox of twentieth-century science. Our

Science is our century's art.

HORACE FREELAND JUDSON

Einstein has shown that reality need not obey our sense of certainty or the workings of the human mind.

[1] Paul Davies, *God and the New Physics* (New York: Simon and Schuster, 1983), p. 123. Reprinted by permission.

intuitive feeling that there must be an objective time and space is actually just a projection of a human point of view. We view the world normally as Kant and Newton held. We can accurately calculate the motions of the planets within our solar system using a perspective of three-dimensional space and uniform time. We can send our robot spacecraft to the outer planets and beyond. Our equations work. But that they work in this domain does not prove that the concepts we assume in applying the equations are valid for other domains.

As we will see in our discussion next of quantum physics, this realization is only the beginning. As in a long romantic relationship, one's partner may eventually reveal a totally unexpected set of personality traits. After Newton we thought we knew what the universe was like. Little did we suspect the unnerving surprises it had in store for us in the twentieth century.

Concept Summary

In the previous chapter we saw how Newtonianism is forced to *assume* that space and time are absolute—that each event has one objective spatial and temporal location. In this chapter we have seen that part of Einstein's discovery involved a philosophical insight, that Newtonianism involved this assumption, and that scientific knowledge of space and time would require empirical tests.

It seems paradoxical that we cannot safely assume and know upon reflection alone that time flows on uniformly throughout the universe as our common sense dictates. It never occurs to me that if I leave my home at 8:00 and arrive at my office at 8:30 that I am assuming that it is not then 9:00 at home. In the eighteenth century, in spite of the tremendous practical success of Newtonian physics, the philosopher David Hume showed that the application of another major concept, causality, was also paradoxical. No matter how successful a science, such as Newtonian physics, had been in predicting the conjunction of events, it could not be known that one event actually caused the other.

For the Newtonian philosopher and physicist Immanuel Kant, this was a disturbing epistemological challenge. As the sophists had challenged the ancient Greeks centuries before, Hume was challenging all Newtonians to demonstrate how this new science provided a durable knowledge, to show how the "truths" of this new science would not be overthrown by future experiences. Kant concluded that this was possible only through a radical inversion of our normal thinking about objectivity. Rather than seeing such fundamental concepts as causality, space, and time (and the basic principles of mathematics used to describe applications of these concepts) as reflections or representations of a known objective reality external to the human mind, such concepts must now be seen as fundamental categories of understanding or conceptual "filters" by which external reality must be experienced. For Kant, Hume may have shown that it is impossible to be certain that Newtonian concepts mirror the things-in-themselves (the real universe independent of human perceivers and knowers), but we can be certain nevertheless in knowing that all future experience will conform to these concepts, because these concepts make perception possible; they are fundamental concepts that we "impose" upon the world.

Kant's influence was great, but he prepared the way for Einstein, and much of modern thinking, most by being wrong. By attempting to demonstrate that the fundamental concepts of Newtonianism were irrefutable, he drew attention, as did Newton, in his attempt to rationalize the absoluteness of space and time, to the fragileness and anthropocentrism of these assumptions. If reality could not be shown to conform to common-sense human assumptions, then the door was wide open for much creative thought on other possibilities.

Einstein recognized that one could not assume that two widely separated, initially synchronized clocks kept the same time. He recognized that time must be tested by measuring and comparing what time each clock records. To compare such clocks, one must take into account the speed of light and the relative motions of the reference frames of each clock. Einstein showed that when this is actually done, because the speed of light had been discovered to be the same regardless of its direction and the speed of its source, our intuitive sense that time is something that just clicks along independently of moving objects will be violated when we compare initially synchronized clocks. Simultaneity is relative to a reference frame, and time dilation, the slowing of time relative to another reference frame, is a fact of life. If my home and office were separated by many light-years, not only could I not assume the times to be the same, but my home (depending on the distance and the speed of my travel) could be many thousands of years in the future when I arrive at work.

Although the success of Einstein's theories does not imply that everything is relative or that scientists create reality, it does show that an honest empiricism, one that tests fundamental assumptions, has brought the observer into twentieth-century physics. To some extent what is real does depend on us. Einstein's theories also set the stage for the great paradox of twentieth-century science. We must be careful about what we assume to be objective; what seems to be obviously an objective property of reality may be a subjective projection of a merely human point of view, one applicable to only a limited range of experience. Nothing seems to me more certainly objective and independent of my wishes than the thought that it is the same time at my home now as at my office regardless of the distance between the two. Kant may have shown why I have this sense of certainty in terms of the way the human mind works, but Einstein has shown that reality need not obey my sense of certainty or the workings of the human mind.

Suggested Readings

Relativity for the Million, by Martin Gardner (New York: Macmillan, 1962).

One of the first places to start for an understanding of Einstein's Special and General theories of relativity. As the title implies, this book is intended to be the introduction for the "rest of us," with a direct style and many illustrations. While the book offers simplicity, it is also authoritative. Gardner, a noted scientist and mathematician, has devoted considerable energy to overcoming the murky misrepresentations derived from some of the implications of modern physics.

Introduction to the Special Theory of Relativity, by Claude Kacser (Engelwood Cliffs, N.J.: Prentice-Hall, 1967).

Although it involves considerable mathematics (not much beyond simple algebra, however), this book is still one of the best step-by-step technical introductions to the Special Theory of relativity. Although the first three chapters are sufficient for gaining a feel for, and proof of, time dilation, the book develops the necessary concepts for a preparation for the General Theory.

Other Worlds: A Portrait of Nature in Rebellion—Space, Superspace and the Quantum Universe, by Paul Davies (New York: Simon and Schuster, 1980).

Although this book deals mostly with the astonishing implications of quantum theory (the subject of Chapter 8), there is a very clear, thought-provoking summary chapter on relativity theory, "Things Are Not Always What They Seem." Also see the chapter entitled "Time" in Davies's *God and the New Physics* (New York: Simon and Schuster, 1983).

Einstein's Universe, by Nigel Calder (New York: Viking, 1979).

Although this book is unusual in starting with the General Theory, it is a masterful attempt at presenting Einstein's two theories without mathematics. Each chapter begins with a series of simple, but tantalizing, declarative sentences that summarize what is to follow.

Time's Arrows: Scientific Attitudes Toward Time, by Richard Morris (New York: Simon and Schuster, 1984).

Another book on the modern conception of time for nonspecialists, but Morris places this view within a historical context by comparing it with many other cultural conceptions of time.

Autobiographical Notes, by Albert Einstein, trans. and ed. by Paul Arthur Schilpp (La Salle, Ill.: Open Court, 1979).

Einstein's reflections, written toward the end of his life, on the states of mind that produced his discoveries, from his early years as a "precocious young man" to his struggle to convince physicists not to give up the notion that the goal of science is to describe the "real state of things" via mathematical descriptions (see Chapter 8). For Einstein's thoughts on matters other than physics, see his *Out of My Later Years* (New York: Philosophical Library, 1950).

The Philosophical Impact of Contemporary Physics, by Milic Capek (Princeton, N.J.: Van Nostrand, 1961).

A thorough philosophical discussion of the implications of the transformation from the Newtonian world view to the new concepts of modern physics. See especially chapters 10–12 on space and time. For further philosophical treatments of Einstein's work, see *Albert Einstein: Philosopher-Scientist,* ed. by Paul Arthur Schilpp (Evanston, Ill.: Library of Living Philosophers, 1949), *The Philosophy of Space and Time,* by Hans Reichenbach (New York: Dover, 1958), and *Philosophical Problems of Space and Time,* by Adolf Grunbaum (New York: Knopf, 1963). Grunbaum's book, especially, is not for the novice investigator; it requires not only a background in the philosophy of science but also the utmost patience and persistence. Although an important work, the reader must be prepared to grapple with very dense prose.

CHAPTER 8

Our Time:
2. Quantum Physics and Reality

A new scientific truth does not triumph by convincing its opponents and making them see light, but rather because its opponents eventually die, and a new generation grows up that is familiar with it.

MAX PLANCK

What we observe is not nature itself, but nature exposed to our method of questioning.

WERNER HEISENBERG

Anyone who has not been shocked by quantum physics has not understood it.

NIELS BOHR

The twentieth century is a remarkable story of technological achievement. Within a few decades, electricity, radio, and TV, not to mention lasers, fiber optics, plastics, and computers, have all become an everyday facet of our lives. I take for granted that I can turn on a TV set in Hawaii and receive, almost instantaneously, a program that originated in Atlanta or New York. Only a short time ago exchanging information between New York and Hawaii required the same time it takes to send a spacecraft to Mars today. Except for power failures, few people in the developed world know what it was like for most of the human species every night, throughout 99 percent of our history, to face the blackness of space and its sea of stars alone without the reassuring lights of civilization. We live in a special time. Never before has such intense, radical technological change taken place. In this chapter we will see that there has been another radical change: Something very strange happened to reality along the way.

In a popular TV show the scientist Carl Sagan sits in the elegant dining room of Cambridge University waiting to be served an apple pie. He tells his audience, "If you wish to make an apple pie from scratch, you must first invent the universe." With this he begins to tell the modern scientific story of physical matter, our view of physical objects such as apple pies, tables, rocks, trees, birds, and even people. He discusses how modern science stretches our imagination to the limit in portraying the dimensions of atoms and the large numbers routinely used by science. (Remember: about 10 billion atoms can fit within a single printed letter on this page.) Sagan explains that every part of ourselves and most of the physical matter on Earth were formed billions of years ago in the ferocious interior of gigantic stars. "The calcium in our bones, the iron in our blood, the gold in our teeth, the complex molecules in our apple pies, all came from the stars. . . . We are made of star stuff."

He continues with the discovery that everything consists of molecules and these in turn are hooked-together atoms, which are subatomic particles and mostly empty space. He characterizes a typical atom as having "a kind of cloud" of electrons on the outside, "clouds of moving fluff."

One ought to be ashamed to make use of the wonders of science embodied in a radio set, while appreciating them as little as a cow appreciates the botanical marvels in the plant she munches.

ALBERT EINSTEIN

What Is an Electron?
Of Particles and Waves

Our story will begin here because it is the electron, and our knowledge of it, that has been responsible for so much of the technology that we take for granted today. Without the electron there would be no electricity, no electric

O amazement of things—even the least particle!

WALT WHITMAN

lights, no TV, and no radio. We would not have supermarket doors that open automatically or computers to play video games and do word processing and spreadsheets for business. But what exactly *is* an electron? In the early moments of the twentieth century, scientists found themselves asking this very question. The discovery of radiation and the atom promised to open up a strange new world of knowledge, understanding, and power.

At first physicists assumed that the atom was like a miniature solar system. At the center was a nucleus consisting of particles glued together somehow, and circling this nucleus were the swiftly moving electrons, like little particle planets. This model did not last long. Although we still use a version of this model today to have some visual handle on what the atom looks like, scientists discovered fairly quickly that mathematical calculations based on this model predicted that the electron would crash into the nucleus in an instant.

Physicists also discovered that electrons could be stripped from the atoms and made into beams of radiation. This was a great breakthrough, because scientists could manipulate these beams and begin to deduce from the beams' behavior the nature of the electron itself. A similar channel of investigation was taking place in attempting to understand the nature of light. From this another remarkable discovery was made: Beams of electrons behaved very much like beams of light.

We saw in Chapter 7 that the speed of light was considered a paradox by many at the turn of the century. By this time the nature of light was also highly controversial and something of a paradox. Under some conditions light seemed to behave as if it consisted of very small particles of matter (now called photons). Under other conditions, however, light showed clear signs of being a wave of energy, a disturbance of a medium, the intensity of which could be measured. To understand how this is a problem, we must first clearly understand that a particle and a wave are very different phenomena.

A *particle* is a piece of matter, like a baseball, that at any given time has a definite size, speed, and location. It can be in only one place at a time. A baseball thrown in Hawaii cannot be in New York at the same time. Furthermore, we assume, ontologically speaking, that we may discover in this marvelous universe some very strange objects but that, regardless of how strange they are, if they are objects, then they will have a definite location at any given definite time. Any object such as a baseball cannot be in two places at the same time.

A *wave,* on the other hand, is a very different kind of thing. In fact, it is appropriate not to refer to it as a thing at all, but rather as an event or phenomenon. *Things* by definition have a definite localized size at a given definite time. Waves do not. Imagine dropping a pebble into a still pond of water. At first there is a small splash, and then circular waves move away from the spot where we dropped the pebble. The wave spreads out; it does not stay in one place, but can be in many places at the same time. Also, it is the medium of the water that transmits the energy of the dropped pebble. The wave is simply a disturbance of the medium. It does not have an existence

There is one simplification at least. Electrons behave in this respect in exactly the same way as photons; they are both screwy, but in exactly the same way.

RICHARD FEYNMAN,
*THE CHARACTER OF
PHYSICAL LAW*

of its own like the smile of the Cheshire cat in *Alice in Wonderland*. Without the water in the pond there would be no waves.

On the north shore of the island of Oahu in the State of Hawaii, every winter large waves pound the shoreline. These waves are caused by the seasonal winter storms migrating northeast of the state in the jet stream on their way to make life miserable for people in the Pacific Northwest, and eventually much of the rest of the continental United States. The winds from the migrating storms cause a significant disturbance in the sea and a series of undulations are transmitted many miles until finally, reaching the reef on the north shore of Oahu, spectacular waves of 20 feet or higher break and push forward a mountain of water and foam toward the beach. On the cliffs overlooking Waimea Bay you can watch a gigantic half circle of water march relentlessly toward the beach and then simultaneously, across a quarter mile area, surge onto the beach. It is a spectacular sight. Tourists travel many thousands of miles to see it, and single-intentioned surfers wait in anticipation all year, hoping to be the first to ride the biggest wave on record and survive.

It would be a strange event indeed, if one day while watching wave after wave break, we saw one wave flow in its normal way toward the beach and then, just as the wave was about to touch the first fingers of vulnerable sand, the entire half circle of water collapsed instantly to a single unpredictable point on the beach and exploded! The wave would have turned into a massive particle located at one place, rather than spread out as waves normally are. Imagine wave after wave doing this, with the location of the collapse being unpredictable each time. Strange indeed this would be, but something like this is what electrons and photons seem to do!

One can't believe impossible things.

LEWIS CARROLL, *ALICE IN WONDERLAND*

Thought Experiments

In the following pages we are going to retrace the same baffling steps that physicists have taken in the past 70 years. Their goal was simply to understand the nature of subatomic objects such as the electron and the photon.[1] The result was a revolution in thought so radical that even Einstein could not accept it. But we will be using a method Einstein would have approved

---

[1] The science of the subatomic realm is called quantum physics or quantum mechanics. The word "quantum" refers to the fact that energy at the microscopic realm comes in packets, or *quanta*; energy is said to be "discrete" rather than continuous. The best way of understanding the implications of discrete motion is to understand the most famous phrase in this science, the "quantum jump." As we will soon see, this does not

of, what are called "thought" experiments. Instead of looking at the actual technical experiments, we will imagine a series of composite pictures that remain true to the actual experimental findings.[2]

Imagine first a lead box impenetrable except for two microscopic slits on one side. On the inside of the box on the side opposite the slits is a photographic film. Imagine that on the outside facing the two slits we have a source of radiation, beams of electrons or light, and that we aim this radiation at the face of the box with the two slits. By looking at the kind of exposure that results on the photographic film, we can deduce what kind of radiation is penetrating the box. For instance, if the radiation consists of beams of particles, then only those particles that happen to be aligned with the two slits will pass through into the box, and the result should be a "particle effect": The photographic film should show a diffused piling up of little hits adjacent to the two slits.

On the other hand, if the radiation is a wave, then a much different effect should result. We should see a "wave effect," roughly what we would see if we dropped two stones into a still pond of water at the same time. Two circular undulations would collide into each other and interfere with each other. In our example, a wave would split in two as it enters the two slits, and then the two new waves would begin to spread out again, eventually colliding with each other as in our pond example. This should cause an "interference effect," a wave picture, on the photographic film. Instead of a piling effect adjacent to the two slits, the radiation would spread throughout the length of the photographic film, producing alternating bands of exposure. Some of the wave crests would meet and accentuate each other, and some would meet the troughs of other waves and cancel each other. This is similar to a wave approaching the beach and a backwash wave meeting it and producing a bigger wave, or a crest meeting a trough of another wave and cancelling each other. The exposed bands on the photographic film would be the result of the crests meeting. Such a resourceful experimental process is what Einstein had in mind with his clock analogy discussed in Chapter 7. We may not be able to see the invisible electron, but we can infer a reasonable representation of what it is by observing the effects it has on macroscopic objects.

---

refer to a continuous, quick motion of an object, but rather a discontinuous, instantaneous movement from one place to another. In other words, quantum objects seem to be able to move from place to place without being anywhere in between. They seem to "pop" in and out of existence.

[2] Technically these are known as the Photoelectric effect, Compton effect, Young and Davisson-Germer diffraction wave experiments, Stern-Gerlach interferometer experiments, Bell's inequality theorem, and the Aspect experiments.

When similar experiments are done, the result is remarkable. The photographic film always shows an interference effect indicating a wave (see Figure 8-1a). Unfortunately, the radiation produces this same effect in passing through a vacuum. How can a wave exist without a medium to disturb? Also, when we look closely at the exposure of the film, the exposed areas show piles of little hits, as if millions of particles hit the film, each blackening only a single grain of film in unpredictable locations (see Figure 8-1d). Remember that if the radiation is a wave, then as it reaches the film, it should be spread out along the entire length of the film like a wave breaking on a beach. But how can it hit at only one unpredictable place? This is as ridiculous as the possibility of watching a wave move toward a beach and seeing the entire wave collapse at a single point on the beach!

Obviously, more experiments are necessary. Baffling results are common in science. So let's close one of the slits and see what happens. Perhaps the particles are so small that as they penetrate the slits they ricochet all over the interior of the chamber, bouncing off each other in a wild, unpredictable manner that eventually produces the illusion of an interference effect. After all, the electron is about 10,000 times smaller in mass than an average atom. Thus, in closing one of the slits, we would lessen this wild ricocheting, and a piling effect should result adjacent to the single slit.

Sometimes nature cooperates. If we alternate in opening and closing each slit, then the result appears consistent with the particle hypothesis—a double piling effect adjacent to each slit (Figure 8-1b).

But wait. To be sure that we are dealing with a particle, let's return to our original experimental setup with two slits open simultaneously. This time, however, we will lower the intensity of the radiation. In other words, another way to lessen the possibility of the ricocheting effect, and at the same time rule out the wave hypothesis is to filter the radiation to such a point that only a single "measure" of radiation passes through into the chamber at a time. If we assume that the radiation consists of particles, then if only a single particle is passing through at a time, *it can go through only one slit or the other.* If it is a particle, then it cannot be in two places at the same time and thus will not ricochet off other particles or itself.

Conducting this experiment will take a long time for an exposure to develop because millions of particle hits will be required to make an exposure, and each particle has three choices: to penetrate the chamber through either of the two slits or be repulsed by the lead barrier. Nevertheless, we should eventually observe a particle effect—a double piling effect of hits adjacent to the two slits (Figure 8-1b).

Alas, nature fails to cooperate. The result is an interference effect, exactly as in our first case (Figure 8-1a and 8-1d)! Now we are really in trouble. Why should we get a wave effect with two slits open, even though the exposure is the result of cumulative unpredictable hits, and a particle effect with only one slit open? With two slits open the radiation is acting as though

In the universe great acts are made up of small deeds.

LAO TSU, *TAO TE CHING*

FIGURE 8-1

If light is a wave (a), then a series of alternating light and dark fringes should appear on a photographic film after the light is made to pass through a pair of slits. This can be explained by "picturing" a set of circular waves starting from each slit. Where the waves intersect, they are reinforced, producing maximum light intensity and the banding effect on the film. The banding effect appears regardless of the intensity of the light. If light consisted of tiny particles (b), then a single pair of bright stripes should appear on the film. If alternate slits are opened and closed, this picture is recorded. If a detection device is added to the arrangement of both slits open (c), then a particle effect is recorded. Close inspection (d) of the photographic film showing a wave effect also shows a gradual piling of individual particle hits.

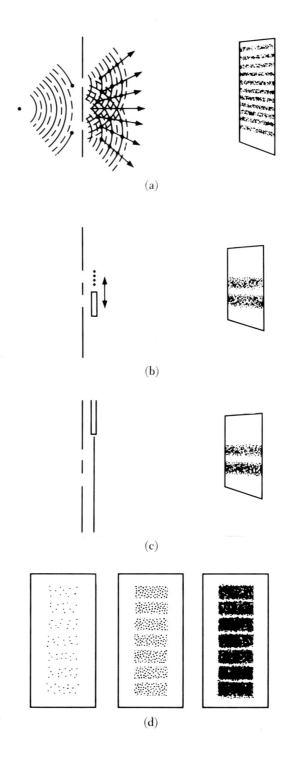

(a)

(b)

(c)

(d)

it penetrates the chamber at two places simultaneously, something only a wave can do. With one slit, however, the radiation is more localized,[1] as we would expect from particles, which can be in only one place at a time.

It is easy for the beginner to lose sight of the philosophical significance of these results. Many of the great names of science, however, who were working at this baffling microscopic realm, also had a background in philosophy, and it was immediately apparent to them that there was something major at stake here. Since the time of the ancient Greeks and the fledgling beginnings of scientific exploration, we have assumed that we are dealing with one world, one consistent reality. That is, even though we expect the world to be baffling at times, with strange and new details of discoveries, we also expect that whatever these details are, they stay the same independent of our knowing. They are "out there" waiting for us to discover, and they are what they are regardless of our knowledge or ignorance. We assume as Newton did that the world does not depend on us or how we choose to make our observations of it. We do not expect something to be a particle on Mondays, Wednesdays, Fridays, and Sundays and a wave on Tuesdays, Thursdays, and Saturdays—especially when these phenomena are entirely different types of phenomena. What kind of a world would it be for us if dogs were dogs on Mondays but turned into cats on Tuesdays? It would surely make dealing with the world and taking care of pets very difficult.

Because this notion is so important, let's try one more example. What we want to know is, does the radiation pass through both open slits simultaneously or only through one? Consider then the following experimental arrangement: both slits open, one measure of radiation entering the chamber at a time, but with an added feature—a detection device inside the chamber that will reveal whether or not the radiation is passing through both slits as a wave would or only one or the other of the slits as a particle would (Figure 8-1c). Because the situation is almost identical to the case in which an interference effect was recorded, we would expect to see the detection device react as though a wave were surging through both openings simultaneously. On the other hand, if the radiation consists of particles, then only one instance of detection should be recorded at a time. Remarkably, the latter is the case—only one instance of detection is recorded at a time—and the photographic result is now consistent with the arrangement with only one slit or the other open (Fig. 8-1b)!

As physicists in this century conducted further experiments with subatomic phenomena, they found that all subatomic phenomena display this same ambiguity, which has come to be known as *wave-particle duality*. This ambiguity

The "paradox" is only a conflict between reality and your feeling of what reality "ought to be."

RICHARD FEYNMAN, *THE CHARACTER OF PHYSICAL LAW*

---

[1] Actually a *diffraction pattern,* a diffused piling effect, results, which is also a wave effect. So the wave effect shows particle characteristics, the individual hits on the film, and the particle effect shows wave characteristics, the diffraction pattern.

was not easy to accept. One of the most fundamental principles of science seemed to be mocked by these results: the notion that we are dealing with, and can know the details of, an objective world. To use Einstein's cosmic clock analogy, we expect that the internal mechanism stays the same regardless of our hypotheses and beliefs about what the internal mechanism is. We do not expect the internal mechanism to change as we change our experimental attempts to know the internal mechanism. It is perhaps one of the greatest achievements of this century that in spite of this shock, a very successful mathematics was developed that not only allowed physicists to predict the results of the above experiments but also produced one of the greatest scientific and technological success stories in recorded history. In 1926 the physicist Erwin Schrödinger gave us the Schrödinger wave equation, and the science of physics has never been quite the same. The equation "explains" the preceding results but with a high epistemological and ontological price.

As we would expect from the name, the equation literally portrays the radiation as a wave, but a very strange wave. According to the equation, in our two-slits-open configuration as soon as the radiation leaves its preparation point, it begins to spread out in a strange multidimensional "hyperspace." As it encounters the slits it splits, as any real wave would, passing into the chamber and interfering with itself. As the radiation touches the photographic film, however, all of it *collapses* to a single unpredictable point! We can never predict at what point the radiation will be received, but we can always, with a remarkable consistency, predict the probability of where it will strike and the overall statistical pattern, not only for this particular arrangement but for all the others as well.

Few physicists, however, accept this literal interpretation; most have been taught to think of the equation as a calculation device. The special mathematical function used is thought to represent only a "probability function"— that is, given initial conditions, the probability of finding a hit, or a pattern of hits, at a particular location. Thus, the only waves that exist are said to be "probability waves."

But wait. What happened to reality? What is a probability wave? What is an electron? What is a photon? Are these questions no longer meaningful? Let's look at one more example.

In Figure 8-2a imagine a light source directed at a half-silvered mirror, a mirror covered with a very light reflective coating. Such a mirror functions as a beam-splitter, and if we assume that light consists of little particles called photons, then the physical properties of the half-silvered mirror should cause each photon to pass through the mirror or be reflected at an angle. Thus, if we set up photon detectors at the appropriate angles, at points A and B, individual detections at A or B should result. With this experimental arrangement, the mathematics predicts that over a sufficient period of time 50 percent of the light will be received at A and 50 percent will be received at B. Furthermore, if the intensity is lowered through filtering, such that only a single photon approaches the mirror at a time, then only a single whole

If we take quantum theory seriously as a picture of what's really going on, each measurement does more than disturb: it profoundly reshapes the very fabric of reality.

NICK HERBERT, *QUANTUM REALITY: BEYOND THE NEW PHYSICS*

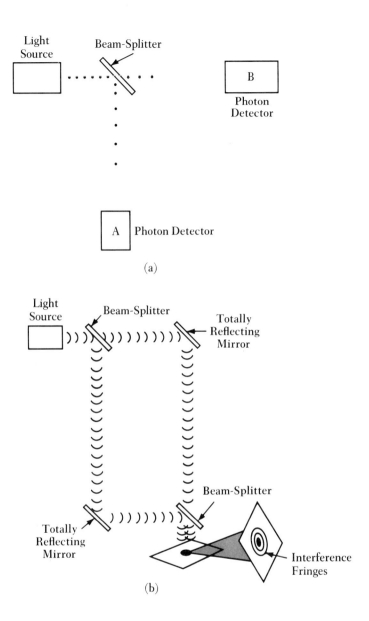

Light
Source

Beam-Splitter

B

Photon
Detector

A | Photon Detector

(a)

Light
Source

Beam-Splitter

Totally
Reflecting
Mirror

Beam-Splitter

Totally
Reflecting
Mirror

Interference
Fringes

(b)

FIGURE 8-2

In (a) light is made to pass through a beam-splitter. If the intensity of the light is lowered to a single photon "particle" at a time, then each should either pass through and be detected at B or be reflected and be detected at A. Detections at A and B should not be recorded simultaneously. In (b) the detectors are replaced with totally reflecting mirrors. If light is a wave, even if the intensity of the light is lowered to a single photon, it will be split and show an interference effect when the beams are recombined. How can the same radiation be consistent with a "particle" picture in (a) and a "wave" picture in (b)? If the radiation is in both channels (b), why is the detection of photons not simultaneous in (a)?

photon should be detected at a time. Detections at A and B should never be recorded simultaneously.

When such an experiment is actually conducted, it is a relatively simple matter to show that one whole unit of energy is detected at either A or B, confirming again that we are dealing with a single object, a little thing that can be at only one place at a time. If a photon is a particle, then it will pass through the mirror and be detected at A *or* be reflected and detected at B.

However, the Schrödinger equation is a wave equation. Although we must be careful using such language, according to this equation, the half-silvered mirror splits a wave packet into two "hyperspatial, virtual/real, probability" waves. At the exact moment that the energy reaches the detectors, some sort of strange decision is made, and the entire unit of energy is received at only one point, at either A or B! The wave packet collapses. If a whole unit is received at A, then the energy that was approaching B has *jumped* over to A. In addition, the equation predicts this will happen even if the two detectors are separated by many light-years, even if one detector is much closer to the half-silvered mirror than the other. The latter case implies that the energy that is approaching the one that is closer, say B, waits(?!) until the energy approaches A and then either jumps to A or the energy that was approaching A goes "backward in time" and collapses at B. The mathematics always works, but what it describes literally seems impossible. Like Alice in Wonderland, we cannot believe in impossible things, can we? According to the mathematics, there is an *instantaneous* collapse of potentiality located multidimensionally to a three-dimensional localized spot. Strange indeed. So physicists explain that we must not think of the split wave packets as real, but only as a description of the probability of where photons will go.

But wait. Now comes the crucial question. Does the light really pass through both channels? Quantum jumping aside, we can at least test for the radiation passing through both channels. Consider the following arrangement (Figure 8-2b). This time we will create an interferometer by placing totally reflecting mirrors at the points where detectors A and B were. Thus, if the light beam is really split by the half-silvered mirror, the totally reflecting mirrors will now reflect the split beams of light. If we aim these totally reflecting mirrors so that the beams will meet again, then we can take a picture of the waves interfering with each other, just as we did in the two-slit experiment. With this arrangement, interference fringes result similar to that found in the two-slit experiment. The interference effect can be produced by having one of the totally reflecting mirrors slightly farther away than the other, so that the light waves will arrive out of phase. The beams are recombined by another half-silvered mirror and transmitted to a chamber with a photographic plate.

If the intensity of the light is reduced to one photon at a time, then the interference effect can only be accounted for by assuming that the photon really splits into two wave packets and then recombines. In fact, if we pick up an ordinary playing card and block one of the paths, there is no interference picture. Instead, a diffused piling exposure is created, similar to the particle picture we received when only one slit was open. If the radiation is a wave, then we can understand the interference picture. If the radiation is a particle, then we can understand the fact that only one detector at a time receives one whole unit of energy. The result of one arrangement indicates that a wave of some kind is really passing through both channels simultaneously. The result of the other makes sense, if we assume that the radiation is passing through only one channel at a time. If the radiation is passing through both channels at the same time, why do the detectors not trigger simultaneously? How does the radiation passing through one channel get over to the other detector? How could this possibly happen if the detectors are far enough away that any transmission of a signal between them would require a speed greater than the speed of light? It's time for a little philosophy.

The Copenhagen Interpretation

Nature at the subatomic level apparently does not conform to normal logic. Since the time of the ancient Greeks, Western logic, through Aristotle's law of excluded middle, has demanded an "either-or" in our relationship with the universe. Either light is a particle or it is not a particle. Either light is a wave or it is not a wave. Either the light splits and goes through both channels or it does not. If it goes through both channels, then it should be detected at both channels. It is not detected at both channels, yet it does go through both channels. If it goes through both channels, why is only one whole unit of energy detected at one detector? How do two halves spatially separated become one whole unit instantaneously?

There is no logical inconsistency within the mathematics that explain these phenomena. In the particle effect case, the mathematics allows us to predict that approximately 50 percent of the time detector A will record a unit of energy and 50 percent of the time detector B will record a unit of energy. In the interferometer arrangement, the mathematics predicts an interference effect and even allows a straightforward calculation of the wave length of light by measuring the interference fringes. The problem is more in our reaction to the results of these experiments and the success of the mathematics. We want to know what kind of a "thing" is producing these results. What is going on out there that enables the mathematics to be successful? Our minds desire a complete understanding. What is real? What is the truth?

These questions reflect our natural tendency to want to go deeper, to find the basic, hidden causes of all things. Western civilization and its science since the ancient Greeks has assumed an ontology: The cosmos consists of one distinct, complete reality full of details. We have also assumed an epistemology: The details, whatever they might be, can be known, and knowing these details does not affect what the details actually are independent of the knower. This is consistent with our common sense and what each of us experiences everyday: a world undisturbed by human thoughts, wishes, and desires, full of things, spatially separated from each other, and interacting with each other through distinct recognizable forces. If someone has a tangerine tree in his yard, he might wish that it was an apple tree, but it will

In a sense [for the Copenhagen Interpretation], the observer picks what happens. One of the unsolved questions is whether the observer's mind or will somehow determines the choice, or whether it is simply a case of sticking in a thumb and pulling out a plum at random.

DIETRICK E. THOMSEN

Atoms are not things.

WERNER HEISENBERG

still be a tangerine tree. Similarly, we do not think of someone thinking cancer into existence or wishing it away. We think of the cancer being "out there," something beyond our mental control, like trees. We can cut down trees and operate on cancer, but they are distinct realities that we "discover" with our thinking, not something that we "create" with our thinking.

So it is natural for us to think of the electron as an independent thing. It shows signs of being a particle, so we begin to think of it as if it is really a particle independent of our observations of it. But it also shows signs of being a wave, and it cannot be both a wave and a particle at the same time, no more than a tree can be both a tangerine and apple tree at the same time.

In the 1920s a few philosophically minded physicists, led by the Nobel Prize–winning physicist Niels Bohr, realized that nature was trying to tell us something very important. Once again nature was using paradox to convey a fundamental error in the assumptions we were making and the way we were asking our questions. According to Bohr, and what is known as the *Copenhagen interpretation*,[1] the results of these encounters with subatomic phenomena amount to a major *epistemological discovery*. Descriptions such as "particle," "wave," "position," "mass," and "spin," are human concepts. These concepts involving assumptions of space and time work for us at a normal macroscopic level and will always be indispensable for describing the results of our physical experiments. But nature is now making it clear to us that we have reached a barrier in our attempt to describe it in terms of human concepts derived from ordinary experience.

Wave-particle duality is nature's way of informing us that we have no right to impose our human concepts on the subatomic level. Just as Einstein had discovered that we have no right to impose our normal assumptions of space and time to all levels of reality, so quantum physics reveals that we have no right to impose our most basic thoughts about the nature of reality on the subatomic realm. The idea of an extended thing sitting in a three-dimensional space, waiting for us to discover it, is revealed as another human projection, a limited image of reality, more of an echo of the way our minds work than of reality itself. According to Bohr, nature reveals this to us by showing that we can have only *complementary* views of reality. If we set up an experimental arrangement that allows for a wave manifestation of subatomic phenomena, then that is what we will observe. If we set up an experimental arrangement to view subatomic phenomena as particles, then that is what we will observe. According to Heisenberg, another major contributor to the Copenhagen interpretation, what we observe in our experiments is not nature itself but nature exposed to our methods of questioning nature. In short, an electron is not a thing until we observe it!

Causality may be considered as a mode of perception by which we reduce our sense impressions to order.

NIELS BOHR

For Bohr, many fruitful explorations still exist, relationships yet to be described and mathematical trails yet to be followed, but the search for a "hidden" reality will not be one of them.

---

[1] So-called because much of the work by Bohr, Werner Heisenberg, and others was done in Copenhagen, Denmark.

According to Bohr, this is a necessary, "pragmatic" response. Experiments must be conducted in human terms, in laboratories full of macroscopic equipment in three dimensions. Our laboratory equipment must be capable of measurements that are understandable through the conceptual reference frame of human beings. This barrier, however, should not be seen as an end to science or as an imposed state of ignorance. It is a discovery, a momentous discovery about ourselves and the nature of science. To discover that complementary views of reality exist, rather than only one unified view, is as important as Einstein's discovery that the reference frame of an observer is crucial for measuring space and time. Rather than limiting science, Bohr viewed this new knowledge as liberating the sciences from the tyranny of thinking that each science must explain itself in terms of a more basic science such as physics and chemistry. Biology, for instance, could very well be a complementary perspective on living things, not totally reducible to physics and chemistry.

When Einstein has criticized quantum theory he has done so from the basis of dogmatic realism.

WERNER HEISENBERG, *PHYSICS AND PHILOSOPHY*

A Debate:
Bohr versus Einstein

Ironically, the main resistance to the Copenhagen interpretation came from Albert Einstein and a few of his followers. Einstein objected very much to the idea that we had stumbled upon a barrier to knowing what is real. Philosophically, Einstein was a realist who believed that the goal of science was to conjecture boldly about the nature of reality from the details of our observations. He acknowledged that as we continue to probe nature for her secrets, we would encounter more and more exotic features, most of which we could never directly observe because of the nature of our observational limitations. He believed, however, that the human mind could always fathom at least the most likely hypothesis about the nature of the reality causing the events we do observe. Thus, although Einstein introduced a revolutionary view of space and time, one that destroyed the classical or Newtonian conceptions of absolute space and time, he nevertheless remained a classical physicist faithful to the concept of reality Descartes stated centuries earlier: "There is nothing so far removed from us to be beyond our reach or so hidden that we cannot discover it."

As we noted previously, for Einstein, nature was like a mysterious clock. We are limited to observing only the exterior features of this clock. We may never be able to see directly inside and know for certain how the clock works, but by observing and thinking about the movement of the hands long enough, the human mind will provide a likely answer to how the clock works. For

We believe in the possibility of a theory which is able to give a complete description of reality, the laws of which establish relations between the things themselves and not merely between their probabilities. . . . God does not play dice.

ALBERT EINSTEIN

Einstein a clockwork for the universe exists and can be known. For Bohr, to assume that a clockwork exists independent of our observations is only another human philosophical bias, another example in a long line of assumptions that experience validates at a certain level, but which experience at another level now demonstrates cannot be considered "the way things are."

Although Bohr thought quantum physics to be in part an important epistemological discovery, and the barrier between the human mind and reality primarily pragmatic, the Copenhagen interpretation does raise the question of whether this epistemological discovery is also an ontological one. For Einstein, Bohr's interpretation was much too close to, and seemed to imply, a traditional ontology—an ontology historically very much opposed to the major goal of scientific method. If an electron is not a thing until it is observed by some instrument, does this not imply that reality depends on our observations and hence, ultimately, the thoughts we use to frame the world? Does this not imply that reality is created by human thoughts?

Metaphysical idealism is an old and widespread belief stating that the physical world as we experience it is basically an illusion; the perception of a world of material things separated in space is said to be only an appearance. Individual things exist only insofar as we have an idea of them. If there were no human observer or recording instrument of any kind in a forest, then a falling tree would make no sound. In fact, there would be no tree to fall and no forest. When I walk out of a room, I assume that the physical room and all its contents are still there. But according to the idealist, the room ceases to exist if no one is there to have a thought of the room.

Most scientists have always viewed this metaphysics with disdain, as more of a symptom of despair of the sometimes harsh realities of the physical world, as primarily a religious view associated with those who find the physical universe threatening and who desire a more perfect but duller world. Does quantum physics validate this philosophy? How embarrassing for Western science if it does. Imagine that after thousands of years of struggling to know the details of Democritus's atom, Western science shipwrecks into a religious philosophy it thought it had left behind at a more primitive time!

Thus, Einstein viewed quantum physics as an incomplete theory: We simply do not know enough yet. Because we cannot produce a consistent picture of subatomic phenomena, we obviously do not know exactly what these things are yet or enough about the mysterious forces governing their motions and manifestations. "God does not play dice with the universe," according to Einstein. He has created one universe and does not choose to have it manifest itself as waves at one moment and as particles at another for no reason.

Bohr and Einstein had several public debates over what was the proper interpretation of the results of quantum physics. These were fascinating discussions between two intellectual giants, but little was resolved at the time. The vast majority of physicists heeded Bohr's advice that there was a pragmatic limitation inherent in our measuring devices. Physicists should be interested primarily in being able to predict experimental results and not in the question of what is real. They were persuaded, with the help of a philosophical tradition that began with Hume, that the question of what is real is primarily an unanswerable philosophical question. Physics must concern itself primarily with complex experimental arrangements and the derivation of the complex mathematical formulas needed to predict the "constant conjunctions" of appearances first discussed by Hume. On the other hand, motivated by the goal of finding a hidden reality, physicists have also pursued Einstein's dream of a unified picture of reality, of seeking a theory that enables us to understand at a fundamental level the mysterious forces of nature.

Physics is still proceeding in two, perhaps complementary, perhaps schizophrenic, directions—one following Einstein's dream and the other developing a series of experiments to confirm Bohr's theory of complementarity. Using a particle approach and a model of subatomic objects consisting of different types of quarks, many physicists have become confident that they are approaching an understanding of the basic clockwork of the universe. These physicists believe that what appear at a certain level to be different forces in nature are actually different manifestations of a *superforce,* a force that existed for only a brief moment under the superhot, superenergetic conditions of the first microseconds of the universe. Because of the present relatively mild conditions of the universe, this force is hidden from us, and because it is hidden we are left with the many paradoxical results of quantum appearances. Thus, throughout the world the race is on, with billions of dollars being spent, to set up the wild conditions that will, it is thought, finally coax nature into revealing her true self. And there are those who, like Paul Davies in his book *Superforce,* have declared confidently, not unlike Helmholtz did for Newtonian physics, "that for the first time in history we have within our grasp a complete scientific theory of the whole universe in which no physical object or system lies outside a small set of basic scientific principles."

First they told us the world was flat. Then they told us it was round. Now they are telling us it isn't even there!

IRVING OYLE

Bell's Discovery

But does nature have a "true" self? Following Bohr, experiments have been conducted that are consistent with the view that it does not, that in our relationship with the universe we can have only different pictures of its clockwork—actually, to be more precise, that *a precise clockwork does not exist until we attempt to picture it!* For many years following the Bohr-Einstein debates it was thought that the issue between them must forever be relegated to the realm of inconclusive philosophical perspectives. No one knew of a conceivable experiment that could disconfirm either one. Bohr could argue that the present experimental data are most consistent with his theory of complementarity, but he could not prove that some day we would not discover some

Bell's theorem is easy to understand but hard to believe.

NICK HERBERT, *QUANTUM REALITY: BEYOND THE NEW PHYSICS*

bizarre "hidden" reality that explained how an electron could manifest itself as a wave in one situation and a particle in another. Similarly, the followers of Einstein could argue that if we think, and search, long enough, someday we will find this hidden reality. No one thought an experiment could be devised that would eliminate the possibility of a hidden reality.

In 1964 physicist John Bell discovered that it was theoretically possible to test whether or not quantum physics was a complete theory. By tinkering with the mathematics, he discovered that an experiment could be devised to confirm or disconfirm hidden processes, or "variables" as physicists refer to them.

Before we describe this discovery and its application in crucial experiments, let us review first why quantum measurements are so puzzling. The essence of all the puzzles, according to the physicist Henry Stapp, is "How do energy and information get around so fast?" In the interferometer experiment we can demonstrate that a wave is passing through *both* channels. But when we modify the experiment to detect the radiation in each channel, we detect only *one* whole unit of energy at a time per channel, implying not only that the radiation consists of particles, and therefore not waves, but also that the radiation is not in both channels. In the particle detection experiment the Schrödinger equation describes a wave-splitting process with a "probability" wave in both channels and then an instantaneous collapse of a potential existence to one localized "actual" spot, to either detector A or B. The Copenhagen interpretation deals with this puzzle by claiming it is inappropriate to think of the radiation as some kind of real thing before we measure it. The radiation "becomes" something only after we measure it. (It is always a particle after we measure it, even though some measurements suggest the particle had wavelike properties between measurements.) Reality, specific attributes possessed by things, according to the Copenhagen interpretation, can only be discussed in terms of an "entire experimental arrangement."

According to Bohr, the problem of quantum measurement can be interpreted as a pragmatic epistemological discovery and does not necessarily imply an idealist metaphysics. Concepts such as "particle" and "wave" are human concepts, and we have discovered that nature will not allow us to picture it consistently with these concepts. Insofar as we must always conduct our experiments through a human framework, with human concepts, there is an epistemological barrier that no future scientific discovery will change. For Bohr, the success of quantum theory represents a "treasure chest" of scientific and philosophical discoveries. The Copenhagen interpretation should not be viewed as advocating a dogmatic end to research and discovery but rather a dramatic discovery that continues a trend first started by Copernicus and sustained most recently by the startling discoveries of Einstein: The universe is not required to conform to human concepts.

In a fundamental way Bell's discovery allowed physicists to test Bohr's claimed epistemological discovery. We could now see whether or not the subatomic realm had a true self independent of our measurements.

No elementary phenomenon is a phenomenon until it is a recorded phenomenon.

JOHN ARCHIBALD WHEELER

Quantum Jogging

Because the actual experiments and Bell's discovery are somewhat complex, let us try an analogy first. Suppose we have a large group of runners. Half of the runners are tall and half are short. Suppose that each of the short runners and each of the tall runners has a twin. Each of the twins will begin running at the same point, but will run a course in the opposite direction to a finish line that is the same distance from the original point of departure. Suppose also that each runner will run the course at the same speed, and that the spacing between the times when each runner leaves is such that no runner will be able to overtake the runner immediately preceding him. No tall runners will overtake short runners or vice versa. Imagine then a continuous stream of runners leaving the original point and running in opposite directions. We might have something like this: Two short runners leave the starting point one after the other simultaneous with their respective twins, then two tall, then two short again, then one tall, and one short after that, then two tall, and so on. Suppose that overall the pattern is random. Suppose further that the contingencies of the course and physical training of each runner are such that many of the runners will not finish. Suppose also that each twin has a very strong desire to finish, such that any one twin will want to finish if and only if the other twin finishes.

Now we are ready to carry out the implications of our thought experiment. In spite of the strong desire of each twin to finish if and only if the other does, our common sense would predict that finishing together is not likely. Suppose one of the short runners pulls a muscle just before the finish line. How likely would it be that the twin, running on an independent track, separated by a considerable distance, either knows this and decides to stop running or pulls a muscle also and does not finish? In other words, if we were to observe the runners finishing and established a mathematical correlation of completion, we would *not* expect it to be very high. Suppose that about 90 percent of the tall and short runners did not finish; it would not be likely that every time a short or tall runner finished or did not finish, the respective twin finished or did not finish as well. If we found the random result at one finish line to be T, T, S, T, S, S, T, S, we would not expect this result to be highly correlated or equal to the result at the other finish line. We would expect an *inequality* in the results.

There is one possibility, however, where the results could be highly correlated. Suppose each runner carried an electronic beeper, such that whenever a runner knew he could not finish, he would signal the twin not to finish. In other words, if the runners could communicate, a very high correlation could be established.

Suppose though that we change our thought experiment a little. This time we will control, at one finish line, which runners finish and which ones do

The hope that new experiments will lead us back to objective events in time and space is about as well founded as the hope of discovering the end of the world in the unexplored regions of the Antarctic.

WERNER HEISENBERG

not. Suppose at a point immediately before one of the finish lines we set up a fork in the course, such that the short runners must take one path and the tall runners must take the other. Suppose further that we have control over an electronic switch that closes the path by throwing up a barrier for either the short or tall runners. By randomly changing the switch we can change which path is open and which type of runner finishes. It is important to be able to do this after the runners have already left. Otherwise the runners could know ahead of time what kind of course they must run and adjust their actions accordingly. Suppose that the barriers are so close to the finish line, and we are able to switch the barriers so rapidly, that there is no time for each twin to signal the other whether he is going to finish or not. Now clearly there could not possibly be a very high correlation. It would be a strange result, indeed, if even most of the time when a tall runner finished, his twin also finished, and most of the time when one did not, his twin did not, and likewise for the short runners.

We assume that the local conditions at a barrier cannot instantaneously influence the local conditions at the other finish line. This *locality assumption* is an inherent part of our normal view of reality. We assume that the runners are independent individuals who will face independent conditions at independent places. What Bell showed is that if this assumption is correct and also applies to the subatomic realm, then the results we obtain in the subatomic realm with particles should reflect the same kind of inequality in correlation we expect to find in our macroscopic realm of short and tall runners.

Quantum theory, on the other hand, predicts an entirely different situation for subatomic particles. Because it is incorrect to refer to subatomic particles as having any definite state with a definite place until a measurement takes place, an analogous runner's example to what happens in the subatomic realm would be the following: Our runners do not exist as definite runners until they are observed to finish, and a measurement at one finish line will instantaneously produce a correlated set of characteristics at the other finish line! From a quantum perspective, the locality assumption is denied; it is incorrect to think of our runners as real independent entities, in real independent places, experiencing real, local, independent circumstances. Instead, between the time we see them leave and finish, our runners are a "superposition of states" of existence. They are neither tall, nor short, nor fast nor slow, but all these potential states at once.

If quantum theory is true, then an analogous experiment in the subatomic realm should result in a significant violation of Bell's inequality deduction because it is incorrect to think of subatomic particles as independent things with definite properties until a measurement takes place. If experiments are devised where "twin" particles are created and fly off in opposite directions like our runners, then quantum theory predicts that there will be a high correlation of the particle states when they are measured at a quantum finish line because a measurement of one particle instantaneously collapses a wave function of potential states, a wave function that was created at the time of the twin particle creation.

I think I can safely say that nobody understands quantum mechanics. . . . Do not keep saying to yourself . . . "But how can it be like that?" because you will get "down the drain," into a blind alley from which nobody has yet escaped. Nobody knows how it can be like that.

RICHARD FEYNMAN,
THE CHARACTER OF PHYSICAL LAW

The Aspect Experiment

Perhaps because the locality assumption is so obvious, or perhaps because the technological tools were not sufficiently developed to conduct the proper experiments, or both, recognition of the significance of Bell's work was slow in coming. A decade after Bell published his work, intense discussion and experimental work finally began. The most recent experimental results, based on a design that physicists acknowledged ahead of time would be the crucial experiment, show decisively that in the subatomic realm Bell's inequality is violated and the predictions of quantum theory are correct. The results show that the measurement of a subatomic particle at one finish line instantaneously determines the state of its twin at another finish line, regardless of how far apart the two finish lines are.

In the realm of subatomic particles, our runners are replaced by mathematical objects with attributes such as "charge," "spin," "velocity," and "momentum." We naturally tend to think of these attributes in the same way we think of the attributes of our runners. Just as we think of each runner as a real, independent body with definite characteristics such as being short or tall, fast or slow, we are more comfortable thinking of a particle having a real location or a real spin. Quantum physics, however, seldom allows us to be comfortable. Consider quantum spin. What kind of real attribute requires a subatomic particle to turn around twice before it shows its original face! Imagine looking at a position on the Earth from the Moon, say New York, and watching the Earth spin around twice before New York is visible again.

As bizarre as quantum attributes are, quantum physicists have learned how to deal with them mathematically and even set up experiments that create twin particles with opposite spin. The most recent, and most conclusive, we will call the Aspect experiment.[1] Using polarization, a property that can be thought of as similar to spin, physicists tested Bell's inequality prediction.[2] Atoms were excited to produce twin photons of light that sped away with opposite polarization. Methods were developed to test the states of the photons at their respective finish lines. In many respects this experiment was analogous to our thought experiment with the barriers and electronic switch. Bell's inequality theorem was violated: The spins of particles at distant finish lines were highly correlated. (In this experiment the main interest was in

Nature loves to hide.

HERACLITUS

My own suspicion is that the universe is not only queerer than we suppose, but queerer than we can suppose. . . . I suspect that there are more things in heaven and Earth than are dreamed of, or can be dreamed of, in any philosophy. That is . . . why I have no philosophy myself, and must be my excuse for dreaming.

J.B.S. HALDANE, "POSSIBLE WORLDS"

---

[1] After the French physicist Alain Aspect, who was the leader of a team that conducted this crucial experiment. The results were published in an unassuming three-page paper "Experimental Tests of Realistic Local Theories via Bell's Theorem," *Physics Review Letters*, Aug. 17, 1981.

[2] Polarization is what makes Polaroid lenses and dark glasses possible. A Polaroid lens allows photons of light with only a particular spin orientation to pass through. Those without this orientation are blocked, thus selectively lessening the intensity of light that passes through.

how often the photons at different finish lines would be blocked.) Because there was an analogous switching device,[1] there was no possibility that a signal could be sent at a normal cosmological speed (the speed of light) causing the particle's spin to be correlated. In summary, the result was as fantastic as our hypothetical, unlikely thought experiment concerning runners, where we find to our amazement, in spite of all of our precautions, that most of the time a tall runner finishes or does not finish, so does the twin, and the same is true for short runners. There is now little doubt that a violation of Bell's inequality is a fact of life. If there is a "hidden reality" with forces influencing the results of our paradoxical measurements, these forces must travel faster than the speed of light. They must be instantaneous.

Note that the violation of Bell's inequality is a "factual" demonstration that at least one assumption of Einstein's realism must be false, what we referred to earlier as the locality assumption. To accept the totality of Einstein's realism, we must assume that the local conditions at one finish line could influence the local conditions at the other finish line *only if* the two locations are linked by a causal chain whose transmission of effects does not exceed the speed of light. In other words, if reality consists of separate objects, then one object cannot influence another object unless some sort of signal or influence travels from one object to the other during some amount of time. If the movement of one object "instantaneously" influences the movement of another object, then they are not really separate objects. In addition, if someone is standing on one side of a dark room with a flashlight, the flashlight must be turned on before an object can be illuminated on the other side of the room. Recall from Chapter 7 that some very strange results are possible if the speed of light can be exceeded. Our mother astronaut could return to Earth and be involved in a fatal automobile accident before her child was conceived and before leaving for her space voyage. Thus, for many reasons, a hidden force traveling faster than the speed of light is ruled out as a possible explanation for the puzzling results of quantum experiments. The Aspect experiment shows that we must reject the totality of Einstein's realism, but not necessarily all possible versions of realism. The entire universe at the subatomic level could be one object.

The results of the Aspect experiment and the violation of Bell's inequality are also consistent with the Copenhagen interpretation: Quantum objects should not be considered things until a measurement takes place. Unfortunately, the implications of this interpretation for the nature of reality are philosophically disturbing for most physicists when they bother to think about them. Thus, most physicists accept the pragmatic aspect of the Copenhagen interpretation and ignore the reality question. The reality question is something for "the philosophers" to worry about. This response is often portrayed

Physics tells us much less about the physical world than we thought it did.

BERTRAND RUSSELL

---

[1] Switching devices were activated by high-frequency waves at a rate 100 million times per second. Because the finish lines were 10 meters apart, no signal could be sent at the speed of light.

as a sophisticated, modern point of view: physics should not be concerned with futile philosophical questions, but keep to the business of predicting results and applying quantum mathematics to novel situations such as computer technology, fiber optics, chemistry, and even biology. By any standard this approach has been very successful. However, is this instrumentalist approach any different from the reaction of past scientists to Ptolemy's epicycles, Copernicus's circles around invisible points, or Newton's gravity?

Quantum Ignorance and Reality

For many, the reality question beckons still. The history of physics, and science in general, shows that the traditional pursuit of "the way things really are" is not just an idle ivory tower game. A quest for a deep understanding has been valuable not only for its own sake but for the purpose of maximum practical application as well. The history of science has demonstrated repeatedly that when we understand the way things are at an invisible level, we are better able to understand, control, and predict the visible world in which we live. Until quantum physics, the vast span of scientific endeavor has vindicated Einstein's simple vision: The better we have been able to understand the invisible mechanism of the cosmic clock, the better we have been able to understand the motions of its visible hands. We may not be able to see Kepler's ellipses or Newton's gravity in the starry night, but an understanding of these veiled realities has enabled us to embrace the night sky— to predict, to control, to see, to explore—in a manner undreamt of by the ancients who so patiently and relentlessly watched this surface reality. Other examples abound: The understanding of the molecular and atomic constitution of matter has enabled us to deal with the surface experiences of heat, temperature, and pressure; by understanding a deeper level of reality, we have been able to create objects that do not exist in nature, such as plastics; and now, by understanding the invisible structure of DNA, we are close to controlling the development of life itself, with many practical applications in agriculture and medicine.

Is it over? The Copenhagen interpretation implies a strange kind of ignorance—call it *quantum ignorance*. According to Bohr, it is a mistake to search for a hidden, deeper mechanism that will explain the results of quantum measurements, because *between measurements there is nothing there to know*, that is, *nothing there can be conceptualized in human terms*. This is nature's way of educating us, of revealing its ultimate message: "Picture me with your human pictures if you must, but do not take your pictures too seriously."

For those sympathetic with Einstein, there must be something more; the results of quantum experiments must be only an example of what can be called *classical ignorance*. There must be something there that we are "dis-

No language which lends itself to visualizability can describe the quantum jumps.

MAX BORN

turbing" when we interact with it in attempting to measure it. We are ignorant of why quantum events happen as they do only because we do not know all the forces acting on subatomic particles, just as we cannot predict each throw of the dice in a dice game, because there are too many minute factors involved and any attempt on our part to measure these minute factors in the act would disturb the results. In the case of dice there are other ways of demonstrating the existence of these factors, and thus we have every reason to believe that they are there, even if we cannot control them.

Bell's theorem and the consequent experiments do not rule out some kind of realism, that some kind of hidden force or reality is at work in the subatomic realm. They do demonstrate, however, that these forces, if they exist, must be very strange forces. Unlike any previously discovered forces, they must be capable of propagating instantaneously regardless of distance. If our finish lines for subatomic particles were billions of miles away, then the violation of inequality would be the same. If one of our finish lines was located in the vicinity of the star Betelgeuse, 540 light-years distant, and the other on Earth, then quantum physics predicts that there would be no difference in the results. The results of Bell's theorem and the Aspect experiment show not only that quantum theory is a complete theory but also that any interpretation of quantum physics must incorporate the fact of instantaneous action.

So what kind of a reality do we live in? Notice that even the language of this question is misleading. To ask what kind of a reality we live in suggests that there is one reality independent of human beings and our attempt to know and measure this reality. Human language has evolved in a context of ordinary macroscopic reality. So how can we even begin to describe the subatomic realm? If it is a mistake to think of the electron as a thing with a definite place, with a definite velocity, until "it" is actually observed with a measurement, then it is difficult, to say the least, to understand how an "it" can exist without a location prior to a measurement that then gives it a location.

The concept is less difficult mathematically but no less strange. Mathematically, quantum physics allows a distinction between the *static* properties of the electron, such as "charge" and "mass," and the *dynamic* properties, such as "position" and "velocity." In this way most physicists believe that they can avoid versions of idealism, such as that of the eighteenth-century Irish philosopher and bishop George Berkeley, who taught that physical matter possessed reality only insofar as it was perceived by a mind. Put more dramatically, Berkeley believed that only mind or consciousness exists. For Berkeley, the entire physical universe is only an idea in the mind of God. Here is how Nick Herbert in his *Quantum Reality* describes the reaction of one group of Copenhagenists:

No believer in observer-created reality, even the most extreme, goes as far as Berkeley. Every physicist upholds the absolute existence of matter—electrons, photons and the like—as well as certain of matter's static attributes. . . . Electrons certainly exist—with the same mass and charge whether you look or not—but it is a mistake to imagine them in particular locations or traveling in a particular direction unless you actually happen to see one doing so.[1]

In other words, almost all physicists are convinced that something is out there, even though they are convinced that whatever it is, it will not conform to classical attempts to describe reality. But how can there be some "thing" without there being an idependent "place" for this thing to be? When we think of things like ordinary runners or elementary particles, we assume that they must have independent, objective attributes. What would be left if we took away from a tall runner his "tallness," his "speed," and his individual identity of being in one "place"? What kind of a runner could exist who was both short and tall, fast and slow, and neither short nor tall, fast nor slow? What kind of a runner could exist who only became a tall runner after we observed him at the finish line? In the quantum realm such "things" are commonplace. Whatever they are, quantum objects are not ordinary things.

A Paradigm for the Twenty-First Century?

According to Nobel laureate Richard Feynman, we "can safely say that nobody understands quantum mechanics." Consider, however, the following provocative possible paradigm for our time.

In Chapter 6 we saw that the paradigm of Newtonianism involved a combination of epistemological and metaphysical assumptions: What was real did not depend on us, and reality is reducible to small independent particles of physical matter and empty space; thoughts, ideas, colors, emotions were all considered to be secondary realities, not real, but rather the result of the movement and interactions of particles. This view, which we will call *metaphysical reductionism,* is seriously contradicted by the science of the twentieth century, particularly by the Copenhagen interpretation. What is real does seem to depend on us and our method of questioning nature. As the physicist E. P. Wigner has claimed, a measurement cannot legitimately be said to have taken place until it is acknowledged by the conscious awareness of a human

The observer is never entirely replaced by instruments; for if he were, he could obviously obtain no knowledge whatsoever. . . . They must be read! The observer's senses have to step in eventually. The most careful record, when not inspected, tells us nothing.

ERWIN SCHRÖDINGER, *MIND AND MATTER*

---

[1] Nick Herbert, *Quantum Reality: Beyond the New Physics* (Garden City, N.Y.: Anchor Press/Doubleday, 1985), p. 168.

being. Far from being a secondary reality, consciousness has a much greater significance in quantum theory. We confront the world with the filters of our human thoughts about the world, and nature conforms to these thoughts to some extent. A reality becomes manifest based upon the thoughts behind one of our experiments. We do not measure reality as Newton and all classical physicists believed; we measure the "relationship" between reality and our thoughts.

In the quantum realm we cannot pin down a consistent reality, and nature teaches us in the process not to take our thoughts about reality too seriously, on the one hand, and to take them very seriously, on the other hand. We should not think of our human concepts of "particle" and "wave" as reflecting an independent reality, but we have been forced to recognize the creative power of human concepts. The mathematics of quantum theory pictures not a precise clock with definite parts but a strange indefinite cosmic substance capable of manifesting an infinite number of fleeting faces. Quantum theory pictures the particles that make up everything that we touch and feel not as little, hard, definite, independent things, but as a tangle of possibilities entangled with every other tangle of possibilities throughout the universe. As with the particles in the Aspect experiment, the particles in my body may be connected in some way with the particles of your body, and these in turn with particles in a distant sun, in a distant galaxy, billions of light-years away.

Neorealism

There is little disagreement today among physicists and philosophers of science that the metaphysical reductionism of the seventeenth, eighteenth, and nineteenth centuries has been destroyed by the science of the twentieth century. But there is no consensus on a replacement. The results of relativity and quantum theory have sent physicists and philosophers of science scurrying in many different philosophical directions. Although most physicists have accepted the practical dictates of the Copenhagen interpretation, David Bohm, among others, has refused to abandon entirely the realism of Einstein, opting instead for a radical *neorealism*. For Bohm, the Aspect experiment does not disprove a "hidden" reality but only one that consists of separate things! A universe of "undivided wholeness" is consistent with all the experimental results. A real universe exists independent of our observations of it, but it is not like the room that I am in now: a bowl of space with apparently independent objects separated into different locations. "Underneath," so to speak, from a perspective of a multidimensional hyperspace or superspace, this appearance of separateness can be seen to melt like ink dots in water.

Mathematical equations that literally describe a hyperspace, a multi-dimensional space, which scientists often cryptically refer to as "configuration" or "phase" space, are common in the mathematics of modern physics. As we have noted, most physicists have been taught during their university educations to think of these as only mathematical devices because it makes no sense to use ordinary language or pictures in an attempt to ascribe a reality to such bizarre number juggling. Bohm, however, following the epistemological lead of Einstein, suggests that what works in our equations may point to an underlying reality.

Consider the following analogy from Bohm's *Wholeness and the Implicate Order.* Imagine a fishbowl with fish slowly swimming round and round, occasionally darting here and there, changing direction unpredictably. Imagine two TV cameras filming the activity of the fish from different points of view. Imagine finally that in another room a person is sitting watching two TV sets receiving the transmissions from the two cameras. This person at first might think that he is watching two different fishbowls and fish movements, except that he would notice an amazing correlation in the movements of the two sets of fish. Every time one of the fish in one TV screen unpredictably changes direction by darting to the left or right, a fish in the other screen changes directions also. After watching this activity for a while, this person should be able to deduce that the separate images are different perspectives of one reality. According to Bohm, this is what the long road of scientific endeavor, culminating in the experiments of quantum physics, has revealed to us: Our normal world of separate objects is but separate images of one underlying reality. We set up our three-dimensional experiments and then wonder how particles separated by light-years can be correlated, but from the standpoint of hyperspace the particles are right "next" to each other, so to speak; the two apparently separated particles are the same particle, just as the two apparently separated fish are the same fish.

There is the immense "sea" of energy . . . a multidimensional implicate order, . . . the entire universe of matter as we generally observe it is to be treated as a comparatively small pattern of excitation. This excitation pattern is relatively autonomous and gives rise to approximately recurrent, stable and separable projections into a three-dimensional explicate order of manifestation, which is more or less equivalent to that of space as we commonly experience it.

DAVID BOHM

Flat Land and Hyperspace

Because of our Kantian-Newtonian filters, it is difficult, if not impossible, for us to imagine what a multidimensional hyperspace is like.[1] We can, however, get an idea of what existence in a higher dimension is like by comparing our three-dimensional existence with a hypothetical two-dimensional existence called Flat Land.

---

[1] Some attempts at unifying all the known physical forces into a super-force use mathematical devices that refer to 11 dimensions.

Imagine a world that is flat like a piece of cardboard upon which flat, two-dimensional creatures live. Imagine that on this world there are flat, two-dimensional houses and flat, two-dimensional creatures that look like triangles, squares, and circles. Because they are two dimensional, these peculiar characters can go about their two-dimensional business by moving forward or backward and left or right, but "up" and "down" have no meaning in this world. Relative to this world, we would find that three-dimensional creatures like ourselves have supernatural powers. We could peer into their houses from above and watch what they are doing; we could cause strange events to happen at great distances simultaneously; we could cause correlated behavior in objects that seem separated to our flatlanders. We could even cause strange objects to appear out of nowhere.

Suppose we picked up an ordinary salad fork from our three-dimensional world and poked it in and out of this two-dimensional world. A flatland creature observing this event from his two-dimensional world would see only four mysterious dots appear from nowhere, move around in a coordinated manner, and then vanish as mysteriously as they appeared. If we picked up one of these two-dimensional creatures and pulled him up into our three-dimensional world, then he would have a mystical experience; he would experience a reality for which he has no language to describe. If we then placed him back onto his two-dimensional world, perhaps where a number of his friends are discussing his mysterious disappearance, he would appear to them to materialize out of nowhere. If he attempted to explain to his friends what he had experienced, then he would undoubtedly sound like a crazy fool, much like the enlightened man in Plato's cave.

According to Bohm, our observations of electrons and other subatomic phenomena in our three-dimensional laboratories with three-dimensional equipment are not the result of an act of creation of consciousness but rather an interfacing of a multidimensional reality with a three-dimensional one. Just as our flatlanders experienced mysterious, unpredictable events that were explainable from the point of view of another dimension, so the behavior of electrons and other subatomic phenomena are understandable from the point of view of an overlaying, but concealed, "implicate" hyperspace. Just as the actions of the four correlated dots produced by the three-dimensional fork are seen to be one reality, so our entire world of apparent separated particles that seem to make up separate objects is but a manifestation of one undivided hyperspatial whole.

The major virtue of such an interpretation of the mathematics and experimental results of quantum physics is that the realism of our normal three-dimensional world is preserved. When we walk out of a room, the room is still "there" in a sense. From a hyperspatial perspective, more than a three-dimensional room may be there, but the three-dimensional room is still there for any three-dimensional creature to see. We do not create the room from nothingness.

Many Worlds

There is no logical necessity for believing in one universe any more than there is for believing that Earth is the center of existence. Another interpretation of quantum physics that attempts to preserve the general philosophical position of realism is known as the Many Worlds interpretation. This interpretation preserves realism with a vengeance. In the 1950s Hugh Everett III, then a graduate student at Princeton University, decided to see what would happen if the mathematical equations of quantum physics were consistently taken literally. To see how this would work, let's return to our previous experiments.

Recall the experiment attempting to prove that single particles of light pass through only one channel (Figure 8-2a). Our result of detecting only one whole unit of energy at detector A or B was consistent with this interpretation. This interpretation was not, however, consistent with the outcome of the experiment with totally reflecting mirrors replacing detectors A and B. The Schrödinger equation depicts waves of some sort passing through both channels, and the experiment with totally reflecting mirrors demonstrates that light, as a wave that splits into two waves, is in both channels. According to the Many Worlds interpretation, there is a simple, but shocking, explanation for the first result. The Schrödinger equation depicts the radiation in both channels as real; the reason we only observe it at one detector or the other is because when a measurement is made, the world splits into two equally real worlds! When the radiation is detected at A, it has also been detected at B. We do not detect it at B because B is an event taking place in another world.

According to this interpretation, all the possibilities delineated by the Schrödinger equation are real. In making an observation of a particular possibility, we are not collapsing a wave packet or creating a reality from a number of possibilities. Rather, like a road with many forks, we are choosing a world to travel on from many possible worlds. All the alternate worlds are paths in hyperspace; they are equally real, but we are forever cut off from them. In every observation we are choosing a branch of reality. If the Copenhagen interpretation implies that nothing is real independent of observation, then the Many Worlds interpretation implies that everything is real. We do not create a universe with an act of observation; we choose a universe that is already there as a possible path. According to John Gribbin, an enthusiastic supporter of this interpretation, "By the act of observation we have selected a 'real' history out of the many realities, and once someone has seen a tree in our world it stays there even when nobody is looking at it."

In the two-slit experiment, when an attempt is made to see if the photons are passing through both slits (Figure 8-1c), we found the radiation passing through only one slit or the other. According to Gribbin, in his book *In Search*

By the act of observation we have selected a "real" history out of the many realities, and once someone has seen a tree in our world it stays there even when nobody is looking at it.

JOHN GRIBBIN, *IN SEARCH OF SCHRÖDINGER'S CAT*

of Schrödinger's Cat, here is the proper interpretation of what the electron is doing:

> Faced with a choice at the quantum level, not only the particle itself but the entire universe splits into two versions. In one universe, the particle goes through . . . [one hole], in the other it goes through . . . [the other hole]. In each universe there is an observer who sees the particle go through just one hole. And forever afterward the two universes are completely separate and noninteracting—which is why there is no interference on the screen of the experiment.[1]

Physics is neither epistemologically nor ontologically neutral.

F.S.C. NORTHROP

This means, however, that just as there are many routes to the future, there are many versions of "us" that will follow these paths. Because every observation splits the path we are on into alternate universes again and again, there are literally billions of alternate paths through hyperspace. These alternate worlds, however, are not parallel to us, as in so much science fiction, but like our three-dimensional view of two-dimensional Flat Land, they are at right angles. Somewhere in this hyperspace there is a world where the South won the American Civil War, a world where the Spanish Armada defeated the British, a world where John F. Kennedy was not assassinated, and a world where World War III has happened and the human species is extinct. You might take a rest here and contemplate the implications of this interpretation. While you are at it, also contemplate the fact that human beings can think such thoughts and that some physicists take this interpretation very seriously as the only way out of the quantum paradoxes. There are no limits to our gestures of understanding with the universe.

The Participatory Universe

As a paradigm for our time, some scientists have found it less shocking to carry out the implications of the Copenhagen interpretation than to believe that each moment we are splitting into 10^{100} equally real copies of ourselves. John Wheeler, a distinguished American physicist, has argued that we must

---

[1] John Gribbin, *In Search of Schrödinger's Cat: Quantum Physics and Reality* (N.Y.: Bantam Books, 1984), p. 241. The title is taken from a paradox first discussed by Schrödinger. If a cat is placed in a special box with a

abandon the basic tenet of traditional realism—that the universe is in some sense sitting out there for us to uncover. In its place, according to Wheeler, we must boldly embrace the concept of a "participatory universe." (However, Wheeler has been highly critical of those who would use this abandonment of realism as an excuse for believing in the occult or mysticism. See the next section.)

Adherents of this view claim that all vestiges of traditional realism must be abandoned. Both Bohm's neorealism and the Many Worlds interpretation are but symptoms of our inability to give up a traditional metaphysics. There is no clocklike world in any sense sitting out there for our observational benefit alone. We do not observe "the real world"; we participate with reality by creating a reality for us. More precisely, we do not create reality; we select a concrete reality from out of an intermingled dance of intangible possibilities. (In the Many Worlds interpretation, all the possibilities are concrete.)

This concept is not as difficult to understand as it may seem. Wherever you are right now there are many hidden, potential manifestations of energy that all of us have come to take for granted in the twentieth century. There are many potential channels of electromagnetic information. Although we cannot see them or feel them, there are many AM, FM, and TV signals passing by us at any given moment. They are both here and not here. To make these signals of information manifest, to make them concrete, we must "tune them in"; we must have a device such as a radio or TV set to collapse the indefinite electromagnetic waves into concrete electronic digits of information. The human mind is like a radio receiver stuck on one channel. When we set up our three-dimensional laboratory equipment, when we peer into our high-tech telescopes and see galaxies millions of light-years away, we participate with the infinite by manifesting one of its faces. It is not a mask; it is definitely there. But only when we observe it; just as radio music is music only when we tune it in.

Our confrontation with the microcosmos has taught us this: The results of our experiments are due to our being on one channel, but the microcosmos is kind enough to reveal to us, through mathematics and observational paradoxes, that there are many other channels. It has taught us that when we go out on a crisp, clear night and peer through a pair of binoculars at the Andromeda galaxy and receive the light that in our normal mode of thinking is 2 million years old, we are instantly creating a 2-million-year-old past. The universe, in a sense, is here because we are here. But we should not get too uppity about this; the universe will do just fine without us. There is still a

Physics, too, is only an interpretation of the universe, an arrangement of it (to suit us, if I may be so bold!), rather than a clarification.

FRIEDRICH NIETZSCHE,
BEYOND GOOD AND EVIL

My mind, in an undisciplined way, detects the cosmic within the nitty-gritty and the trivial within the infinite.

HAROLD MOROWITZ

deadly vial of poison and a quantum device is used to trigger its release, then until we open the box to measure the state of the cat, the cat is both alive and dead. Like the electron, the cat is represented by a superposition of states. The Many Worlds interpretation solves this paradox by claiming that in one world the cat is alive and in another it is dead.

kind of a past even if I am not looking, just as there is potential music in my room, even if my radio is off.

Mysticism and the Convergence Thesis

One more interpretation of quantum physics deserves some comment. It is a very controversial interpretation, in part because it has attracted a faddish and cultlike following, which claims that the results of modern science have validated a particular religious orientation. The possibility of such a development is one of the reasons scientists are often reluctant to communicate with the general public. An idea, however, cannot be responsible for its misuse by uncritical followers, and the misuse of an idea does not prove the idea false.

For the purpose of identification let's refer to this final interpretation as the *convergence thesis.* Essentially, this view argues that our confrontation with the quantum has demonstrated that Western science, founded upon the logic and philosophy of the ancient Greeks, has, after traveling a much different philosophical path, converged with the philosophy of the East, especially the mystical philosophies of Hinduism and Buddhism. This view was popularized in the 1970s by Fritjof Capra in *The Tao of Physics* and by Gary Zukav in *The Dancing Wu Li Masters.* According to Capra, "What Buddhists have realized through their mystical experience of nature has now been rediscovered through the experiments and mathematical theories of modern science." Zukav says, "Hindu mythology is virtually a large scale projection into the psychological realm of microscopic scientific discoveries."

For many thousands of years, it is argued, the mystics have had a cosmological, ontological, and epistemological view of things that the Western world is just beginning to understand. Cosmologically, Western science has understood only recently that the universe is extremely old. In 1965 the temperature of the universe was measured for the first time, resulting in our present estimate of the age of the universe as 15 billion years old. In the ancient literature of the East one does not, of course, find such precise figures. Instead there are analogies such as the following. Imagine an immortal eagle flying over the Himalayas only once every 1,000 years; it carries a feather in its beak and each time it passes, it lightly brushes the tops of the gigantic mountain peaks. The amount of time it would take the eagle to completely erode the mighty Himalayas is said to be the age of the present manifestation of the universe. Such a conception of time, which predates modern science by thousands of years, is thought to be remarkable, especially when it is compared to the slow realization of Western science and religion to the possibility of a less humanlike time scale.

For if those who hold that there must be a physical basis for everything hold that these mystical views are nonsense, we may ask— What then is the physical basis of nonsense? . . . In a world of ether and electrons we might perhaps encounter nonsense; *we could not encounter* damned nonsense.

ARTHUR EDDINGTON, *THE NATURE OF THE PHYSICAL WORLD*

Ontologically, Eastern mysticism is also consistent with the results of quantum physics. The mystics have always rejected the idea of a hidden clocklike mechanism, sitting out there, independent of human observation. The number one truth is that reality does not consist of separate things, but is an indescribable, interconnected oneness. Each object of our normal experience is seen to be but a brief disturbance of a universal ocean of existence. *Maya* is the illusion that the phenomenal world of separate objects and people is the only reality. For the mystics this manifestation is real, but it is a fleeting reality; it is a mistake, although a natural one, to believe that maya represents a fundamental reality. Each person, each physical object, from the perspective of eternity is like a brief, disturbed drop of water from an unbounded ocean. The goal of enlightenment is to understand this—more precisely, to experience this: to see intuitively that the distinction between me and the universe is a false dichotomy. The distinction between consciousness and physical matter, between mind and body, is the result of an unenlightened perspective.

Epistemologically, our so-called knowledge of the world is actually only a projection or creation of thoughts. Reality is ambiguous. It requires thoughts for distinctions to become manifest. We have seen that in the realm of the quantum, dynamic particle attributes such as "spin," "location," and "velocity" are best thought of as relational or phenomenal realities. It is a mistake to think of these properties as sitting out there; rather, they are the result of experimental arrangements and ultimately the thoughts of experimenters. Quantum particles have a partial appearance of individuality, but experiments show that the true nature of the quantum lies beyond description in human terms. Our filters produce the manifestations we see, and the result is just incomplete enough to point to another kind of reality, an ambiguous reality of "not this, not that."

For the mystic, the paradoxes of quantum physics are just another symptom of humankind's attempt to describe what can only be experienced. We are like a man with a torch surrounded by darkness. The man wants to experience the darkness, but keeps running senselessly at the darkness with his torch still in hand. He does not realize that he must drop the torch and plunge into the darkness. The proliferation of philosophical interpretations of quantum physics is a symptom of the shipwreck of a traditional Western way of understanding, of our inability to "let go" of our Western torch—our traditional logic, epistemology, and ontology. It is also a symptom of our inability to let go of our egocentricity, our persistent attempt to define everything in purely human terms, as if we were somehow special and separate from the rest of the universe. Like a nervous, self-centered teenager at a party, concerned only with what others think of him, our entire field of vision and understanding is narrowly defined in terms of "me." Because of our fear of letting go, we are missing much that is right in front of us.

According to this interpretation, the mathematics is complete just as it is. What the Schrödinger equation depicts for microscopic objects is also true for any macroscopic object. The universe is not full of separate objects,

[We must] continue to insist on the centuries-long tradition of science, in which we exclude all mysticism and insist on the rule of reason. And let no one use . . . [quantum] experiment to claim that information can be transmitted faster than light, or to postulate any so-called "quantum-interconnectedness" between separate consciousness. Both are baseless. Both are mysticism. Both are moonshine.

JOHN ARCHIBALD WHEELER

people, and places. Rather, it is an unbounded field of entangled possibilities. Because of the level of our conscious awareness, we fail to realize that duality, ambiguity, and interdependence are the rule rather than the exception. Mathematics may be one of the closest ways we can come to representing this in terms of a human language. All languages, however, are ultimately inadequate. Myths, stories, analogies, pictures, mathematical equations—all such symbol systems can just point to what can only be fully understood through a visionary experience.

In the episode entitled "The Edge of Forever" in the "Cosmos" television series, Carl Sagan visits India, and by way of introducing some of the bizarre ideas of modern physics, he acknowledges that of all the world's philosophies and religions those originating in India are remarkably consistent with contemporary scenarios of space, time, and existence. However, adamantly skeptical of the knowledge value of a nonrational mystical intuition, he concludes that although these religious ideas are worthy of our deep respect, this is obviously a "coincidence." Using natural selection as a model, Sagan proposes that this consistency is "no doubt an accident" because given enough time and possible proposals, given enough creative responses to the great mystery of existence, some ideas will fit the truth just right.

Other critics of the convergence thesis have not been as charitable. They argue that it is just plain silly to interpret an ancient belief system, founded upon certain psychological needs and within a historical context, in terms of any modern perspective. It is obvious, they argue, how the Hindu and Buddhist beliefs could soothe people living under extreme conditions. If our day-to-day reality is but a fleeting manifestation, then the vicious misfortune and meaningless suffering of this world are not real. For these critics, the methodology of psychological need as an origin of these ideas implies there is no connection. By revealing the obvious psychological motivation for a set of beliefs, it is argued, one can question the truth of these beliefs. To further suggest that there is any connection between these beliefs and the results of rigorous experimental science is ludicrous.

Both of these arguments are flawed. If the ideas of Hinduism and Buddhism are simply the result of a lot of guessing, and the serendipitous contingency of evolutionary processes the appropriate model, then shouldn't all the guessing that takes place over time be consistent with a macroscopic environment, not a microscopic environment with which a primitive people have no experience? And even if it is true that a belief system serves a set of psychological needs, does this prove the belief system false? Many scientists are also surely motivated for many reasons to hold the beliefs they do: a philosophical perspective, the need for certainty, or the need for security (be it a government grant or tenure at a prestigious university). That scientists have biases and motivations to believe what they do does not prove that what they finally believe is false.

Both of these arguments, however, do reveal a sobering point. The philosophical consistency between Hinduism and Buddhism and the results of modern science does not prove much by itself. Historically, we have seen many

In our teaching we have an obligation to help our students to think about the uncertainties and ambiguities of nature as they are found at the interface between the known and the conjectural, but we have also . . . the higher responsibility to help them function on this side of that interface. On this side— well back from the exciting and esoteric frontier where Einstein and Bohr still wrestle to a draw—our students are presented with obstacles to clear thinking and daily assaults against science and against the integrity and reasoning of the people who do it.

CHARLES STORES, a master science teacher

instances of a philosophy or a religious view being consistent with the science of a time, and a consequent rush to claim that the new science validates a religion or a philosophy. For both Copernicus and Kepler, the heliocentric system of the planets was consistent with their Neoplatonism and the idea that the sun was the "material domicile" of God. Similarly, for Bruno the heliocentric system was consistent with a larger universe and a greater God. For Newton a universe based upon the laws of universal gravitation was consistent with a conception of God as a master craftsman, a creator of an almost perfect machine who left a few defects to give Himself something to do. For some of the initial supporters of Darwin, natural selection was interpreted as a vindication of a philosophy of inevitable progress based upon a capitalistic economic system.

Perhaps the more pertinent question, applicable to all the interpretations of quantum physics, is not which offered paradigm is the truth, but which one will give us the most mileage? Which one, if followed as a guide, will be the most fruitful in stimulating the imagination of the next generation of scientists in devising new ideas, mathematical relationships, and experiments? In this chapter we have not given much attention to the area of modern physics that recently has gotten the most notoriety. In spite of the overwhelming success of the experimental demonstration that a traditional metaphysics of reductionism is inadequate, most physicists, concerned with the daily demands of obtaining research grants and Nobel Prizes, have simply filed such demonstrations away and continued with the Einsteinian quest, searching for more and more exotic particles, new "things" that will prove the supersymmetry theories, unifying all the known forces of nature and catapulting our understanding to the first microseconds of the universe and perhaps beyond.

Yet in spite of Nobel Prizes for the discovery of some of these new particles and public pronouncements that the end of the Einsteinian quest is near, one senses that all is not well with this approach. Physicists themselves complain that the proliferation of particles necessary to explain everything is too complex to be consistent with a simple universe. One senses many ad hoc approaches and a situation not unlike followers of Ptolemy adding epicycle after epicycle to make the data fit. Some experiments reveal serious anomalies. Particles that allegedly consist of a "bag" of quarks are not supposed to pass through each other, but in some cases, if the spins are just right, they do! One senses that nature is not yet ready to succumb completely to our latest gestures of understanding.

Every past success at understanding has produced new mysteries. Why should it be any different now? Perhaps the results of quantum physics discussed in this chapter are revealing to us a great discovery after all. This great partner we call the universe is not a static personality, but grows and is formed by us, as we are by it. There is every reason to believe that our romance will continue, that there are many mysteries left for a new generation of physicists. Although there have been many pretenders since the time of Kepler, no one has yet read the mind of God.

I believe that certain erroneous developments in particle theory . . . are caused by a misconception by some physicists that it is possible to avoid philosophical arguments altogether. Starting with poor philosophy, they pose the wrong questions. It is only a slight exaggeration to say that good physics has at times been spoiled by poor philosophy.

WERNER HEISENBERG

Concept Summary

One of the major scientific questions of this century seems so simple. What are electrons, photons, and other subatomic objects that have made the amazing technological revolution of this century possible? After more than 70 years of asking this question, no consensus on an answer has been reached, and widely divergent views on the nature of reality and the role of science in dealing with reality have resulted.

Experiments with subatomic phenomena show effects that are difficult to reconcile with our normal view of an objective world. Particles of matter are independent objects located in one place at a time. Waves can spread out and/ or split and be in many places at the same time. All experiments with subatomic phenomena show wave-particle duality; rather than a definitive, objective world, reality seems to be ambiguous at the quantum level.

Although a very successful mathematics was developed enabling physicists to interact with, explore, and extend applications of subatomic reality, the interpretation of the mathematics is a philosophical muddle. One of the most influential interpretations of quantum physics was that of Niels Bohr and what has come to be called the Copenhagen interpretation. According to this interpretation, the ambiguity and complementarity of quantum experimentation reveal a startlingly pragmatic, epistemological discovery: Our macroscopic experiments must be conducted from the point of view of a human conceptual reference frame, but nature at the quantum level need not, and apparently does not, conform to macroscopic concepts. Accordingly, what we measure in our quantum experiments are the results of our relationship with nature, not nature itself. Between moments of preparation and measurement of quantum events, there is nothing definitive to know, understand, or measure, because nature has revealed to us that there is nothing there that can be conceptualized in human terms. Subatomic phenomena such as photons and electrons become definitive objects only after measurements are made with macroscopic equipment.

Because Einstein was convinced that the goal of science is to reveal the clockwork mechanism of nature itself, not just the probabilistic results of our relationship or experimental tinkering with nature, he objected to this interpretation. Quantum theory could not be a complete theory. For Einstein there was an underlying reality that we did not understand yet; for Bohr, the goal of knowing an underlying, definitive reality was an antiquated philosophical relic of classical physics. For Einstein, the Copenhagen interpretation implied defeatism at its best and classical idealism at its worse. For Bohr, many fruitful explorations still exist, relationships yet to be described and mathematical trails yet to be followed, but the search for a "hidden" reality will not be one of them.

The work of Bell and Aspect has resulted in a strong experimental confirmation that in the quantum realm it is wrong to think of quantum phenomena as independent hidden entities influenced by independent local circumstances. Like Newton confronting the problem of gravity, most physicists in the twentieth century have been trained to adopt a pragmatic or instrumentalist stance to these results—science is supposed to describe the objective properties of experiments, not speculate on a hidden reality between measurements. But the compelling need for a philosophical understanding has produced numerous proposals.

Some physicists have argued that the success of quantum theory shows that far from being a secondary quality, consciousness produces a definitive, relationship-reality from an ambiguous, featureless whole. David Bohm has suggested that an interpretation of radical neorealism is still possible, one that describes a multidimensional hyperspace of implicative wholeness "behind" the explicative or definitive reality of our common-sense world. Others have argued that a rigorous interpretation of the mathematics of quantum theory reveals that the divergent results of interactions with the quantum realm can be explained in terms of real, branching, or splitting universes. Still others have claimed that we must abandon traditional realism altogether. The world cannot be pictured as "sitting out there" for us to uncover; our "participation" with an intermingled dance of possibilities yields a concrete reality. Finally, some have even claimed that quantum theory represents a convergence of Western and Eastern philosophies; modern science has uncovered the same indescribable, interconnected oneness discussed by mystics for centuries. For the mystic the result of quantum experiment is like a man with a torch surrounded by darkness. The man wants to experience the darkness, but keeps running senselessly at the darkness with his torch still in hand. Most scientists believe this is total nonsense.

One thing seems clear. There are plenty of mysteries left to stimulate the next generation of physicists.

Suggested Readings

Quantum Reality: Beyond the New Physics, by Nick Herbert (Garden City, N.Y.: Anchor Press, 1985).

Although many books have attempted to convey to a generalist audience the philosophical excitement and perplexities inherent in the development of quantum physics, this book is highly recommended for its readable style, objectivity, and boldness. It presents each of the major interpretations of quantum physics fairly and is written by a physicist willing to discuss issues of reality in a nonmathematical language (something most physicists have been taught not to do). It also incorporates a historical perspective with important recent work by Bell and Aspect. For other introductory presentations for the nonspecialist, see *Taking the Quantum Leap: The New Physics for Nonscientists* by Fred Alan Wolf (San Francisco: Harper and Row, 1981) and *In Search of Schrödinger's Cat: Quantum Physics and Reality* by John Gribbin (New York: Bantam Books, 1984), both of which advocate a particular philosophical perspective.

The Dancing Wu Li Masters: An Overview of the New Physics, by Gary Zukav (New York: Morrow, 1979), and *The Tao of Physics: An Exploration of the Parallels Between Modern Physics and Eastern Mysticism,* by Fritjof Capra (Berkeley, Calif.: Shambhala, 1975).

Although these books are also intended to be introductory, both are controversial, as noted in this chapter, in advocating the convergence thesis. If nothing else, both books show how developments in the quantum domain have caused the Western mind to reach beyond its cultural tradition for some philosophical help and guidance in constructing a new image of reality. Also see *Einstein's Space and Van Gogh's Sky: Physical Reality and Beyond,* by Lawrence L. LeShan (a psychologist) and Henry Margenau (a physicist) (New York: Macmillan, 1982) for an attempt to frame a new view of reality and mind using Eastern philosophy as a guide. The authors even discuss parapsychology and extrasensory perception within this context.

Atomic Physics and Human Knowledge, by Niels Henrik David Bohr (New York: Wiley, 1958), *Physics and Philosophy: The Revolution in Modern Science,* by Werner Heisenberg (New York: Harper & Row, 1958), and *Mind and Matter,* by Erwin Schrödinger (Cambridge, England: Cambridge University Press, 1958).

In these books three of the major players in the development of quantum physics give their interpretations of what this development means. Also see Heisenberg's *Philosophical Problems of Quantum Physics* (Woodbridge, Conn.: Ox Bow Press, 1979), and *Across the Frontiers* (New York: Harper & Row, 1974).

The Philosophy of Quantum Mechanics: The Interpretations of Quantum Mechanics in Historical Perspective, by Max Jammer (New York: Wiley, 1974).

An essential, complete scholarly resource for anyone ready to get serious about understanding the different schools of thought, and their historical origin, that have arisen in response to quantum physics. With some higher-order mathematics, the book covers from the 1920s up through the significance of Bell's work and the

Many Worlds interpretation. Includes a very nice development of the Copenhagen interpretation and the Bohr-Einstein debates.

The Shaky Game: Einstein, Reality, and the Quantum Theory, by Arthur Fine (Chicago: University of Chicago Press, 1986).

The book's title is taken from Einstein's concern that the Copenhagen interpretation implies playing a "risky game" with reality, that physics was abandoning its role of determining the independent physical states of a natural world. The author argues that Einstein was misunderstood and that recent developments in quantum physics (Aspect and Bell) may be incompatible with a "reductive" and classical realism, but are not necessarily incompatible with a "minimal" realism or what the author calls a "natural ontological attitude." Although the author's attempt to semantically navigate around the implications of the Aspect experiment is suspect, the book summarizes the philosophical issues well.

Quantum Theory and the Schism in Physics, by Karl Raimund Popper, ed. by William Warren Bartley (Totowa, N.J.: Rowman and Littlefield, 1982).

This compilation of writings and thoughts on quantum physics starts in the 1920s. It represents an attempt by one of the major philosophical figures of the twentieth century to counter the "subjectivism," and what the author calls "the great quantum muddle," produced by Heisenberg's and Bohr's Copenhagen interpretation. Popper argues that a proper understanding of quantum physics, involving a "propensity" particle interpretation, in which there are no waves and only objective probabilities of (admittedly queer) particles, can return science to its rightful enterprise of relentlessly getting us closer to the truth. Because this interpretation would ultimately have us return to thinking of electrons, photons, and protons as independent real things capable of precise locations, a position apparently refuted by the Aspect experiment, the author has been accused of violating his own epistemology (see Suggested Readings in Chapter 2) and imposing a dogmatic metaphysics upon science. For an especially scathing criticism of Popper's interpretation, see Paul Feyerabend's "On a Recent Critique of Complementarity," *Philosophy of Science* 35 (1968): 309–331 and 36 (1969): 82–105, and "Trivializing Knowledge: A Review of Popper's *Postscript,*" *Inquiry—An Interdisciplinary Journal of Philosophy* 29, no. 1 (1986): 93–119.

Wholeness and the Implicate Order, by David Bohm (London: Routledge & Kegan Paul, 1980).

An important work. Bohm is most noted for his thus far unsuccessful attempt at creating a "hidden variable" interpretation of quantum physics that will lead to novel, testable predictions. But as noted in Chapter 8, Bohm's book gives us a glimpse of a possible, creative neorealism that preserves the undefinable, indescribable, and immeasurable nature of quantum reality. It also comments on Eastern mysticism and the ultimate philosophical questions generated by modern science. For Bohm's thoughts on the latter, also see his discussions with the noted mystic Jiddu Krishnamurti in their *Truth and Actuality* (San Francisco: Harper & Row, 1980). For two other interesting attempts to reestablish some kind of realism in quantum physics, see Bernard d'Espagnat's *In Search of Reality* (New York: Springer-Verlag, 1983) and Alastair I. M. Rae's *Quantum Physics, Illusion or Reality?* (Cambridge, England: Cambridge University Press, 1986).

Quantum Theory and Reality, ed. by Mario Bunge (New York: Springer, 1967), *Paradigms and Paradoxes: The Philosophical Challenges of the Quantum Domain,* ed. by Robert Garland Colodny and Arthur Fine (Pittsburgh: University of Pittsburgh Press, 1972), and *From Quarks to Quasars: Philosophical Problems of Modern Physics,* ed. by Robert Garland Colodny and Alberto Coffa (Pittsburgh: University of Pittsburgh Press, 1986).

Books of readings by philosophers of science. The first one contains the original article by Karl Popper, revised and expanded in his *Schism* book, in which he characterizes the Copenhagen interpretation as causing "the great quantum muddle," and the last two have excellent articles on Einstein. See "Quantum Mechanics without 'The Observer'" by Popper in the Bunge book, "The Nature of Quantum Mechanical Reality: Einstein Versus Bohr" by C. A. Hooker in *Paradigms,* and "Einstein and the Quantum: Fifty Years of Struggle" by John Stachel in *From Quarks to Quasars.*

Articles to Knock Your Socks Off

"Are Superluminal Connections Necessary?", Henry Pierce Stapp. *Nuovo Cimento* 40B, no. 1 (1977): 191–204.

"Demonstrating Single Photon Interference," Arthur L. Robinson. *Science* 231 (Feb. 14, 1986): 671–672.

"Einstein Was Wrong," Paul Davies. *Science Digest,* (April 1982): 40–41.

"Princeton University Dean of Engineering Justifies Psychic Research." *Science News* 116, no. 21 (Nov. 24, 1979): 358–359.

"Testing Superposition in Quantum Mechanics," Arthur L. Robinson. *Science* 231 (March 21, 1986): 1370–1372.

"The Copenhagen Interpretation," Henry Pierce Stapp. *American Journal of Physics* 40 (Aug. 1972): 1098–1116.

"The Quantum Theory and Reality," Bernard d'Espagnat. *Scientific American* 241 (Nov. 1979): 158–181.

CHAPTER 9

Darwin's Universe Revisited:

Intelligence, Communication, and Extraterrestrial Life

We have little more personal stake in cosmic destiny than do sunflowers or butterflies. The transfiguration of the universe lies some 50 to 100 billion years in the future; snap your fingers twice and you will have consumed a greater fraction of your life than all human history is to such a span. . . . We owe our lives to universal processes . . . and as invited guests we might do better to learn about them than to complain about them. If the prospect of a dying universe causes us anguish, it does so only because we can forecast it, and we have as yet not the slightest idea why such forecasts are possible for us. . . . Why should nature, whether hostile or benign, be in any way intelligible to us? All the mysteries of science are but palace guards to that mystery.

TIMOTHY FERRIS

I don't think there's one unique real universe. . . . Even the laws of physics themselves may be somewhat observer dependent.

STEPHEN HAWKING

Natural selection does only one thing: it produces organisms better adapted to the local environment. It contains no built-in "self-perfecting" principle that guarantees a particular outcome, such as intelligence. . . . The evolutionary process is more subtle than the operation of some law of nature which unfailingly generates complex intelligent creatures.

EDWARD C. OLSON

The Egocentric Illusion

Nature is a relentless teacher. The results of quantum physics may be difficult to understand, but from a broad historical perspective a pattern emerges that makes intelligible, at least at a philosophical level, the results of modern science. Every age suffers from the same philosophical stumbling block: We consistently underestimate the role our humanness plays in observing and learning. History shows that we succumb easily to what Brian Stableford calls the "egocentric illusion" in his *The Mysteries of Modern Science:* the unconscious tendency to think of human concepts and perceptions as central or privileged in depicting the nature of reality and the structure of the cosmos.

From the beginning of our struggle to know, we have acted, in one sense or another, as if we were the center of everything. Every age has been able to recognize the egocentricity of the past generation but incapable of recognizing its own version. With quantum physics the pattern continues; nature will not allow human awareness and its principal analytic tool, science, to continue to the next stage until it discards another vestige of anthropocentric thinking. We have learned that the laws of nature may not be observationally independent as science has always assumed. This possibility has a profound implication for the major topic of this chapter: the existence of extraterrestrial life and the possibility of communicating with such life. To understand what it means to say that the laws of nature are *observationally dependent*, let's first review where we have been by sketching the egocentric pattern revealed by the history of science.

It was natural for early people to believe that they were the center of all things. Objective perceptions gave abundant evidence for believing this. Every morning the Sun rises in the east; every night the Moon, planets, and stars all seem to move around us. As we have seen, Aristotle, Ptolemy, and their followers provided a sophisticated astronomical, philosophical, and physical explanation for these common-sense observations. The Earth could not possibly move because if it did there would be an incredible east wind and objects would fly away when thrown into the air. Objects have a natural tendency to move down toward the center of the Earth, according to Aristotle, and this movement could not be accounted for if the Earth moved. The motion of the planets could be made predictable, according to Ptolemy's model, if the Earth was the center and all the celestial bodies revolved around it. Later, Tycho added the idea that the Sun could not be the center, because this would imply that the stars are too far away—a waste of space. Surely an economical, fastidious God would not have created such a situation. It was a tidy universe; almost everything made sense. Almost everything.

But there were puzzles. How could the planets physically revolve on an epicycle around an invisible point? What were those strange motions Galileo was observing through his telescope? Why did Venus show phases like the Moon? And most important of all, why were so many circles needed? If God was the creator of a mathematically harmonious universe, could He not have created a more elegant mathematical system than that modeled by the crazy complex circles of Ptolemy?

For Copernicus, Galileo, and Kepler, it was obvious that He could. So by the seventeenth century we discovered, somewhat painfully, that our home is not the physical center of it all. But our Sun was still considered special. For Copernicus and Kepler, the central importance of our God in this mathematically elegant universe must be matched by the central importance of our Sun. The entire cosmos, even the now very distant stars, must revolve around our Sun.

The universe was no longer a cozy backyard universe. In the eyes of de Cusa and Bruno, this was no problem; it was simply consistent with a greater God. Newton's theory of gravitation would soon not only provide an answer for how the Earth could spin without an uncontrollable east wind and destructive centrifugal forces but also would explain how the orbits of the planets would have to be exactly as outlined by Kepler. We were, however, still the center in that we were obviously the special creatures of a benevolent God. No other creature was allowed to read the mind of God, to determine the special laws of nature and the harmonious mathematics that made it work. One either believed with Descartes that God planted these ideas in our minds or that we possessed, according to the British philosopher John Locke, a special empirical ability to discover God's floor plan through experience. The theories of how we came to know God's plan may have been different, but the result was the same: We were special.

For Newtonians our knowledge and our point of view still had a privileged status. As the ancient Greeks believed, there was one world and it was full of details for us to know. The details of forces and material particles played out the dance of existence independent of our human measurements, but the perspective of space and time in which the dance took place was still a human perspective. The space and time of this new universe was still Earthlike. The universe was now much larger, but the successful Earthlike spatial geometry of three dimensions was thought everywhere to be the same. Similarly, time flowed everywhere as it does on Earth in terms of a uniform, absolute direction of past, present, and future. God's time was our time. What we observed to be true here on Earth—the laws of nature, space and time—were true throughout the universe. It all seemed so clear that Kant declared that no future experience could ever be inconsistent with this viewpoint, Leibniz announced that we are obviously living in the best of all possible worlds, and Laplace alleged that God's omniscience (if such a supreme intelligence existed) could be given a scientific definition. Everything made sense—almost everything.

But what was all that space for? Men of the Enlightenment such as Christian Huygens began to speculate that we were not alone and not the only

Man has undergone agonizing decentralization. He has waged a steady struggle against *decentralization, but at the same time— paradoxically—his accumulating knowledge has gradually forced him to abandon all illusions about his centrality.*

JAMES CHRISTIAN

creatures that could read the mind of God. There must be many planetary systems; otherwise God would have "wasted" a sun. There would have to be inhabitants of these planets, of course; otherwise God would have wasted a perfectly good planet. Even though Huygens acknowledged, in the nonegocentric spirit of his age, that these creatures could be physically and culturally different from ourselves, they would still have writing and geometry. It did not yet occur to many that this vast sea of stars might be there for no purpose or for one not even remotely related to human concerns. Humankind was still part of a well-designed, purposeful universe, and geometry was now God's gift to all His intelligent creatures, whether terrestrial or extraterrestrial.

The steady realization of the implications of natural selection, supported by the fossil record, began to unveil evolution as a tree with no special branch and no special creature. Millions of creatures had lived and died since the beginning, all as wonderful as the human species in their special adaptable way. The human species could now be seen as a relative newcomer, "a flash in the pan," with no guarantee of any special status. Soon some scientists and philosophers would begin to think the unthinkable—that our "intelligence" may not be a special adaptable trait that will guarantee a special survival.

This realization was not immediate. Darwinism was first supported uncritically by humanists and social Darwinists, who, despite their different political perspectives, saw in evolution a philosophical tool with which to overcome the otherworldliness of religion and replace such preoccupations with humankind and worldly business. For the humanists and the social Darwinists, evolution became mistakenly synonymous with their presupposed brand of "progress" and was used to justify their business and social philosophies.

While the full implications of Darwinism were being discussed, astronomers were unveiling another shocking picture of our physical place in the universe. Although the idea of a universe with no center had been proposed by Democritus as early as the fifth century B.C., and again in the fifteenth and sixteenth centuries by de Cusa and Bruno, it was not until the first quarter of the twentieth century that astronomical methods could provide concrete evidence for a radical new cosmology. Although Arabian astronomers recorded the brightest and nearest galaxies (the Clouds of Magellan and the Andromeda galaxy) as early as the eleventh century (with ideal viewing conditions, they are visible to the naked eye) and after the development of the telescope systematic cataloging of mysterious spiral nebulae took place from the seventeenth century on, until the twentieth century the universe was thought to be equivalent to the Milky Way. Then as quantum physicists were learning how to make beams of radiation in their laboratories and realizing that a classical representation of reality would not work in understanding the electron, the American astronomers Harlow Shapley, Edwin Hubble, and Milton Humason discovered unmistakable evidence that the Milky Way was just one of millions of galaxies. Together, these astronomers established that the spiral nebulae were other "island universes," and that, except for a few near galaxies, all the other galaxies were rushing away from our galaxy. By ana-

In some remote corner of the universe, poured out and glittering in innumerable solar systems, there once was a star on which clever animals invented knowledge. That was the haughtiest and most mendacious minute of "world history"—yet only a minute. After nature had drawn a few breaths the star grew cold, and the clever animals had to die. . . . There have been eternities when [human intellect] did not exist; and when it is done for again, nothing will have happened.

FRIEDRICH NIETZSCHE,
"ON TRUTH AND LIE IN AN
EXTRA-MORAL SENSE"

lyzing the light from these galaxies, Hubble and Humason established a mathematical relationship between the distance and the speed at which each galaxy is receding from our galaxy. Soon galaxies were measured to be hundreds of millions of light-years away and receding at speeds close to half the speed of light.

If most of the other galaxies were moving away from us, then perhaps we were the center after all. Einstein's General Theory of relativity shattered this comforting thought. By applying the insights gained with his Special Theory to problems of gravitation, a new geometry and physics of "stretched" space was used to understand how all the galaxies were receding from ours, and each other as well. As de Cusa had noted centuries before, it was possible to be living in a universe with no center at all. The work of Einstein and Hubble gave reasonable evidence that it was true. The universe was now thought to be like an expanding balloon. On its surface were not only all the galaxies but space as well, expanding along with the galaxies. It is impossible to visualize the universe using this analogy—remember that there is no space inside or outside the balloon—but a mathematical description is now commonplace, and the astronomical evidence gathered throughout this century supports it.

Einstein also shattered the idea that the way we experience space and time on Earth is the same throughout the universe. With no privileged observational location from which to view space and time, different observers can view different histories. The observer for the first time was implicated in what is real; what is real depends to some extent on the existence of a conscious perspective. The belief that reality could in part depend on observation gained scientific respectability, and the floodgates of the strange were opened.

The Special and General theories combined to provide fertile ground for incredible scenarios. Mothers can become younger than their sons, and the universe itself can contradict human logic by being finite but unbounded.[1] Places, such as black holes, can exist where space and time freeze into eternity, and the physical universe itself with its space and time included can be thought of as emerging from a mathematical point, a singularity with space and time inside.

In Einstein, however, we see still perhaps another vestige of the egocentric illusion. Although strange, the overriding goal of this new cosmology is to save the absoluteness of the laws of nature. Our knowledge is still privileged. We are still describing an objective reality, the "things in themselves," not just a human view. With Einstein we are still assuming that the universe is like a grand mysterious clock with hidden distinct parts, ultimately intelligible to us. Our job is to observe the external motions of the clock and deduce the

[1] Unlike de Cusa who arrived at a centerless universe as a deduction from an infinite universe, Einstein makes us work harder. The universe can have no center, but can be finite.

Understanding the world for a man is reducing it to the human, stamping it with his seal.

ALBERT CAMUS

Every man takes the limits of his own field of vision for the limits of the world.

ARTHUR SCHOPENHAUER

hidden details. The world still consists of separate things and parts; our job is to figure out how the parts all fit and how the forces that bind the parts interact and make them move.

In quantum physics we receive what may be the ultimate egocentric shock. In spite of our progress, since the time of the ancient Greeks the myth of total objectivity has prevailed. We have assumed that our perceiving, measuring, and knowing something can be dualistically separated from that something. We have assumed that our experience of the outer world is somehow privileged and more real, somehow superior in the knowing process compared to our subjective experiences. Although different philosophers had suggested this possibility before, and the entire culture of the East had taken it for granted for centuries, we now begin to realize, experimentally and mathematically, that the conscious awareness of the observer is intimately bound up with the results of measuring, observing, and knowing. The idea of an observer must be replaced with the idea of a participant: We do not objectively observe nature, we participate with it. We do not re-present what is already out there independent of our tools of representation. Our involvement with the universe creates or makes manifest, to some extent, a reality for us from a flexible, faceless, potential reality.

Although, as we have seen, there are other interpretations, the results of the most recent experiments in the subatomic realm at least show that such talk of consciousness-created reality is no longer philosophical gibberish. The thought of ourselves as little privileged gods who can stand back and take it all in, as if we do not touch the world at all with our presence, is revealed to be just another human thought. Quantum physics clearly reveals this assumption in the shipwreck of this assumption. We cannot think of electrons, photons, and quarks as independent things with independent characteristics until we participate, through observation, with nature, and our participation will always carry with it the imprint of ourselves.

Can this concept be extended one step farther? Are the *laws of nature* themselves also dependent existences, not something out there that we discover but rather, as with the spin of an electron, *something made manifest* upon our participation with nature? "Everywhere we look the laws of nature are the same," boasts Carl Sagan. Perhaps this is only because *we* are doing the looking. According to Niels Bohr, the mathematical laws of quantum physics are best understood as describing not an independent existence but rather the relationship between the macroscopic reality of physicists and the quantum realm. It is a mistake to think that we can go "deeper," to think that we will discover some new mathematics that will allow us to measure this strange reality. There is no deeper reality. There is nothing there to measure—not that there is absolutely nothing there, but nothing that humans can conceptualize. Thus, the concepts that we create, and the laws of nature we discover on the basis of these concepts, cannot be thought of as pointing to or representing independent cosmic laws. To think otherwise is as egocentric as concluding that we are the physical center of the universe because a practical

We cannot doubt the existence of an ultimate reality. It is the universe forever masked. We are a part of an aspect of it, and the masks figured by us are the universe observing and understanding itself from a human point of view.

EDWARD HARRISON

world view can be constructed from the apparent motions of the celestial bodies.

For the people of the Middle Ages it was consoling to believe that they were the center of the universe. In the twentieth century it is consoling to believe that our perspective of the universe is privileged, that the laws we discover, that work for us, are the laws that any intelligent form of life would eventually discover. The most recent experiments in physics, however, cast doubt on whether this will be believable in the twenty-first century. Quantum physics may well be the Copernican inversion of our time, and we have only begun to reflect on its ramifications. As we will see next, the possibility of the observational dependence of the laws of nature may mean that we are not only alone in a biological sense but alone with our thoughts of the universe as well.

Extraterrestrial Life

Sometimes I think we're alone. Sometimes I think we're not. In either case, the thought is staggering.

BUCKMINSTER FULLER

The egocentric pattern continues. We find it consoling to be part of a purposeful, God-crafted mathematical universe. It would be consoling if our universe is not like a deserted island, but is full of life. And we would prefer a universe full of intelligent creatures, many of which have discovered the same objective laws of nature that we have. Besides validating our science, such a universe would allow for meaningful communication by using the laws of nature as a universal standard. And we would prefer any visitors to be a benevolent, superior life form, one that is watching us now, waiting to step in and stop our self-destruction.

The important questions of whether or not life exists elsewhere and whether communication with any form of extraterrestrial life is possible necessarily involve all that we think we know. Involved is what we think we know about the cosmological structure of the universe—its size, age, and layout in terms of galaxies and stars, the number and type of these things, the distances that separate them, the chemicals they consist of, and how they have evolved and will evolve. Also involved, of course, is what we think we know about life and intelligent understanding of the universe. Do we know what life is? Have we agreed on a definition of life? Do we know the essentials behind the origin of life? If life has arisen on other planets, do we live in a Darwinian universe? Is the process of natural selection universal? Is intelligence rare or inevitable? Do we know what intelligence is?

The science of *exobiology,* the study of extraterrestrial life, has scant data to work with. We have, in fact, not yet discovered any subject for it to study. With the *Viking* project we have tested for microbial life on Mars, but the results were inconclusive. Recently, with an infrared telescope orbiting the

Earth, large amounts of dust and debris have been detected circling distant stars. But there is currently no consensus whether any star other than our Sun has planets. This question is so important to us that it is the one area of scientific endeavor in which scientists seem willing to speculate uncontrollably. As the chemist Robert Shapiro and the physicist Gerald Feinberg have noted in *Life Beyond Earth*:

> The question of the extent of life in the Universe . . . affects the framework of values through which we perceive the purpose of our lives, the goals of humanity, and the place of mankind in the Universe. . . . If our civilization continues on its present course, we will be faced with many choices involving our intentions toward the rest of the Universe. Very different sets of goals may occur to us if we learn, on the one hand, that life is extremely rare and we are essentially alone, or on the other, that the Universe is alive with fascinating and diverse living beings. The question concerning extraterrestrial life is one which we will want answered before we come to more final decisions on our goals and the meaning of our existence.[1]

By combining then all that we think we know, a popular scientific speculation on the extent of life elsewhere suggests that although the origin of life is a relatively improbable event because of relatively few hospitable biological locations, given that the basic chemical building blocks for life exist throughout the universe, and given the enormous number of stars and galaxies, life probably exists elsewhere. And some intelligent life forms capable of forming technological civilizations are also likely to exist. Just within our own galaxy there are between 8 to 24 billion yellow dwarf stars like our Sun. If just 20 percent of these have planetary systems, and only 1 percent of these have life, this would still leave millions of chances for intelligence to evolve.

But often this view is immediately qualified by pointing out that it is unlikely that we have been visited and that other intelligent forms of life are unlikely to resemble the human species.

Assume for the moment that other intelligent creatures exist in our galaxy and that they have evolved to the point of understanding how to use electricity, light, and electromagnetic communication in general. How would they know of our existence? The human species has developed the capability of electromagnetic communication only within this century. Just within the past few decades have we begun to "leak" the radio signals of our existence into deep

The idea that we shall be welcomed as new members into the galactic community is as unlikely as the idea that the oyster will be welcomed as a new member into the human community. We're probably not even edible.

JOHN BALL

---

[1] Robert Shapiro and Gerald Feinberg, *Life Beyond Earth* (New York: Morrow, 1980), pp. 20, 214.

space.[1] Our signals are traveling at the speed of light, but there are only 200 stars within 30 light-years of Earth. Within 500 light-years there are 1,000,000 stars, but it will be the twenty-fifth century when our signals reach all these stars. The greater the distance, the greater the sensitivity of receiving equipment needed to detect these signals.[2] Thus, given also that most UFO occurrences lack the kind of evidence the scientific method requires, most scientists are very skeptical about the Earth being visited at this early stage of our technological adolescence.

Furthermore, we must remember that we are living in Darwin's universe. Once life begins it is subject to the law of natural selection. That life on Earth began shortly after the Earth formed could mean that, given the right conditions, life may be common elsewhere. However, just as the weather conditions experienced by a tree over the course of its life can significantly affect its shape and the number and length of its branches, so enumerable conditions on Earth had to be just right to produce the life we see today.

Billions of coincidental events were necessary on Earth to produce the human species. Overcrowded conditions in primitive water basins, coupled with the right mutations, eventually produced land animals. If a primitive fish had not developed a lung, then there would be no amphibians, reptiles, or mammals today. Two hundred million years ago the giant continent Pangaea began to split apart, drastically changing the climate and isolating different animals on different continents. Sixty million years ago the dinosaurs became extinct, allowing for the proliferation of the mammals. At about the same time flowering plants and fruit trees evolved. This new method of plant reproduction provided a rich food source not only for insects but also for any creature that could adapt to life in trees. Thus, instead of perhaps reptilian monkeylike creatures, the primates evolved with stereoscopic vision to better judge distances between branches and grasping hands with which to navigate in this new environment.

What nature giveth, she also taketh away. As the continents continued to drift, the climate changed and large sections of lush forest began to disappear some 10 to 15 million years ago. In the place of the forests large sections of

---

If there are any gods whose chief concern is man, they cannot be very important gods.

ARTHUR C. CLARKE

---

[1] Much of this leakage is from the powerful military warning radars used by the Soviet Union and the United States, and TV broadcasts. Because of the length of the electromagnetic wave used, much of the AM broadcasting is reflected back to Earth by our atmosphere.

[2] This, of course, does not preclude some superior intelligence developing an innovative quantum physical technology. At least one scientist (Jack Sarfatti) has suggested that an "instantaneous communication device" could be constructed using a principle similar to the instantaneous quantum correlations obtained with particles in the Aspect experiment. Other scientists, however, have scoffed at this suggestion, pointing out that even if there are instantaneous correlations in particle spins over vast distances, no information could be transported in this way because the results at either finish line are random. See the runner's analogy in Chapter 8.

grassland appeared, and many species of apelike creatures either adapted to this new environment, remained in isolated pockets of forest, or became extinct. Eventually, the relentless sorting of natural selection produced a relatively physically weak creature, who for some reason, much debated by evolutionary biologists, began to walk and run upright. Its grasping hands, which originally evolved for an entirely different purpose, were put to use to create tools for food gathering and defense against the terrifying, physically superior creatures that were everywhere. Its disadvantageous situation forced it to be very clever and cooperate; otherwise it would have perished.

Evolution is irreversible and nonrepeatable. The lesson of natural selection is clear: There was no inevitable progressive march from a flippered fish to a four-legged terrestrial walker to a four-legged tree swinger, and finally to a two-legged runner. Evolution on Earth has been a jerky, messy affair; were life to start over again on Earth, no creature would evolve into its exact present form.

Natural selection, like a diversified stock portfolio, is in the business of producing diversity to enhance life's chances against the unpredictable contingencies of a fussy, heedless environment. Starting from a common ancestor, perhaps a single cell, perhaps a naked strand of DNA, life on Earth has diverged into an astonishing variety of ways for dealing with the many environmental niches on this planet. There are microbes that can exist above the boiling point and below the freezing point of water; some eat oil; some are even purple because they process a different wave length of light than do normal bacteria. Some insects farm; others build elaborate skyscrapers complete with air-conditioning; many hear with their legs, and some have antifreeze in their blood. In the great depths of the ocean there are light-bulb creatures producing their own light through bioluminescence. There are plants that catch and eat insects, fish with lungs, flying frogs and squirrels, animals with armor, mammals that lay eggs, birds that talk, and birds that build little nestlike houses and even paint the interiors. There are even animals that express feelings through sound and "see" through objects. And there have been creatures impossible to imagine without nature's fossil record for guidance: birds 12 feet tall, anteaters as big as a horse, a reptile that looked and acted like a dolphin, and the great dinosaurs, some as large as a six-story building.

If natural selection can produce this much diversity on a single planet, what could it do given trillions of planets, revolving around billions of stars, within millions of galaxies? Here is how science writer Gene Bylinsky summarizes this theme:

> From all we know now about evolution of life on Earth and the evolution in interstellar space of molecules that make up life, to assume that we represent the only life in the universe would be to return to the egocentricity of the Dark Ages, which placed the Earth at the center of the cosmos and proclaimed man life's crowning achievement.

Both competition and cooperation are observed in nature. Natural selection is neither egotistic nor altruistic. It is, rather, opportunistic.

THEODOSIUS DOBZHANSKY

The variety of animal life on Earth, with more than a million species, which walk, crawl, hop, swim, fly, burrow, squirm, stay fixed to one spot, and range in size from malarial parasites one eighth [sic] thousandth of an inch long to whales more than one hundred feet in length, represents only a fraction of life-forms that have, or could have lived on Earth.

Multiplied by billions of life-bearing planets circling other suns, the total number of cosmic species staggers the imagination.[1]

We must constantly remember that this diversity strongly suggests that there is no ultimate goal of evolution. Nothing will ever be repeated exactly the same because nowhere will there be exactly the same conditions with the exact same sequence of circumstances. Based on the best evidence available to us, we can conclude that we are the only human beings ever to have evolved or will evolve in this vast space and time. We are unique and alone in this sense.

Intelligence and Evolutionary Convergence

Evolution is capable of playing the same game with a different cast of players.

So why should we expect other intelligent technological civilizations to exist? If human beings are unlikely, why is intelligence likely? The answer, according to advocates of the existence of extraterrestrial intelligence, is that we must also remember that there are two aspects to the process of natural selection: (1) the *randomness* of mixing and mutating genes and (2) the *determinism* of environmental selection. Randomness produces *divergence,* but environmental selection, given similar environments, tends (from time to time) to produce *convergence.* In other words, because the environment tends to select these forms of life that are best adapted, it molds life into similar channels given similar environments and optimum ways of life within these environments. Intelligence, it is argued, is more likely to be a convergent property than a divergent property of life. Evolution tends to repeat in broad outline the best survival strategies. Intelligence is a highly adaptable trait, so it is likely to occur again given the right circumstances. The laws of nature represent the environment that is common to every place in the universe. Thus, given the astronomical odds, it is likely that creatures will evolve (more than once) that will "resonate" with these laws.

---

[1]Gene Bylinsky, *Life in Darwin's Universe* (Garden City, N.Y.: Doubleday, 1981), p. 70.

To understand how this might be possible, let's look at some examples of convergence on Earth.

Given the density of the Earth's atmosphere, it is not surprising that the development of wings occurred several times—in insects, reptiles (pterodactyls), birds, and bats. Being able to fly is an obvious survival trait. North American tourists visiting Hawaii are often surprised to find what they think are North American hummingbirds living on these tropical islands. What they are seeing, however, is not a bird at all but an insect, the hummingbird moth. It looks and acts very much like a hummingbird. This moth and the hummingbird obviously have no recent common ancestor, but their body structure and way of life have converged.

Similarly, given the density of water, it is not surprising to find so many creatures that have developed streamlined fishlike bodies. The dolphin and the shark are different in many ways. The dolphin is a mammal with a four-chambered heart; it breathes with lungs and has mammary glands and even some hair. The shark has the two-chambered heart of more primitive creatures, breathes with gills, and does not take care of its young. But both have the same basic shape, because it is the best shape for swimming fast and catching food in an aquatic medium. The ichthyosaur, an extinct reptile, was an even more striking example. It looked and probably acted very much like a modern dolphin—breathing with lungs and giving birth to live young. But the ichthyosaur evolved from a lizardlike land creature, whereas a dolphin evolved from a doglike land creature.

Finally, given the life-style of a land hunter, it is not surprising to find wolflike, catlike, and doglike shapes developing many times, in many unrelated creatures. In South America and Australia at various times there existed the marsupial (pouched mammal) mirror images of the North American wolf and cat. Although both groups are mammals, their common ancestry is very remote.

Thus, evolution is capable of producing a complex mixture of differences and similarities. On other planets where life evolves, it is unlikely that we will find animal and plant life exactly like that on Earth; too many circumstances must be repeated. But on other planets, so the argument goes, given similar environments, there should be some recognizable similarities of structure and behavior.[1] Because the laws of nature will be the same, no matter what the particular circumstances, natural selection will tend to pick, given countless chances, some solutions of survival as optimal.

However, optimal selections are not direct or inevitable. Computer simulations show that the optimal solution of an upright, intelligent reptile was

If this [humankind's extinction] happens I venture to hope that we shall not have destroyed the rat, an animal of considerable enterprise which stands as good a chance as any . . . of evolving towards intelligence.

J.B.S. HALDANE, "MAN'S DESTINY"

[1] Some may be very strange though. For instance, on a planet with a very thick atmosphere, one approaching the density of water, we might find fishlike and jellyfishlike creatures swimming in the air! On another planet perhaps marsupial humanlike creatures would live, favoring the evolution of a larger head and brain.

possible on Earth. Because of the same radical environmental change that led to the extinction of the dinosaurs, it did not happen. On another planet, especially one with a desert climate favoring reptilelike creatures, this optimal solution might exist. On Earth, if just a few circumstances had been different in the evolution of a two-legged runner from a four-legged tree swinger, there might have been no intelligent creature on this planet at present.

To summarize, given that the laws of nature are the same everywhere, an optimal survival solution would be for creatures to evolve that learn to be aware of these laws in a cognitive way and learn to manipulate their physical circumstances accordingly, rather than simply reacting instinctively to the results of these laws. Life will always take advantage of environmental opportunities, and understanding the laws of nature is the ultimate in environmental opportunities. But this optimal survival solution will not be inevitable because of the many chance elements involved in evolution. Thus, intelligence as a convergent property, as an obvious optimal solution to the problem of survival, may not be inevitable, but given the astronomical odds, its occurrence is probable in many locations even within our own galaxy.

Because convergence is so important for the possibility of other intelligent creatures, let's look at one more example. Let's return for a moment to our playful example used in Chapter 3, the basketball environment. There we noted that if our cultural environment changed radically, such that everyone must play basketball to survive, we should not be puzzled if tall people were the norm for the human species after a period of time. Members of the black race might be more numerous, not because of any inherent genetic superiority, but because of the relationship among prejudice, social isolation in ghettos, lack of job opportunities, a lot of free time, and basketball as one adaptation. Many tall, talented black adolescents would have an unintentional head start, just as the first mammals had a lucky start when the environment changed unfavorably 60 million years ago for the dinosaurs. The terrorized, hyperactive tree shrew living in its oppressed night ghetto was suddenly free to take advantage of new environmental niches.

Note now that our basketball analogy must, of course, be more complicated than was portrayed in Chapter 3. *Evolution is capable of playing the same game with a different cast of players.* Not only would tall, socially disadvantaged black adolescents have a head start but so would tall, socially disadvantaged white adolescents, who also had little opportunity to do anything else and a lot of free time. So might this new niche be taken advantage of by short, agile adolescents, who could jump three or four feet off the ground. Hence, similar circumstances could produce divergence. But it would also tend to produce convergence: Over a period of time, regardless of the different cast of characters and circumstances, good basketball players would tend to be favored.

The word *tend* is important. Evolution does not guarantee that optimal solutions will always emerge. A disease could strike all the tall players—perhaps a strange bone disease that coupled with gravity renders all tall players extinct. Short, mediocre players, if the timing was right and they were still alive, would rush in to fill the vacated niche. If there were no competition from better players, they could stay mediocre. Alive today are "living fossils," primitive forms of life that have inhabited narrow niches for hundreds of millions of years. Their survival is due not to a superior design but to their luck in being in a niche where competition and predators are few. Similarly, as the tree sloth demonstrates, a new species is not necessarily a biological improvement.

Sagan's Cosmic Rosetta Stone

One of the most popular advocates of this convergence-resonance theme has been the astronomer and science writer Carl Sagan. He has rationally and poetically attempted to draw the attention of the general public and the world's political and business leaders to the acute practical value of the cosmic perspective. He has courageously attempted to communicate the abstract and esoteric thoughts of scientists to a world that is often too busy to romance the universe and see the relevance of the intellectual embrace of science to their daily lives. Like the man in Plato's cave, Sagan has risked ridicule in attempting to broaden our horizons.

The ideal universe for us is one very much like the universe we inhabit. And I would guess that this is not really much of a coincidence.

CARL SAGAN

In *The Cosmic Connection,* *The Dragons of Eden,* and *Cosmos,* Sagan has advocated the convergence-resonance theme. Mindful of the importance of divergence in understanding the development and preciousness of human life and the possible development of life and intelligence elsewhere, he has nevertheless done much to popularize the notion that intelligent extraterrestrial life probably exists and that communication with such life is possible. According to Sagan, what makes such communication possible is the universality of the laws of nature and the capability of understanding these laws as an optimal survival trait.

Sagan sees the basic principles of science and mathematics as an interstellar or cosmic Rosetta stone. Just as the French archaeologist and linguist Jean-François Champollion finally deciphered the ancient Egyptian hieroglyphic writing by comparing the hieroglyphics and Greek written on a stone discovered in 1799 near Rosetta, Egypt, so the principles of mathematics and science provide the common key that will enable us to communicate with, or at least discover the existence of, a very different extraterrestrial culture.

It is a compelling and consoling thought that in this vast lonely space and time there might be others at least a little like us with which we could exchange insights on the meaning of existence. We have acted on this possibility in designing the message contained on the *Voyager* spacecraft. To play the record described in Chapter 6, an intelligent extraterrestrial Champollion must be able to decipher a scientific code. Such a creature must know about hydrogen atoms and be able to count starting with the number one. We have also spent some time and resources "listening" with our radio technology for unmistakable mathematical footprints indicating the existence of other intelligent life. According to Sagan, a reception of the series of prime numbers, numbers that can be divided only by themselves and one, would prove the existence of an intelligent sender. (We have also used the 1,000-foot diameter Arecibo radio telescope in Puerto Rico to send a message to another galaxy.) Such an event would mark a turning point in human history as significant as the Copernican revolution and hopefully, according to Sagan, would lead to a more enlightened and saner world.

It's an encouraging thought, but will it ever happen? We must confront the possibility of considerable egocentricity in this theory.

First of all, consider the conception of life that this theory assumes. All life that we know of, life on Earth, is carbon and water based. The carbon atom is an excellent molecular connector. That is, its particular atomic properties allow for the creation of long molecular chains such as the molecule of DNA. Most scientists believe that without some way of storing and transferring vast amounts of information—information necessary for the development of a living organism from a single cell to a complex creature with trillions of cooperating cells—life would be impossible. DNA serves this purpose for all life on Earth, and without the carbon atom such a long chain of stored information is impossible. Many scientists believe that no other atomic element is capable of entering into such long molecular chains. Life, so the argument continues, also needs a solvent—some liquid medium for energetic reactions to take place. Water, consisting of hydrogen, the most abundant element in the universe, is not only an excellent solvent but also remains a liquid over a relatively large temperature range. Thus, Sagan unashamedly calls himself a carbon-water chauvinist and argues that a reasonable person has no other alternative than to believe that if life exists elsewhere it will have the same fundamental basis as life on Earth. And with the same fundamental basis, the beliefs of the convergence-resonance theory follow as reasonable.

But it is a big universe, and if science has taught us anything, it is that there are surprises in every corner. That life has taken a particular form on Earth is no guarantee that this form is universal. In *Life Beyond Earth*, Feinberg and Shapiro have referred to carbon-water chauvinists as "carbaquists" and argue that our particular form of DNA-based life on Earth is more likely to be a special case. By assuming a different definition of life, emphasizing organization and information processing rather than particular elements, they have argued reasonably, if not persuasively for most biologists, that carbon and water need not be essential ingredients of life. Given radically different conditions than those on Earth, chemical reactions and molecular combinations that would be rare or impossible on Earth could produce the necessary forms of organization for life.

In low-temperature environments, Feinberg and Shapiro argue, life could flourish in liquids such as ammonia or even oil, and chains of nitrogen atoms,

In our time this search [for extraterrestrial life] will eventually change our laws, our religions, our philosophies, our arts, our recreations, as well as our sciences. Space, the mirror, waits for life to come look for itself there.

RAY BRADBURY

explosive on Earth, would be stable enough to form the long molecular information chains analogous to DNA. Similarly, in high-temperature environments where the carbon combinations of Earth-based life would be destroyed, a silicate-based life could exist. Given such possibilities, what would be the implications for intelligence and communication? Would we have anything in common at all with such creatures? Would they develop mathematics and science? Although the possibility of alternate chemistries will increase the chances of life elsewhere in the universe greatly, it also increases the likelihood that our particular way of relating to the universe is a divergent property of evolution. Perhaps mathematics and science exist nowhere but here on Earth.

A second, and perhaps more serious, consideration is the obvious circularity in believing that intelligence is a convergent evolutionary property. That our species is alive and possesses a certain characteristic hardly by itself proves that this characteristic is an optimal survival trait. That we have the ability to reason and think naturally leads us to conclude that the ability to reason and think is a positive thing. That we relate to the universe through the analytic tools of science and mathematics leads us to conclude that scientific evidence supports the view that science would be the preferred method of relating to the universe by any advanced creature.

Recently, Marvin Minsky, one of the world's foremost computer experts, has argued that if advanced extraterrestrial creatures exist, no matter how strange they are, they will reason the same way we do. Assuming that any form of life will be an information-processing machine of some sort, Minsky used an advanced computer to simulate "all" possible thought processes, all possible ways of processing information from an external environment. The computer showed that the only simulations of information processing that responded to input in an interesting, creative way were machines that performed a kind of counting operation. In other words, the only kind of coherent processing of input—input mimicking information from an external environment—was accomplished by machines that could do arithmetic. The machines that could not count stopped thinking altogether or ran in nonproductive circles.[1] Although Minsky's result is interesting, aside from assuming that all forms of life are essentially machines, this is hardly convincing, considering that a computer, whose thinking is based on counting, was used to prove that counting is the only way to think.

> *[Civilization] is a highly complicated invention which has probably been made only once. If it perished it might never be made again. . . . But . . . it is a poor thing. And if it is to be improved there is no hope save in science.*
>
> J.B.S. HALDANE, "MAN'S DESTINY"

[1] For Minsky's work in this and related matters see his comments and participation in *Communication with Extraterrestrial Intelligence,* edited by Carl Sagan (Cambridge, Mass.: MIT Press, 1973), the proceedings of an important international conference; his article, "Communication with Alien Intelligence," *Byte* 10, no. 4 (April 1985): 126–138; the "Space" section by Edward Regis, Jr., *Omni* 8, no. 6 (March 1986): 18, 82; Minsky's article "Why People Think Computers Can't," *Technology Review* 86 (Nov.–Dec. 1983):67–70, 80–81; and his recent book, *The Society of Mind* (New York: Simon and Schuster, 1986).

Looking back over the geological record it would seem that Nature made nearly every possible mistake before she reached her greatest achievement Man—or perhaps some would say her worst mistake of all. . . . At last she tried a being of no great size, almost defenseless, defective in at least one of the more important sense-organs; one gift she bestowed to save him from threatened extinction—a certain stirring, a restlessness, in the organ called the brain.

ARTHUR STANLEY EDDINGTON, *SCIENCE AND THE UNSEEN WORLD*

Before continuing, let's consider what "intelligence" means in the convergent-resonance argument: Human nature has evolved the powerful, and potentially flexible, capability of a cognitive awareness of the environment. That is, we have the ability to figure things out, to use symbols and abstractions to map how the environment works, to "re-present" artificially the things-in-themselves, and to reflect in a self-conscious way on the details of our representations with a sense of history and a possible future. This capability has enabled our species to respond in a flexible way to environmental change; we are not limited to reacting with the wrong physical skills to a radical environmental change—a situation that causes extinction—but can predict and plan with a sense of purpose. In short, human nature implements Lamarckian evolution at a cultural level. If the environment changes radically, then people can respond by acquiring new ways of living, and we can pass these acquisitions on to future generations.

However, comparing human nature with other forms of life on Earth shows that many creatures are more directly superior, physically speaking, in terms of information gathering. Many creatures have better eyesight and hearing, can directly experience electromagnetic fields, or can subtly detect chemicals; some have sonar. Because a bat directly senses where it is going by bouncing sound waves off objects, it has no use for a flight traffic controller using a mathematically programmed computer (and the cognitive skills this implies) to determine its flight path. An insect with its small brain does not need a sophisticated science laboratory to detect the temperatures, sounds, and scents it requires for survival; it directly senses this information—temperature with its feet, sound with special organs on its legs, and scents from miles away with antennae. A blue crab can detect amino acids (indicative of the creatures it eats) with its antennae. Sharks are so sensitive to magnetic fields that they can locate a wounded fish by detecting the electrical signals its struggling motion produces.

To a large extent, the human species has learned to duplicate these modes of information gathering, possessing indirect methods for obtaining different slices of reality. We are an evolutionary irony. Our physical inferiority coupled with the other contingencies of our evolution—as mammals we require a long period of parental care, allowing for a slow but durable learning process, and our apelike ancestors' environment required an upright posture for survival, freeing our hands for tool creation and use—have led to an *indirect* power and control over, and understanding of, our environment. However, the lesson from evolution on Earth seems to be that a *direct* sensing of the environment is the norm, the convergent evolutionary property, rather than an indirect intelligence. And if it is a convergent evolutionary property on Earth, it follows that if advanced creatures evolve on other worlds, they too will most likely use the direct sensing method displayed by most of Earth's creatures.

From this point of view, our indirect method is the result of one peculiar branch of Earth's evolution. Thus, even if the laws of nature are the same everywhere, an assumption we must examine more closely in a moment, cognitive intelligence would seem to be an unlikely evolutionary response. It is more probable that creatures elsewhere would converge around the direct methods so many of Earth's creatures display. If the direct method is a convergent evolutionary property, then there may be little need to develop the indirect methods of science and mathematics.

Consider the story of the physicist who was invited to play softball with a group of colleagues. He had never played before and his first attempt was disastrous. He could not hit a single pitched ball, and his attempts to throw and catch were simply dangerous. So embarrassing was this first attempt that he vowed to be an expert at this silly game by the next time he was asked to play. So he purchased a ball, glove, and bat and submitted each to a rigorous mathematical analysis. By the next time he was invited to play he knew everything there was to know "cognitively" about the physics of catching, hitting, and throwing, but to his consternation and repeated embarrassment, he still could not do any of these things. The philosopher Nicholas Rescher once commented, "Expecting extraterrestrials to be doing natural science as we do on Earth is like expecting a newly discovered desert-island race to be speaking grammatical English." Expecting science and mathematics to be optimal survival solutions may be as unlikely as believing that the best way to learn softball is from a physics textbook. The universe may allow for many optimal ways to play the game of survival.

In the sixteenth century it was difficult to believe that the Sun was the center of planetary motion because this implied that God had wasted a lot of space. In the seventeenth century it was difficult to believe that other planets did not have inhabitants with writing and geometry; otherwise God would have wasted a planet. In the twentieth century it has been difficult to believe that the universe could waste all this space on nonintelligent creatures. Renaissance scientists sought to understand the "music of the spheres," the elegant mathematical laws originating in the mind of God, proving in the process the exceptional nature of humankind's place in the universe. Perhaps twentieth-century scientists listen for mathematical signals from space hoping to validate our way of knowing, to prove that we are special and that our minds resonate with the way things are.

Finally, we must face the possible message of quantum physics. In claiming that the laws of nature are the same everywhere, the convergence-resonance hypothesis assumes that we can "re-present" reality. We have seen, however, that quantum physics implies that the laws of nature may not be the same everywhere, no more than space and time are the same everywhere. Because there may be no "things-in-themselves," conscious awareness of an environment implies participation, not objective observation. Thus, like a radio receiver our particular form of consciousness has evolved on one channel; it is only one of possibly many forms of conscious awareness capable of collapsing an ambiguous reality into a concrete reality. Thus, aside from the fact that the origin of life elsewhere may be improbable, aside from the possibility of different chemistries for life, aside from the many difficult, improbable paths

Expecting science and mathematics to be optimal survival solutions may be as unlikely as believing that the best way to learn softball is from a physics textbook.

life must take to evolve intelligence, aside from the possibility that a direct intelligence is more likely than an indirect cognitive one, we must also consider very seriously that even if extraterrestrial intelligent creatures evolve, they could be on a completely different conscious channel. Our logic, mathematics, and natural science could have no parallel anywhere in the universe. We could be totally alone with our thoughts.

The Ontological Status of Mathematics

To conclude this chapter, let's explore one final point. In Chapters 4 and 5 we raised the question of why mathematics works so well. For the advocates of the convergence-resonance theory, the workability of mathematics is very important. We have seen that Renaissance scientists believed that mathematics represented a direct channel to the mind of God and His secret, underlying plan for the universe. For the advocates of the convergence-resonance theme, it works so well that it is still easy to believe that mathematics represents a direct channel to understanding the secrets of the universe. Even in quantum physics it is mathematics that reveals that our mathematical representations may not be absolute.

A mathematical truth is timeless, it does not come into being when we discover it. Yet its discovery is a very real event, it may be an emotion like a great gift from a fairy.

ERWIN SCHRÖDINGER

In some science fiction movies, when contact is made finally with extraterrestrial creatures, an expert mathematician is summoned to communicate with them. As we have seen, many assumptions lie behind this innocent Hollywood notion. There is also a philosophical history that deserves some comment. Recall that Plato, struggling to answer the relativism of the sophists, found in mathematics the apparent certainty needed, and for various reasons, he also concluded that mathematical truths were not of the physical world. For Plato, mathematical truths existed in an eternal nonmaterial realm, another dimension from the material world of our everyday experience—a realm of truth that we could tap with our thinking and then apply to the confusing, shadowy realm of daily life, saving the phenomena and establishing order in the process. For Plato, our mathematical applications can never be totally accurate, but aside from being the best we can do, mathematics at least demonstrates that we are capable of knowing absolute truth. Plato was an idealist who believed that ideas are more real than the things they are supposed to represent. The number "2" is not a physical thing, but it is a very powerful concept that can be applied to an infinite number of physical things. For Plato, the entire universe could be destroyed, but 2 + 2 would still be 4.

For Aristotle, Plato's most famous student, this mystical interpretation of absolute truth had too many problems to be defended. Aside from seeing the technical problem of how our minds could "participate" with this eternal realm, Aristotle was a realist who was more comfortable believing that the truths of logic and mathematics were in the physical world. In a sense for Aristotle, "2 + 2 = 4" is a physical fact. Just as a sculptor's statue has a form, so the objects of the world have formal relationships that humans can recognize. And just as the form of the statue could not exist unless there was a physical substance to form, so the truths of mathematics could not exist unless there were physical objects to count. For an Aristotelian, mathematical principles work because they represent formal truths of reality. They are the invisible floor plan, the skeletal structure that supports all existence. In philosophical terms, mathematical (and logical) truths represent the formal structure of Being. Thus, for an Aristotelian, when we apprehend a mathematical truth on Earth, when a child finally understands the difficult abstraction of "2," a truth is known that applies to the entire universe. For an Aristotelian, mathematical principles represent independent objective truths about the physical universe, truths that we discover just like other physical facts.

Thus, for a modern Aristotelian, it is easy to understand why we are able to fly our spacecraft billions of miles from Earth and arrive at precise points at precise times. The mathematical principles we discover on Earth are universal; these principles can be applied anywhere with the same success. The nineteenth-century mathematician Charles Hermite demonstrated the historical pervasiveness of this belief when he said, "I believe that the numbers and functions of analysis are not the arbitrary product of our spirits: I believe that they exist outside of us with the same character of necessity as the objects of objective reality; and we find or discover them and study them as do the physicists, chemists, and zoologists."

The convergence-resonance theory is consistent with the Aristotelian position. As we have seen, in our modern period what was previously a religious harmony between our minds and nature has now developed into an evolutionary resonance.

Kant, however, offers a completely different interpretation of why mathematics works so well. Recall from Chapter 7 that faced with the problem of induction and the skepticism of Hume, Kant concluded that it is impossible to claim with any certainty that our thoughts of nature represent the things-in-themselves. For Kant, a more epistemologically honest position was to believe that mathematical truths were not the formal structure of reality but formal filters by which human beings understand and organize their experience of reality. For Kant, the principles of mathematics were still objective and universal, not relative to a particular culture or period of human history. But the universality of these principles was relative to the human mind. They represented "our way" of filtering reality. In the words of the twentieth-century physicist Sir Arthur Eddington:

Where science has progressed the farthest, the mind has
but regained from nature that which the mind has put into
nature. . . . We have found a strange footprint on the shores
of the unknown. We have devised profound theories, one

after another, to account for its origin. At last, we have succeeded in reconstructing the creature that made the footprint. And Lo! it is our own.

In short, small wonder that everywhere we look, everywhere we go, mathematics works. Everywhere we go, we take our minds with us! Kantianism implies that the principles of mathematics are not truths of the real world but simply a way of conceptualizing or filtering the world. Would other forms of consciousness evolve the same filters? At the very least, Kantianism introduces another unlikely variable into the convergence-resonance scenario.

Another interpretation of the status of mathematics, one that is as old as Protagoras but also very popular today, is *conventionalism*. The conventionalist argues that mathematical truths are not "descriptions" at all. For an Aristotelian, mathematical truths describe objective facts about the physical world; for a Kantian, mathematical truths describe the formal structure of human experience. For a conventionalist, a mathematical system is not a system of truth but a "game," and the principles themselves are not descriptions—they are "rules." There is nothing magical, durable, or absolute about the rules of a game. To play a game some set of rules is needed to prevent chaos and anarchy. But there is nothing absolute about what rules are used. The only thing that may matter is which rules are most convenient for the type of game you want to play.

Consider a basketball game. There is nothing absolute about the dimension of the court, the height of the basket rim, or the distance of the free throw line. No god has decreed that this is the only way basketball can be played. Any rule could be changed tomorrow. Because of the height of most basketball players today, a more interesting game might result if the rim were moved up to 12 feet from the floor, rather than its current 10 feet. It might be more convenient, given particular goals, to change the rules.

Suppose you wanted to play some one-on-one basketball. Suppose at first there is nothing unusual about the style of play of your opponent; he seems to know the game well and follows the normal rules. As the game progresses, suppose you get ahead by 10 points. Suddenly your opponent picks up the ball, pushes you out of the way, runs down the court without dribbling, and does a lay-up. Shocked, you ask him what in the world he thinks he is doing. He replies that this is his rule: When he gets 10 points behind, this style of play becomes legal. What has he done? He has not violated a law of a god. He has not foolishly flaunted an objective law of nature, such as jumping out of a sixth-floor window claiming that the law of gravity does not apply to him. He has simply violated a rule that was freely created by human beings. In this case you would have two choices: You would either decide that you would rather not play with this person or you would adopt his rules. Convenience and order are the only things that matter in playing games.

According to this view, "2 + 2 = 4" does not describe any kind of fact; it is simply a rule humans have made up to organize our experience. It is a convenient rule for most situations, but not all. Two things and two things are not always four things. Two female rabbits and two male rabbits do not remain four rabbits for long. Because of a chemical reaction, two quarts of alcohol and two quarts of water do not exactly equal four quarts. In quantum physics there are many states of existence in which ordinary arithmetic is inaccurate and hence not useful. In relativity theory we have seen that ordinary addition does not work in adding velocities that are high relative to the speed of light. In all these cases it is more convenient to use a mathematics different from ordinary arithmetic. In the case of rabbits it would be better to use the statistical procedures of the population geneticist; in chemical reactions, the transformations of the chemist; and in quantum physics, bizarre imaginary numbers such as the square root of -1.

Just as the world is flexible enough to allow for many different successful cultures, so the universe is flexible enough to allow for many different forms of organization. On some strange planet the environmental conditions might be so strange that by the time an object is counted it splits in two. On such a world it might be more convenient to use the rule "$2 + 2 = 8$." In some South Pacific island communities, "$2 + 2$" is not "4" because their method of counting does not use the number "4." Instead, their numbering system is 1, 2, 3, many! Is this an indication of backwardness, an ignorance of an important feature of nature, or simply a different, successful way of relating to nature, one consistent with the life-style and goals of this culture?

The conventionalist claims that there are many different ways of playing the game of life; the Kantian implies that there may be many different kinds of glasses with which to see the universe. Both views introduce possible divergent variables that make it more unlikely that our science and mathematics represent a cosmic Rosetta stone. For another intelligent form of life to possess mathematics and some day send us a message encoded with prime numbers may be as unlikely as someday receiving a message in English.

But even our conventions must work. The advocates of *conjecturalism,* a final view, acknowledge the enlightened and nonprovincial characteristics of Kantianism and conventionalism, but are uncomfortable with the relativism implied by them. Mathematical ideas, like all ideas, may be free creations of the human mind and thoughts we impose on reality, but not just anything will work. Even in our games the rules we adopt must be "constrained" by the physical circumstances of the external environment. According to this view, mathematical systems are in many ways very much like our physical theories. We create different styles of thought and then try them out on the universe for acceptance. Some work, many do not. This result implies a kind of objectivity once again; there must be something in nature captured by those mathematical theories that work. As with all scientific theories in general, mathematical ideas are informed guesses or conjectures that we create, and to be accepted, they must face the test of experience. Something is out there, and it "kicks back" unmindful of our wishes.

Before Kepler discovered the ellipse and the other laws of planetary motion, he thought he had intuited in a flash of insight God's secret mathematical plan for the planets. There were only six known planets and five perfect geometric solids. Often called the Pythagorean or Platonic solids, scientists considered why there could be only five a beautiful mystery. Each solid is a three-dimensional figure that consists of two-dimensional shapes with some number of equal sides. For instance, a cube has six squares for its faces, and a pyramid has four equilateral triangles for its sides and base. Geometric demonstrations can be constructed that prove that there can be only five such solids. By nesting the planets within the shapes of the five perfect solids, Kepler thought he had discovered not only why there must be only five perfect solids and six planets but also why the planets had the spacing that they did. The theory was so elegant and grand that it must be true. But no matter how hard he tried, no matter how many models of nested perfect solids he constructed, he could not get his theory to save the empirical facts of planetary motion. God may have constructed the universe based on an elegant mathematics, but this conjecture was not it. To his great disappointment, Kepler discovered what the Nobel laureate Abdus Salom more recently has noted, that the aesthetics of man are not necessarily those of the Lord.

Insofar as mathematics is about reality, it is not certain, and insofar as it is certain, it is not about reality.

ALBERT EINSTEIN

Does conjecturalism provide the objective basis the convergence-resonance theory needs? If there is something out there that our successful mathematics captures, then given the enormous number of possible places for life to evolve, is it not likely that some, perhaps many, intelligent forms of life, struggling against a heedless environment to survive, will evolve a system of thought to capture this same reality? As a realist, Einstein supported conjecturalism. We are free to boldly create whatever ideas we want, but the cosmic clock responds regardless of the apparent elegance of our ideas. Einstein was genuinely puzzled by the fact that any of our ideas work, when there are always an infinite number of possible ideas. How do we discover the reasonable ideas from the overwhelming number of conceivable ideas? Einstein was sure that we were not just fooling ourselves. Over time we really do find, in the words of Xenophanes, "ideas that are better." But in an age when it is unpopular to believe that we are God's special creature, deserving special gifts, it was a great mystery to Einstein how we do this.

The conventionalist can still argue that our science and mathematics work not because of any special correspondence between these forms of organization and an objective reality but because reality allows many forms of organization to work to some extent. The difference between these views may be only one of semantics. Some future philosopher of science may be able to show that the differences among all these views, including the epistemological fight between realism and instrumentalism, is not as great as first appears.

From a scientific point of view, however, the conjecturalist seems to have the last say. What is needed at this point is for reality to kick back a little. We do not know the answers to the questions posed in this chapter. It is premature to conclude much at all. The purpose of this chapter has not been to "debunk" the convergence-resonance theory or to ridicule spending our resources on the radio listening for a message using prime numbers. The important philosophical and scientific questions raised in this chapter can only be tested by further extraterrestrial research. Our purpose has been to make clear the different possible scenarios of extraterrestrial existence and communication and the philosophical assumptions and historical background behind the thinking on these possibilities.

During the 1960s and 1970s a flurry of related activitiy took place in attempting to communicate with animals. It was popular to think for a time that animals relatively close to our species, such as dolphins, gorillas, and chimpanzees, could be taught a special language. By learning to communicate with a few special animal emissaries, scientists thought that they could tap into the thoughts and feelings of these creatures, learning in the process another perspective on the meaning of existence. Many now think that early encouraging results were the result of egocentricity. Having animals mimic a language to get food does not mean that anything like creative thought initiates this activity. Although the scientific jury is still out on this prospect, it is important to remind ourselves at this point of the noble thought behind this and extraterrestrial research. We are all part of a grand mystery; the more opinions on its meaning, the more likely we will approach an understanding of it. We yearn to understand our existence, and we wish to have some company in this endeavor.

Unfortunately, the present crossroad that the human species faces cannot wait for conclusive evidence on whether we have any company. The human prospect in many ways is a human predicament. Our problems are great, our future existence unclear, our survival hardly guaranteed. Thus, in considering the very difficult choices we must make, we must consider the full ramifications of the cosmic perspective. For one, we must consider seriously the possibility that our entire existence—our culture and history, our values and mammalian heritage, our unique combination of emotion and cognitive thinking, and messy struggle to know, as well as our mathematics and science— may have no parallel anywhere else in the universe. There could well be many other forms of life in Darwin's universe, yet our species could be but a brief crazy accident, never to be repeated.

I shall tell you a great secret, my friend. Do not wait for the last judgment. It takes place every day.

ALBERT CAMUS

To have arrived on this Earth as the product of a biological accident, only to depart through human arrogance, would be the ultimate irony.

RICHARD LEAKEY

Concept Summary

Science may force us to be intimate with the world, but throughout its triumphant history it has also forced us to reflect on the nature of our intimacy with the world. We have learned that each new perspective revealed by science showed how much a vestige of our presence was previously involved in what we thought we knew about the world. The Earth seemed to be the center of the universe, but this was only because we viewed the universe, until recently, from this perspective. It seemed obvious that God's "nows" would be the same as our "nows," that now on Earth would be now throughout the universe, but after Einstein we know that our intuitive feeling of simultaneity is also anthropocentric. Quantum physics forces us to reflect upon the possibility that although we thought we were uncovering an independent nature, as in any relationship where an interaction is created and must be maintained by the existence of both partners, what we have been learning is the nature of our relationship with nature, not nature itself.

The universe is very large and we wish to know if there are other forms of life with which to share our thoughts. In the seventeenth century the best minds assumed other creatures must exist, because they assumed their God would not want to waste His creation of other planets and suns. And, of course, these creatures must think a little like God, and like us they must have writing and geometry. Today we think we know a lot more about this subject.

From one point of view the odds for extraterrestrial life seem very good. With billions of stars within billions of galaxies, and hence, perhaps trillions of planets, with the right chemicals for the evolution of life scattered throughout the universe, it seems absurd that we would be the only ones to establish a partnership with the universe. Whether or not such a conscious partnership like ours can be duplicated depends on to what extent evolution duplicates anything. Although the message of natural selection is clear in one sense— a process of nonrepeatable diversity means human beings were created only once—uniqueness can be complemented by repeating optimal ways of relating to the environment. Thus, evolution not only diverges it also converges.

Because evolution tends to repeat optimal ways of relating to the environment, a case can be made for the notion that our intelligence "resonates" with nature. Our awareness and consequent flexibility in dealing with the environment, our rational ability and its expression in science and technology, comprise an optimal survival trait. Hence, this trait is likely to be repeated at least sometimes when life evolves elsewhere. Although it is unlikely that other intelligent life has visited us—the distance between stars is too vast and our own means of technological communication too new for others to know about us—extraterrestrial intelligence could produce technology and the means of remote communication. This notion assumes that intelligence has something to be intelligent of, that like the characteristic of flight, which needs a common environment favorable for flight, so intelligence has a universal, objective climate—the laws of nature. If this is true, then the laws of nature also make possible a common means, a cosmic Rosetta stone, for communication with other intelligent forms of life. Hence, it is reasonable to listen with our radio telescopes for an unmistakable sign that we are not alone, most likely a coded mathematical message.

But is this view a twentieth-century egocentric version of the seventeenth century writing-and-geometry argument? Twentieth-century advances in science raise the possibility that the laws of nature are not observationally independent as the convergence-resonance argument must assume. What intelligence is intelligent of is not nature itself but our relationship with nature. And just as every relationship is unique, the divergent forces of evolution could well produce only divergent channels of awareness. We could be totally alone with our thoughts, even if many other forms of life in the universe exist. In particular, our mathematical thoughts, so critical to our way of relating to the universe, may be nothing more than our own footprints.

Alone or not, the result of finding out would be staggering. In the final analysis these issues can only be resolved by the self-corrective process of science. Whether or not there is something out there that a successful mathematics captures, and whether or not another intelligent form of life, struggling with its environment, has evolved a system of thought to capture this same reality, can only be answered by testing such notions as hypotheses, by looking for other footprints. Let's hope we have enough time.

Suggested Readings

Life in Darwin's Universe: Evolution and the Cosmos, by Gene Bylinsky, with illustrations by Wayne McLoughlin (Garden City, N.Y.: Doubleday, 1981).

Although this book is strongly in favor of the carbaquist and convergent proposals, it discusses the alternatives and serves as a good introduction to the sources and issues surrounding contemporary thinking on life elsewhere in the universe. Thought-provoking illustrations.

On Civilized Stars: The Search for Intelligent Life in Outer Space, by Joseph F. Baugher (Englewood Cliffs, N.J.: Prentice-Hall, 1985).

From a course, "Interstellar Communication," taught at the Illinois Institute of Technology. Begins with a dream/nightmare fantasy of our "squandering 4 billion years of painful evolutionary progress," and then examines, using mostly carbaquist and convergent assumptions, the prospects for contact between humanity and extra-terrestrial civilizations. Contains informative appendices and a very thorough bibliography.

Natural Acts: A Sidelong View of Science and Nature, by David Quammen (New York: Schocken Books, 1985), and *Evolution of the Vertebrates: A History of the Backboned Animals Through Time,* by Edwin H. Colbert, 3rd ed. (New York: Wiley, 1980).

A thorough appreciation of the possibility of the evolution of life elsewhere requires a thorough appreciation of life on Earth.

Quammen's book drives home for the intelligent layperson the theme of respect for the variety and quirkiness of the natural world. Colbert, formerly a professor of vertebrate paleontology, is now a curator emeritus of the American Museum of Natural History. For over 30 years his classic textbook has brought to life for students the last 400 million years of the evolution of backboned animals on Earth. The third edition has been updated with the latest in fossil finds and the effects of plate tectonics.

The Mysteries of Modern Science, by Brian Stableford (Totowa, NJ: Littlefield, Adams, 1977).

Every solution to a mystery creates in its wake new mysteries. There is a need, says Stableford, to popularize and explore the mysteries created by twentieth-century science, especially since the attitude of most nonscientists toward science is the result of a nineteenth-century perspective. Contains a chapter on the egocentric illusion, discussing how self-deception and discovery go hand in hand and how science is prone to take Copernicanism too far—assuming that what holds for Earth will hold elsewhere.

Intelligent Life in the Universe, by I. S. Shklovskii and Carl Sagan (San Francisco: Holden-Day, 1966), *The Cosmic Connection: An Extraterrestrial Perspective,* by Carl Sagan (Garden City, N.Y.: Anchor Press, 1973), and *Communication with Extraterrestrial Intelligence (CETI),* ed. by Carl Sagan (Cambridge, Mass.: MIT Press, 1973).

Carl Sagan has actively promoted the idea that it is now more likely than previously thought, based on recent scientific discoveries, that life exists elsewhere and that our own technological development gives us the means to communicate with advanced extraterrestrial intelligence. These three books discuss the possible scenarios needed for life and intelligence to evolve on other planets. The third book is the proceedings of an international conference held in the Soviet Union in 1971. Most of the participants were sympathetic to the carbaquist and convergent themes and the idea that there is an "ecological niche for intelligence."

Life Beyond Earth: The Intelligent Earthling's Guide to Life in the Universe, by Gerald Feinberg and Robert Shapiro (New York: Morrow, 1980).

Although this book does not focus on the question of extraterrestrial intelligence, it does strongly question the carbaquist definition of life and suggests that if life is defined in terms of energy flow and systems capable of interacting with and ordering this flow, maintaining sufficient levels of complexity in the process, then the universe may be teeming with radically different types of life. As such, this book places in a much different perspective our technological slice of life, suggesting that it is only a special case of a much more inclusive way of interacting with the universe.

Extraterrestrials: Where Are They?, ed. by Michael H. Hart and Ben Zuckerman (New York: Pergamon, 1982).

The proceedings of a 1979 "Symposium on the Implications of Our Failure to Observe Extraterrestrials." Although the purpose of this conference was to also examine the possibility of life elsewhere, a significantly different picture emerges from most of the participants from that suggested by Sagan and the convergent evolutionists: It is more likely that we are the first civilization in our entire galaxy. Evidence supports the notion that "the Universe appears to be a gigantic wilderness area untouched by the hand of intelligence," and the notion that we are alone (at least with our unique thought processes) can serve as an antidote to the indifferent and ruthless way most people treat life on Earth. Feinberg and Shapiro were participants at this conference.

For recent articles that question the convergent evolution theme, see Edward C. Olson's "Intelligent Life in Space," *Astronomy* 13 (1985): 6–11, 14–15, 18–22. and Frank J. Tipler's "Extraterrestrial Intelligent Beings Do Not Exist," *Quarterly Journal of the Royal Astronomical Society* 21 (1980): 267–281, and "Additional Remarks on Extraterrestrial Intelligence," *Quarterly Journal* 22 (1981): 279–292.

The Anthropic Cosmological Principle, by John D. Barrow and Frank J. Tipler (Oxford, England: Clarendon Press, 1986).

For a completely different perspective on the role of intelligence in the universe, this controversial book is recommended. Rather than seeing intelligence as a by-product of cosmological, chemical, and biological evolution, the authors discuss the possibility that the many apparent coincidences needed to produce intelligent life are too numerous to be coincidences, that intelligence *must* come into existence and the universe is here because intelligence is here. The authors also argue that we are alone because intelligent life is inevitable and if there were other versions of it, we would have observed their presence by now.

Because this view is similar to ancient teleological and design anthropocentric theories (the universe "strives" to create intelligence just as a planet "strove" to maintain circular motion in the Aristotelian-Ptolemaic cosmology), this book has received a great deal of expected criticism. See Martin Gardner's review in the *New York Review of Books* 33, no. 8 (1986): 22–25. If nothing else, however, the authors' discussion of history, cosmological theory, Earth science, and biochemistry serves as a scholarly compendium of factual information related to the questions of life and intelligence.

CHAPTER 10

The Human Prospect

This sense of the unfathomable beautiful ocean of existence drew me into science. I am awed by the universe, puzzled by it and sometimes angry at a natural order that brings such pain and suffering. Yet any emotion or feeling I have toward the cosmos seems to be reciprocated by neither benevolence nor hostility but just by silence. The universe appears to be a perfectly neutral screen onto which I can project any passion or attitude, and it supports them all.

HEINZ R. PAGELS

Ye daring ones! Ye venturers and adventurers, and whoever of you have embarked with cunning sails on unexplored seas! Ye enjoyers of enigmas!

Solve unto me the enigma that I then beheld, interpret for me the vision of the loneliest one. . . .

O my brethren, I heard a laughter which was no human laughter.

FRIEDRICH NIETZSCHE, *THUS SPAKE ZARATHUSTRA*

Our obligation to survive and flourish is owed not just to ourselves, but also to the cosmos, ancient and vast, from which we spring.

CARL SAGAN

News items:

A man who argued with a bar owner shot and killed the man, raped a woman and forced another woman to decapitate the dead bar owner, police said yesterday.

The suspect was arrested hours later after he fell asleep in a stolen taxicab. Police said his victim's head was on the front seat.

Authorities said that the man may have taken the head to keep police from finding the bullet.

Police who responded to a disturbance at an apartment smelled the "unmistakable" odor of burned human flesh and arrested a mother and her live-in boyfriend after discovering a 4-year-old girl's body in an oven.

Neighbors said they had heard loud religious music, sounds of fighting and screams of "Let me out!" before calling police. When police arrived they found the severely burnt body of the 4-year-old little girl.

A woman's body which was apparently dismembered and then thrown down a trash chute was discovered at a downtown high-rise yesterday.

The body had been further cut up by a trash compactor before it was found in a bin by a worker about 9 a.m., police said.

Although homicide detectives spent all day in the compactor room at the high-rise, the woman's head was not found.

Acts of horror similar to the preceding items are reported almost daily in the newspapers. Perhaps because they are so common, or perhaps because we do not wish to think about them, or both, these stories are often short and placed in the back of the newspaper. When such stories are printed, the first page usually relates more moderate human activities, such as tax reform, the president hosting a dinner for a visiting foreign leader, or a beauty contest winner. Underneath the local story on pieces of a woman's body found in the trash chute was a large picture of a boy and his dog. The boy and his six-month-old basset hound had won first place at a local Kennel Club competition.

Reptiles with Nuclear Weapons

Logic does not exclude madness.

ERICH FROMM

If we are not by nature violent creatures, why do we seem inevitably to create situations that lead to violence?

LIONEL TIGER AND ROBIN FOX, *THE IMPERIAL ANIMAL*

If we bother to read such stories at all, most of us console ourselves with the thought that such acts of violence are obviously the result of abnormal, deranged minds. Consider, however, how some normal, sane graduates of our best schools spend their time and intellectual resources. In 1982 another article in my daily newspaper reported that experts in what is called "conflict management" at the U.S. Defense Department were planning for World War IV. The 1982 *Defense Guidance* report, the Defense Department's official, annual policy statement, detailed principles and plans that would guide our response to the situation after World War III. Billions of dollars would be spent on submarines, bombers, and land-based missiles held in a secret reserve, and a communication system that would enable the United States to emerge from the ashes of World War III with sufficient resources for bargaining with and coercing whatever was left of the leadership of the Soviet Union.

Over $1 billion would be spent to render survivable specially modified Boeing 747s and 707s for a specially chosen civil and military leadership. Because fixed command centers are unlikely to survive the unleashing of the thousands of nuclear weapons likely to be used in World War III, the president, his civilian staff, at least 16 top successors, and top military commanders and some of their spouses would take to the air when World War III appeared inevitable. A "top priority" $18 billion would be spent on special "survivable, reconstitutable and secure" communication equipment to ensure that this surviving leadership would be able to communicate with submarines (perhaps hidden beneath the Arctic ice cap) and other nuclear delivery forces held in secret reserve.

Testimony at congressional hearings and discussions with other defense consultants revealed that manipulating the hardware in this plan was much easier than dealing with software or personal considerations. For instance, a "protocol complication" soon arose in considering which spouses would rate seats on special helicopters that would take the civilian and military command force to the waiting planes. Obviously, the spouses of the president and vice president rate seats, but given the limited space, how far down the line of civilian staff and the 16 successors should we go? Presumably, the Defense Department will soon hire consultants in ethics, psychology, and sociology to solve this delicate problem.

Although the article stated the first principle of this plan, that "the United States would never emerge from a nuclear war without nuclear weapons," it did not discuss its overall rationale. The president, and the administration about to spend tax money on this plan, had advocated for some time that the primary, and perhaps only legitimate, role of government was the defense of its people. Because the report acknowledged that only the select civilian and military command force and reserve nuclear forces will likely survive, it is difficult to understand how average citizens, who will be dead, will be defended by this spending of their tax money.

Undoubtedly, the answer that our political leaders would give for the necessity of such planning would involve the concept of "deterrence." As in the rest of nature, it is important that one present a tough, serious stance in defending territory and interests. Because even the winner risks critical injury in fighting, other behavior patterns should be tried first. Many animals when threatened have ways of changing their physical appearance, making themselves appear more fearsome and implying that any attack would mean doom for the attacker. Displays of confidence, deterrence, intimidation, and elaborate bluffs have evolved as methods to avoid violence. Unfortunately, the human species has also evolved cognitive intelligence. As discussed in Chapter 9, our physical inferiority has led to a powerful indirect control over the environment. We have been able to figure things out, not the least of which is that material things consist of atoms and that enormous amounts of energy are released in splitting the nuclei of uranium. Thus, our bluffs can now be backed up with an enormous destructive force.

Since the end of World War II, the Soviet Union and the United States have accumulated an unprecedented mutual deterrence. A common measurement of destructive firepower is the megaton, a million tons of TNT. The total firepower of World War II was three megatons. Forty million people died in this war—20 million alone in the Soviet Union. Estimates of the current combined firepower of the United States and the Soviet Union approach between 18,000 and 20,000 megatons, or at least 6,000 World War IIs.

The statistics of this destructive force can be cut in many ways. By 1987 the Soviet Union possessed about 9,500 nuclear warheads and the United States close to 10,000. But the total megatonnage was 17,000 for the Soviets and only 3,500 for the United States.[1] These weapons of assured death could also be delivered in different ways. About 71 percent of Soviet bombs could be delivered by land-based missiles, 25 percent by submarines, and 4 percent by bombers. The United States has a more balanced strategy: 25 percent land-based missiles, 50 percent submarines, and 25 percent bombers.

Statistics can be very misleading. Defense Department releases often say that the Soviet Union has the largest submarine fleet in the world. True, but most of these submarines are noisy, technologically backward, and relatively limited in the number of nuclear weapons deliverable. The United States has the *Polaris, Poseidon,* and the very modern *Trident* submarines. A single *Poseidon* can carry 9 megatons or three World War IIs, enough firepower to destroy 200 of the largest Soviet cities. In 1987 the United States had about 30 *Poseidons.* A single *Trident* submarine carries 24 megatons, enough to de-

Nuclear war is incalculable.

FREEMAN DYSON, *WEAPONS AND HOPE*

Perhaps a species that has accumulated . . . tons of explosive per capita has already demonstrated its biological unfitness beyond any further question.

ARTHUR C. CLARKE

---

[1] Since smaller warheads can be delivered more accurately, the comparatively low megatonnage of the United States—over 1,000 World War IIs—was not necessarily a disadvantage.

stroy every major city in the Soviet Union. (A single *Trident 2* missile can carry 12 nuclear warheads.) Thus, using sophisticated MIRV technology (Multiple, Independently Targetable, Reentry Vehicles), the United States can deliver many more bombs (between 5,000 and 6,000) than the large Soviet submarine fleet can. MIRV technology enables the United States to launch several nuclear bombs from a single missile. When the missiles achieve the appropriate altitude, the warheads divide into independent bombs that can then strike different targets.

In spite of recent attempts at limiting short-range nuclear weapons, both the United States and the Soviet Union still possess enough nuclear bombs to target major cities and military sites many times. Debates about which side has an advantage seem pointless—targeters on both sides will run out of targets and victims long before they run out of bombs. Consider Jonathan Schell's description of the impact of a major Soviet attack on all the major U.S. civilian and military targets.

In the first moments of a 10,000-megaton attack on the United States . . . flashes of white light would suddenly il- luminate large areas of the country as thousands of suns, each one brighter than the sun itself, blossomed over cities, suburbs, and towns. In those same moments . . . the vast majority of the people in the regions first targeted would be irradiated, crushed, or burned to death. The thermal pulses could subject more than 600,000 square miles, or one-sixth of the total land mass of the nation, to a minimum level of . . . heat that chars human beings. . . . Tens of millions of people would go up in smoke. As the attack proceeded, as much as three-quarters of the country could be subjected to incendiary levels of heat, and so, wherever there was inflam- mable material, could be set ablaze. In the 10 seconds or so after each bomb hit, as blast waves swept outward from thousands of ground zeros, the physical plant of the United States would be swept away like leaves in a gust of wind. The 600,000 square miles already scorched . . . would now be hit by blast waves . . . and virtually all the habitations, places of work, and other man-made things there—substan- tially the whole human construct in the United States— would be vaporized, blasted, or otherwise pulverized out of existence. Then, as clouds of dust rose from the earth, and mushroom clouds spread overhead . . . day would turn to night. . . . Shortly, fires would spring up in the debris of the cities and in every forest dry enough to burn. These fires would simply burn down the United States. . . .

In any city where three or four bombs had been used—not to mention fifty, or a hundred—flight from one blast would only be flight toward another, and no one could escape alive. Within these regions, each of three of the immediate effects of nuclear weapons—initial radiation, thermal pulse, and blast wave—would alone be enough to kill most people. . . . The ease with which virtually the whole population of the country could be trapped in these zones of universal death is suggested by the fact that the percent of the population that lives in an area of 18,000 square miles could be annihilated with only 300 one-megaton bombs. . . . That would leave . . . 97 percent of the megatonnage in the attacking force, available for other targets. . . .

Needless to say, in these circumstances evacuation before an attack would be an exercise in transporting people from one death to another. . . . In a full-scale attack there would in all likelihood be no surviving communities, and . . . everyone who failed to seal himself off from the outside environment for as long as several months would soon die of radiation sickness. Hence, in the months after a holocaust there would be no activity of any sort, as, in a reversal of the normal state of things, the dead would lie on the surface and the living, if there were any, would be buried underground.[1]

Schell continues his narration of zones of universal death[2] by pointing out that these are only the immediate and local effects of a nuclear war. Assuming, of course, that the United States would unleash a similar amount of destruction on the Soviet Union, and perhaps that other countries would also be targeted by both sides because of various alliances or other strategies, we must consider seriously the long-range irreversible global effects of a nuclear war

---

[1] Jonathan Schell, *The Fate of the Earth* (New York: Avon Books, 1982), pp. 56–60. Copyright © 1982 by Jonathan Schell. Reprinted by permission of Alfred A. Knopf, Inc. Originally appeared in *The New Yorker.*

[2] Because calculations involve many uncertain variables, as well as political considerations, views on the number of U.S. citizens who would die have varied over the years. In 1975 the Federal Emergency Management Agency reported that 156 million would perish immediately due to "direct" blast effects. In 1987, with the administration attempting to get Congress to increase money for civil defense, the agency reduced this figure to 112 million. The National Academy of Science, however, in the same year concluded that "our society, and possibly human life, could not survive."

on the life of planet Earth. What would be the fate of all life on Earth from the enormous amount of radioactive fallout that would result from a 20,000-megaton exchange? What would happen to the world's nuclear power plants? As of 1983, the United States alone had some 80 nuclear power plants. What would happen to the world's ecosystem if the deadly, unimaginably long-lived radiation from these plants was added to the already overwhelming radioactive fallout? Would only insects survive? (Depending on the species, insects can survive between 10 and 1,000 times the dosage of radiation that mammals can.) What would be the result of having the Earth littered with the dead bodies of billions of animals and human beings and the destruction of the insects' natural enemies, the birds? If substantial numbers of people survived, what kind of medical epidemics would ensue? With increased radiation, increased rates of genetic mutation in every species would be inevitable, and the cancer rate would increase.

Because much of the world's vegetation would be destroyed, what would be the consequent effects in terms of land erosion and the glutting of the Earth's seas and waterways with the lost topsoil? Any remaining topsoil would most likely be extremely radioactive, so what would happen to the world's agricultural output? What would happen to the ozone layer that protects all life on Earth from ultraviolet radiation? What would be the effects on the Earth's climate of millions of tons of dust and debris thrown into the atmosphere and the smoke from fires blocking out the Sun? Would a nuclear winter result?

Schell concludes by pointing out that we have reached the point in the arms race where for the first time in the history of human combat there is a "zone of uncertainty" about such drastic human-initiated effects. Once we could safely believe that no matter how ignoble, brutal, pointless, and destructive such a conflict would be, life would recuperate and go on. We can no longer believe this. With this uncertainty we have reached a "zone of extinction." As with the cosmic perspective, nuclear war forces us to think about what human beings are the most reluctant to face, but is most important as a guide to action. We must think about our contingency and our finitude. According to Schell:

As human life and the structure of human existence are
seen in the light of each person's daily life and experience,
they look impressively extensive and solid, but when human
things are seen in the light of the universal power unleashed
onto the earth by nuclear weapons they prove to be limited
and fragile, as though they were nothing more than a mold
or a lichen that appears in certain crevices of the landscape
and can be burned off with relative ease by nuclear fire. . . .

Once we learn that a holocaust *might* lead to extinction
we have no right to gamble, because if we lose, the game

will be over, and neither we nor anyone else will ever get another chance.[1]

We have no right, says Schell, after being produced by so many lucky circumstances, of silencing the music of existence on this planet, of toying with the possibility of reinitiating "the simplicity of nothingness." Yet preparations continue. Perhaps as you read this somewhere in the world someone is using the gift of our evolution, our cognitive intelligence, to design more and better bombs, or perhaps using sophisticated mathematical equations to compute the kill ratio per megaton of a ground-burst explosion versus that of an air-burst.[2] And undoubtedly, somewhere there is a bureaucrat, soon to return to his wife and children, signing the necessary forms demanded of organized society for funding the next increment of nuclear weapons.

How did we get in such a predicament? How did the human species, born from billions of years of chemical and biological evolution, historical heirs of stargazers and the inquisitive minds of the ancient Greeks, of the lovers of mathematical symmetry and contemplators of the harmony of natural law, reach this unforgivable plight? Two essential questions must be answered in addressing the human prospect: What in human nature and its role in the cosmic perspective has produced this zone of extinction? And what, if anything, will allow us to live through what Carl Sagan has optimistically called this period of our "technological adolescence"?

Human beings share many behavior patterns with animals. We have day-to-day routines, are sensitive to social hierarchy, position and status, are obedient to precedent and ritual, and engage in acts of aggression, territoriality, and dominance. Our participation in fads and concern for fashion are related to ancient animal sexual strategies. We routinely act in repetitive, obsessive, unthinking ways while thinking that we are thinking about what we are doing. Our leaders "posture" with a facade of strength on a grand diplomatic and military scale both for their people and the other side. In Chapter 3 we took a brief look at some of the possible social implications of our evolutionary inheritance from reptiles and mammals. At the risk of oversimplifying current research in ethology and neuroethology, and perhaps insulting reptiles, here we must consider seriously the question, To what extent can we be considered no more than complex reptiles who have developed more intricate, but dangerous, ways of implementing the basics of survival? To what extent is our reason driven by ancient reptilian fears?

What a chimera . . . is man! What a novelty, what a monster, what a chaos, what a subject of contradiction, what a prodigy! A judge of all things, feeble worm of the Earth, depository of the truth, cloaca of uncertainty and error, the glory and the shame of the universe!

BLAISE PASCAL

---

[1] Schell, *Fate of the Earth*, pp. 65, 95. Reprinted by permission of Alfred A. Knopf, Inc.

[2] An air-burst covers more territory just as a flashlight will illuminate a wider area if it is held higher off a surface, but a ground-burst creates more deadly radioactive fallout.

Evolution will take its course. And that course has generally been downward. The majority of species have degenerated and become extinct, or . . . worse, gradually lost many of their functions. The ancestors of oysters and barnacles had heads. Snakes have lost their limbs and ostriches and penguins their power of flight. Man may just as easily lose his intelligence.

J.B.S. HALDANE, "MAN'S DESTINY"

All species must have some method of defense. Most higher animals have evolved innate displays of defensive aggression. Only one has combined the naturalness of defensive aggression with a cognitive ability; only one has unlocked the mystery of the atom and its implications for a new understanding of reality and meaning in life and combined the resultant knowledge with the naturalness of defending itself, its clan, and its territory to produce an unprecedented zone of possible extinction for itself and all life on Earth.[1] Could our species be but a brief aberration, a momentary violent experiment, never to be repeated again anywhere else in the universe?

As part of a culture convinced of the rationality of existence and the purposefulness of human life, Aristotle defined man as a "rational animal." Consider though how unbiased extraterrestrial observers might see us. How would they view a culture of creatures who go about the business of obtaining basic necessities, producing offspring, loving and learning life—all the while with explosive devices strapped to each of their backs equal to 15 tons of dynamite, any or all of which could go off at any moment? Consider what they would think if they asked these curious creatures why they lived this self-imposed, dangerous way and the creatures responded that living this way was necessary for defense!

The world's superpowers have produced, for the most part, step by rational step, enough nuclear destructive power equal to 15 tons of dynamite for every person on Earth. The United States and the Soviet Union have each adopted, based upon sound historical and geographical considerations, what their leaders think are the best possible policies of defensive aggression. The United States has adopted a mixed policy of "assured destruction," and "limited first use" of nuclear weapons. Strategically, the United States defends its territory through a policy of "deterrence" (a term that translates to mean "intimidation" in Russian). A destructive force has been amassed that ensures the destruction of every Soviet city many times over. Tactically, the United States has adopted a limited first-use policy of containment, thinking that in this way we will avoid an escalation toward assured destruction. If the Soviet Union with its overwhelming superior conventional forces invades Europe, the United States has made it clear that smaller, tactical nuclear weapons will be used to stop this invasion. Thus, the United States has refused to renounce the first use of nuclear weapons. The Soviet Union on the other hand has renounced the first use of nuclear weapons and has adopted the combined doctrines of "counterforce" and "first strike."

The distinction between "first use" and "first strike" will undoubtedly mean little to the millions, perhaps billions, of people killed in a nuclear war, but

---

[1]We are not arguing that only the human species is violent. Almost all vertebrate species are naturally violent to some extent. But only the human species has discovered the means to extend this violence beyond any conceivable defensive purpose.

psychologically it is an important distinction. Not understanding it and how it relates to human nature could have devastating consequences. Historically, Soviet territory has been invaded many times. From the Mongol hordes to Napoleon and Hitler, history has produced in the Soviet people a mixed psychology of the need to strike first when facing an overwhelming outside threat, the inevitable destructiveness of war regardless of who is first, and an agonizing muddling through of some survivors no matter what the level of destruction. Thus, the Soviets have amassed huge weapons and aimed them at the land-based missiles of the United States. If tactical nuclear weapons are used against them in Europe, or if the initiation of nuclear hostilities appears likely, they have notified the world that they will "preempt" the use of missiles against them by unleashing their entire strategic nuclear force first. In other words, they are going to hurt the other side first as much as possible before they themselves get hurt. The policy of the United States says, "Shoot at us and you will be assuring your own destruction"; the policy of the Soviet Union says, "If you are about to shoot at us, we are going to hurt you as much as we can before you do. Although we are going to die, we are going to take as many of you with us as possible."

Thus, the United States remembers Pearl Harbor and is terrified when it sees the Soviet Union pointing missiles at U.S. missiles and preparing its civilian population for some sort of survival. (Survival implies that assured destruction will not work.) The Soviet Union remembers Hitler and the 20 million of its citizens who died in World War II and feels the consequent paranoia of the results of the policy of containment and being surrounded by hostile forces. Consider what it must be like to be a Soviet survivor of World War II knowing that there are missiles in Great Britain, France, Italy, Turkey, the Mediterranean Sea, maybe Pakistan, and definitely China, aside from those in the United States, all of which at any moment could become a new apocalyptic Napoleon or Hitler. Imagine if missiles in Canada, Mexico, and South America at any moment threatened us with another Pearl Harbor?

Consider what the children are taught in each culture. In the Soviet Union every child knows that only one country has been crazy enough to actually use nuclear weapons on defenseless civilians. Only the United States has been demented enough to sear the flesh from the bones of several hundred thousand people and to produce epidemics of mutation and cancer in the survivors of Hiroshima and Nagasaki. Every Soviet child knows that the United States first developed this ominous threat to the human prospect. Every Soviet child knows that this mindless society, governed by heartless, greedy capitalists has the most powerful military force ever assembled and that through its influence and aid has surrounded them with death. Every Soviet child knows that this impulsive country has invaded their country before and that only the historical strength of the Soviet people and their valiant military forces keep it from happening again.

On the other hand, in the United States every child knows that there is only one true threat to global freedom, that an "evil empire" threatens an end

I'll tell you that whatever happens, and whatever mess they make up yonder, we shall win the battle. . . . War is not a polite recreation, but the vilest thing in life, and we ought to understand that and not play at war. We ought to accept it sternly and solemnly as a fearful necessity.

LEO TOLSTOY'S PRINCE ANDREI, *WAR AND PEACE*

to the development of the unlimited potential and growth of humankind. Every American child knows that only one country has made negotiations in reducing nuclear missiles difficult, that their president was first in proposing and has tried many times to reduce the land-based missiles of both sides but meets with unreasonable demands from the funny-talking, sinister-looking atheists from a strange land.[1] And, of course, every American child knows that God is on his side.

Volumes can be written on the technical strategies, and the historical reasons for these strategies, of both sides. We can debate unendingly the relative merits of counterforce versus assured destruction, examining countless pieces of analysis on the incalculable factors of "collateral damage" (read "dead civilians") and missile survivability, or on such esoteric factors as the "throw weight" of a missile warhead. But the picture is as simple as a sandlot fistfight. Someone developed the bomb. The other side became afraid, so they made the bomb. The first side became afraid, so they made more bombs. The other side made more bombs and aimed them at the first side's bombs. They also had the audacity to plan to try to save a few of their people. The first side became absolutely terrified that their own bombs might be destroyed before they could be used and that the other side was "threatening" to survive! So they eventually elected a leader who spent over $1 trillion in a few years on more defense, complete with a more accurate missile called the *Peacekeeper* and research on the use of nuclear weapons and particle beams in space.

Perhaps this view is too simple; perhaps it is not scientific. But a short review of our cultural and biological history seems to document that the following attitude is an inherent part of our nature: "If you have a nuclear bomb, then I must have a nuclear bomb. If you have 10, then I must have 10. If you have 1,000, then I must have 1,000. And if you are stupid enough to reduce your 1,000 to 500, I am still going to keep my 1,000!" In an atmosphere of reciprocal fear, where each natural act of defense terrorizes the other side, any sign of weakness is an invitation to destruction.

In Chapter 1 we noted that the United States and the Soviet Union are like two men facing each other at a close distance with magnum pistols aimed, cocked, and prepared to shoot the other. We can add to this analogy by pointing out that each has decided that just in case the bullet somehow only wounds his opponent, each has added many hand grenades to his arsenal. Each knows that actually using any of the grenades would most likely result in his own death as well. But each must have this extra defense to prove to the other that his resolve is absolute. What would happen if one of the men

Natural selection has poised us on a knife-edge of uncertainty, destined to tumble, luckily, into knowing or, oftener, unluckily, into trembling.

MELVIN KONNER, *THE TANGLED WING*

---

[1] The *glasnost* atmosphere of the late 1980s softened these attitudes somewhat. But serious problems stemming from these attitudes remain. For instance, the leaders of the United States forget to tell their children that they have more bombs on their submarines than the Soviets have and that the United States has refused to include the missiles of Great Britain and France in negotiations.

became sick of this pointless standoff? What would happen if he lowered his gun? What would happen if, as a gesture of sincerity toward a negotiated ending of their mutual threat to each other, he simply uncocks his pistol? Maybe the other man would feel great relief and lower his gun also. Maybe. Or maybe the other man, who has spent day after day feeling the painful discomfort of anxious uncertainty, would see a chance to eliminate once and for all the source of this uncertainty. Maybe the lowering of his opponent's gun would only increase his fear. Maybe it is a trick. Maybe the safest thing to do is shoot first while the chance exists and ask questions later!

Nuclear freeze and disarmament advocates, who see that together each side has produced a deadly atmosphere of reciprocal fear, in which each new act of defense threatens the other side, often see themselves as the only sane participants in the nuclear debate. Why can't the leaders of the world see what the average person sees? Why can't we all just cry "Enough of this, enough. Let's throw away the backpacks. Not little by little, but all of it now!" On the other hand, the leaders of the world—with their military strategists, political experts, and advisors—see the nuclear freeze people as naive, emotional, and irrational. Unilateral disarmament is an uncocking of the gun; a unilateral freeze is a lowering of the gun. Either move could cause the other side to fire.[1]

This then is how we have arrived at the present crossroad. We have used our total nature: irrational fear to some extent, greed in part[2], but mostly our reason. Like the man who took the logical step of decapitating the head of his victim so the police would not find the bullet, each side, starting with its own premises and circumstances, has taken the next logical step toward Armageddon. The essential paradox of the nuclear age is that we have rationally and naturally escalated our stance of defensive aggression to the point that there is no conceivable just cause in which our tools of defense can be used; one mistake now or in the process of reduction could detonate our backpacks. Is the situation hopeless then? If one of our best characteristics, our ability to think and figure things out, has produced such a wide zone of possible extinction, can there be any hope for the human prospect? Is the combination of a cerebral cortex with its powerful cognitive skills on top of a reptile brain with innate survival strategies a hopeless evolutionary gesture?

The military establishment looks on the peace movement as a collection of ignorant people meddling in a business they do not understand, while the peace movement looks on the military establishment as a collection of misguided people protected by bureaucratic formality from all contact with human realities.

FREEMAN DYSON, *WEAPONS AND HOPE*

---

[1] This is also why some American and Soviet political conservatives feel that arms reduction agreements, such as the 1987 Intermediate Nuclear Forces that only reduced our nuclear backpacks by 4 percent, are actually dangerous. Having an insane number of nuclear weapons is a way of "tricking" ourselves into peace. Without these weapons we make the world safe for conventional war and the process of reduction introduces an element of uncertainty that could make one side think it could attack and get away with it.

[2] Someone makes a lot of money in producing all the bombs, and thus each new weapon has a vested economic and political constituency.

Weapons and Hope

Perhaps we have but one hope. If as a species we are a crazy, mixed-up creature, then at least we can be aware of our craziness. And perhaps this awareness will be more than a witness to our destructive nature. Perhaps the knowledge gained from self-reflection can lead to steps that avoid pushing the buttons of our reptile heritage. For instance, animal research has taught us that crowding leads to abnormal destructive behavior. As creatures possessing such knowledge we can take steps to avoid creating situations that will initiate this behavior in ourselves. We can initiate policies of birth control, establish disciplines of research in future studies, and adopt energy and economic strategies that implement the results of understanding ourselves better. Unlike an animal that is unaware that its evolved characteristics are no longer compatible with its environment, the human species has, through its unique cognitive intelligence, the ability to know of its possible impending doom. Even though our rational ability is in large part responsible for getting us into the mess we are in, it will also be responsible to a large extent for getting us out of it.

Freeman Dyson, a highly respected physicist and military advisor, advocates this position in his book *Weapons and Hope*. At a time when the nuclear debate is divided into two extreme camps—the advocates of "more is safer" versus "enough of this"—Dyson, along with the American diplomat and scholar George Kennan, whose ideas Dyson cites extensively, assumes the stance of "a teller of complicated truths to people who prefer simple illusions." Although appreciating the idealism of such critics as Albert Einstein, who advocated that scientists refuse to cooperate in weapons research, Dyson attempts to paint a politically more aware and creatively more realistic picture of our present predicament. According to Dyson, by using the total arsenal of our cognitive intelligence, we can trick ourselves politically and militarily down a path of hope. By becoming aware of the historical, geographical, and human reasons for the development of the present state of affairs, and by developing what Dyson calls "the weapons of David," the Goliath of massive nuclear destruction could become a vestige of an immature time in the development of the human species. In other words, assuming that the dictates of our reptilian heritage and innate imperatives for survival and defense will always be with us, we need intelligent weapons, weapons that will squelch the relentless fire of our inherent fears, "weapons which are capable of defending territory without destroying it in the process."

According to Dyson, recent revolutionary developments in computer information processing have made war and defense preparations more and more a "contest of information." The side that is better able to "see" with computer surveillance everything the other side is doing has a security advantage in several ways. Not only are they more secure in their ability to defend an attack but this security also breeds the necessary confidence in one's defense needed to avoid shooting first out of fear and uncertainty. Consistent with this contest of information is a different type of weapon. Instead of the Goliath weapons of assured destruction, a country could make itself more secure by developing smart precision guided munitions, or PGMs. Dyson cites the Falklands fight between Argentina and Great Britain as evidence of this. Here a small, elite Argentinean air force was able to penetrate the massive naval forces of the British and inflict great damage using PGMs.[1]

Thus given the right perspective, Dyson argues, revolutionary technology produced by scientists can serve as a basis of hope rather than doom.[2] A continuation of this process, plus negotiation based upon understanding the historical fears of both sides, could lead the world's superpowers to realize that the country that maintains the present status quo of nuclear terror risks being the Spanish Armada of our time. For Dyson there is no implied moral superiority in the development of a PGM strategy; it is simply a realistic gambit based upon an awareness of the present situation, its historical background, and humankind's "damnable" attraction to war. Dyson foresees a possible return to an older style of warfare, a time of the professional soldier shooting at other professional soldiers, a time when civilians were not involved and could go about their business without fear of being used as pawns in a MAD (mutual assured destruction) defense policy, a time when differences between countries could be decided without worrying whether life itself would recuperate.

We need then, according to Dyson, a new debate. We must reexamine what nuclear weapons are for—not from the standpoint that only irrational reasons caused the development of these weapons in the first place or that keeping them implies our doom (for this risks making our doom a self-fulfilling prophecy) but from the standpoint that "tragedy is not our business." Above all we must not underestimate the "sagacity" of the human species, our uncanny ability of lucking out when things look most hopeless, our artistic predilection for making the best of a less-than-ideal situation, our so-far successful ability to cleverly bungle our way through this gift of existence. If Plato could give a homework problem on the motion of the planets that would change the

If we are to avoid destruction, we must first of all understand the human and historical context out of which destruction arises. We must understand what it is in human nature that makes war so damnably attractive.

FREEMAN DYSON, *WEAPONS AND HOPE*

---

[1] In one week in 1987 a teenager, flying a single engine plane, flew through the billion-dollar Soviet Union air defenses and landed in Red Square (practically at the door of the Soviet leadership) and a simple, inexpensive Iraqi *Exocet* missile knocked out a multimillion-dollar U.S. warship in the Persian Gulf.

[2] Dyson makes it clear, however, that he is not promoting a "Star Wars," or Strategic Defense Initiative, as advocated recently by the United States. In his opinion this plan, which would move nuclear war into outer space, using nuclear explosions to generate laser beams of death, is just another continuation of the "more is better" response to our fear. He calls such a plan a "technical futures folly."

world while believing that the planets did not even exist, if Copernicus and Kepler could give us the heliocentric model of planetary motion—a model we now use to travel billions of miles from Earth—while believing that the Sun was the home of God, if Newton could give us the theory of gravity while believing that God would get confused if the dance of existence did not take place according to a human time, if a creature who is about to blow itself up is also close to understanding what happened during the first billionth of a billionth of a second of existence, then much that is hopeful is possible.

Yet there are many who feel that the stance of the scholar warrior is a dangerous anachronism, a relic of what got us into this mess in the first place. What would the other side think of as it watched its adversary place vast amounts of its economic and intellectual resources into the production of new mysterious David weapons? Would it feel safe? Would it be tempted to shoot and get it over with before it is too late? Suppose through some magical process we were able to eliminate all nuclear weapons. What about biological weapons? How long would it be before the forces that produced nuclear weapons produced a new horror?

Carl Sagan is most famous for advocating a different strategy for tricking ourselves toward a more hopeful future. For Sagan the solution is a grandiose idealism, an idealism that, given our present crisis, becomes a pragmatic approach to dealing with the "evolutionary baggage" of our reptilian heritage. More weapons, even smart ones, will only fan the fires of our destructive tendencies. The solution, or at least a large part of the solution, is one of perspective, the cosmic perspective, according to Sagan. At present the perspective adopted by the world's leaders leads to *psychological crowding* in which our differences and narrow self-interests predominate. We know that this type of crowding is guaranteed to produce the worst in us. Innate mechanisms of self-defense, tribal identity, and aggression have evolved as part of a successful survival strategy. A creature that does not instinctively fear strangers and defend itself and its immediate genetic relations is unlikely to survive to discuss peace and love. Also, just as a city evolves by building around its older central portion, so the human brain has evolved by adding a limbic system and a cerebral cortex to the ancient R-complex.

The solution is obviously not a lobotomy but an end to the psychological crowding that pushes the button of the destructive potential of our evolutionary heritage. Out there, said Einstein, the universe beckons like a liberation. Out there, says Sagan, is a stunning remedy to our psychological crowding. From the standpoint of the cosmic perspective, the differences between our clans are trivial. Confronted by the awesomeness of the cosmic perspective, our fear and concern are externalized away from each other and toward this mysterious and beautifully threatening place we call the universe. According to Sagan, it is not unrealistic idealism to advocate spending billions of dollars on a joint Soviet-U.S. venture to Mars. It is not merely an ivory tower dream to spend a small portion of our gross national products on exploring the atmosphere of Jupiter and the oceans of Uranus or to listen for the unmistakable signals that we are not alone, that intelligent extraterrestrial

We awake and cringe, stir and cringe, eat and cringe, strive and cringe, even love and cringe. . . . Unlike other animals, we must knowingly look in the face of death, to find it a maw, insatiable. Aside from that, we are—not metaphorically, but precisely, biologically—like the doe nibbling moist grass in the predawn misty light; chewing, nuzzling a dewy fawn, breathing the foggy air, feeling so much at peace; and suddenly, for no reason, looking about wildly.

MELVIN KONNER, *THE TANGLED WING*

life exists, perhaps also groping for the meaning of existence. Our survival may depend on it. The yearly budgets for NASA in the past decade have been between 6 to 10 billion dollars; the yearly military budget for the United States is now over 300 billion dollars. There is a similar disparity for the Soviets. Sagan asks, Which will defend us more from ourselves?

Meliorism and Nihilism:
What Is the Universe For?

Although they differ in their solutions, both Dyson and Sagan advocate a method that is based upon the same philosophical assumption. Both assume that there is a connection between our cognitive awareness and a better world. In both cases our reason supplies the connective link between the facts of existence and the good life. Both are *meliorists*. A meliorist believes that existence gets better with time and that human beings and their efforts are an essential part in making things better. The meliorist believes that progress is real and an inevitable result of struggle and effort.

The ancient Greeks, primarily through the work of Plato and Aristotle, reflect a strong meliorist philosophy. As we saw in Chapter 4, they thought of the cosmos as a good place, full of purpose. Human life was an essential part of this purpose. Our role was to know. The knowledge and understanding gained from fulfilling our purpose would then be applied to a practical technology. The application of a practical technology would then make life easier, allowing for more leisure, which in turn would provide more time to pursue more knowledge. This beneficial cycle of self-actualization produced the good life, not only in that it produced leisure and protection against the harsh aspects of nature but also more directly because in seeking knowledge we would be doing what we are supposed to do. Fulfilling the purpose that nature has allotted us was its own reward. Happiness was developing one's potential, a constant progressive striving toward some ultimate goal; reaching the goal did not really matter. Happiness and the good life meant making progress; it was better to travel than to arrive.

When the progressive philosophy of the ancient Greeks became synthesized with Christianity in such Renaissance men as Descartes and Locke, concepts such as universal freedom and equality became the moral consequents of greater knowledge and understanding. Thus, in our modern period we are taught to believe that not only is the good life produced from knowledge but so too is the progressive movement toward making all human beings free to develop their potential. We are also taught that this is not just an ivory tower philosophy but real progress: There is less slavery in the world today, a greater equality in the treatment of women, and liberation movements are in full

We are a bit of stellar matter gone wrong. We are physical machinery—puppets that strut and talk and laugh and die as the hand of time pulls the strings beneath. But there is one elementary inescapable answer. We are that which asks the question.

SIR ARTHUR EDDINGTON

The quickening not only of the mind, but also of spirit, is the aim of a liberal arts education. As men and women devoted to intellectual pursuits, we have a happy faith that in the future, as in the past, the liberal arts and sciences will continue to be central to any meaningful understanding of the human condition.

From the PHI BETA KAPPA, A HANDBOOK FOR NEW MEMBERS

force; toward those few unenlightened pockets of oppression and apartheid, the rest of the world stands united in its revulsion.

Yet we must consider other voices and points of view. The same knowledge and understanding that produced this alleged progress has also produced universal zones of extinction. If increased awareness and our mastery over nature are supposed to produce progress, then why has there been an undeniable exponential increase in acts of death and destruction? Why is it that 200 years ago war involved only soldiers, yet in World War II more civilians than soldiers died? Why is it that while the laws of Newton were being applied to produce efficient machines to make life easier, the machine gun was invented and the art of killing became mechanized? (In World War I, the war that was supposed to end all wars, 80 percent of the men who died did not see the man who killed him.) If a progressive moral sensibility is the consequence of greater knowledge and understanding, why did saturation bombing and the destruction of whole cities became appropriate military strategies in World War II? It is no small matter of historical fact that the defenders of individual liberty created, with intelligent bombing, a situation where the air caught fire and 70,000 people died in Hamburg and 80,000 in Dresden—whole cities of people roasted in an enormous, gruesome oven. The news item, noted in the beginning of this chapter, of the little girl roasted in an oven by deranged parents, pales in comparison. Why is it that we consider the parents of the little girl demented, but the planners of the Hamburg and Dresden bombing heroes?

If knowledge produces progress, then consider finally the report, *The World at War*, released in 1982 by the Center for Defense Information. According to this report, at this time 45 nations were at war with millions of soldiers directly engaged in combat and millions of people being killed. The report also noted that the United States was the main arms supplier to these wars; the Soviet Union was a close second. The report notes, "The most striking aspect . . . is the degree to which conflict violence and international tensions have increased in nearly every region of the world." If education is a good thing, why do these words of Bertolt Brecht ring true?

Out of the libraries come the killers.
Mothers stand despondently waiting,
Hugging their children and searching the sky,
Looking for the latest inventions of professors.
Engineers sit hunched over their drawings:
One figure wrong, and the enemy's cities remain
    undestroyed.[1]

---

[1] Bertolt Brecht, "1940," *Poems 1913–1956*, 2nd ed., John Willet and Ralph Manheim (eds.) (London: Methuen, 1980). Reprinted by permission.

Insanity in individuals is something rare—but in groups, parties, nations, and epochs, it is the rule.

FRIEDRICH NIETZSCHE, *BEYOND GOOD AND EVIL*

515

We are here as on a darkling plain Swept with confused alarms of struggle and flight, Where ignorant armies clash by night.

MATTHER ARNOLD, *DOVER BEACH*

Enter the *nihilist,* who believes that reason, cognitive intelligence, our ability to know—call it what you will—does not produce progress; the undeniable escalation of violence on this planet proves that progress is a myth. Rather, all that reason produces, other than better ways for us to destroy ourselves, are dignified metaphysical rationalizations and dishonest psychological consolations for the utter meaninglessness of life and the fact that humankind's unconscious awareness of the meaninglessness of our lives makes us the most dangerous of animals.

Friedrich Wilhelm Nietzsche (1844–1900) will never be known for being a jolly fellow, but according to the twentieth-century French existentialist Albert Camus, he was the most honest and consistent philosopher. Nietzsche lived at a time when the developing industrial revolution, powered by the emergence of the modern scientific mind from the intellectual backwardness of the Middle Ages, made it easy to believe in progress. Everywhere scientists and intellectuals were congratulating themselves on possessing a new energy to conquer nature and produce the good life. It was a "can-do" generation that thought it would inevitably produce a relative utopia. Nietzsche saw a different future. With unmatched poetic pronouncements he rendered a lonely, but bold voice, admonishing his contemporaries for their self-deception.

Nietzsche believed that our consciousness, or cognitive intelligence, is not the connective tissue that will link the facts of existence with utopia, but rather it is a "disease" that only we have been unlucky to catch. The role of our reason is not to produce truth, for according to Nietzsche, echoing the ghostly relativism of Protagoras, there is no such thing. Rather, the role of reason is to create rationalizations to cover up the pointlessness of life and give false purposes for our actions. The role of a philosopher is the same as that of a lawyer: to make a case for one's client. It does not matter whether your client is guilty or innocent. The goal is an artistic challenge, to create a favorable, persuasive picture. Every culture has its hidden agenda of selfish interests. The job of reason, and the philosopher who uses this tool, is to keep the hidden agenda hidden, to cover it up with a sense of high purpose. The clients of Plato and Aristotle were the Greek aristocracy who needed a justification for continued economic and political dominance over ignorant barbarians: A culture that understood that the purpose of life was to develop the potential of the mind deserved economic spoils. The clients of Jesus were the poor and weak masses who needed ideological and physical protection from the bold and powerful. Thus, being equal in the eyes of God meant that one could be punished for being too bold and powerful, and if this did not work as a deterrent to keep the bold and powerful from exploiting the weak, at least the weak could be consoled believing that they alone would go to heaven. The clients of the modern period are the scientists who gain prestige in their conquest of nature, the politicians who gain more centralized power in a technologically confused society, and the middle-class businessmen who reap financial benefit from the rape of nature by supplying the masses with distracting toys. Thus, for Nietzsche science is just as much a religion

It is we, we alone, who have dreamed up the causes, the one-thing-after-anothers, the one-thing-reciprocating-anothers, the relativity, the constraint, the numbers, the laws, the freedom, the "reason why," the purpose. . . . We are creating myths.

FRIEDRICH NIETZSCHE,
GENEALOGY OF MORALS

as Christianity and just as much founded upon a metaphysical rationalization.

Nietzsche is best known for two main ideas: the *Übermensch*, or Superman, and the idea of Eternal Return. The first idea has been very much misunderstood, perhaps because Hitler and the Nazis misused it. According to Nietzsche, only a rare few understand that life is never a matter of science and truth but rather of art and creativity. A few exceptional mortals understand that "all seeing is essentially perspective, and so is all knowing." Only a few understand that "we are much greater artists than we know." Only a few have "broken through" and understand that God is dead, that the objectivity, absoluteness, and intellectual security that Western man has sought in vain since the time of Plato are myths.

For Nietzsche, the world may kick back, but there are so many ways of fashioning a response to reality that, like the quantum physicist, the Superman knows it is meaningless to talk about an objective reality when there is no one around to interpret it. Like the religious man who consoles himself by believing in a God of his own creation, scientists attempt to console us by creating "objectivity." But the secret is out. God is dead, and so are all absolutes. Science is simply another artistic response to reality. Reality upon reality exist right in front of us. It is only a matter of boldness and power to bring them into existence. Nietzsche's Superman is not necessarily always a Napoleon or a Machiavellian prince, although both would be an example of one. Rather, the Superman is anyone who has pierced the rationalizations of culture and understood the terrible freedom this power rewards us with. He knows that life resembles a man holding an enraged, uncontrolled squirming snake. Like the endless threatening directions our freedom can take, the snake writhes and thrashes about. The masses shudder and create myths to pretend that the snake—the Dionysian reptile in us all—is not there; the Superman bites off the head of the snake and swallows whole this repulsive truth!

Of the few meliorists who would read Nietzsche during his lifetime, all saw in his writings the demented rambling of a madman. What about moral imperatives? What would the world be like if people did not believe in reason and morality? Anarchy and chaos would be unleashed. To such protests Nietzsche had ready his most painful answer. Anarchy and chaos such as the world had never known were already being, and would increasingly be, unleashed not by madmen, not by artists, but by the sane rationalists. And this inevitable result, according to Nietzsche, is why our consciousness is a disease. Unlike the lower animals, which experience the pleasurable simplicity of a direct contact with the environment, the human species is doomed with its indirect awareness. With our awareness we sense the future, our deaths, and the neutral silence and mocking laughter of existence as we beg it to give us some meaning. So our reason creates a case for us, which might be harmless except that in creating a higher puspose we attempt to repress the serpent truth of our nature. And like a furiously boiling pot of water, sooner or later the steam must be vented. Nietzsche saw a half century before World War I that sooner or later the façade of reason and progress would collapse under the pressure of its own creation, that there would be a destructive explosion of violence the world had never known before. This nineteenth-century madman predicted the world wars of the twentieth century. (And for a modern follower of Nietzsche, an increasingly technological world and the spread of terrorism is not a coincidence.)

Nietzsche's second idea, Eternal Return, is naive scientifically, but with it he raised an important emotional point about scientific understanding and progress. Nietzsche thought that science features the universe as an infinite sea of interacting particles. If this is so, thought Nietzsche, then it would follow that everything must happen over and over again. Like a card game with an infinite number of hands, every card must reappear eventually in the same combination. Our biological and cultural evolution, every historical event, our individual lives—everything must recur again down to the last detail. For Nietzsche this aimless repetition is just another way of mocking those who believe in progress. If time and the development of events are cyclical rather than linear, then progress at some future time is impossible. Nietzsche cheats the thinker who believes that our meaning lies in the future by taking away the future. We do not make progress; we just travel in a meaningless circle repeating all the same mistakes.[1]

Today we may believe that the universe had a finite beginning, but the haunting point behind this doctrine is as much worth considering in the twenty-first century as it was in the nineteenth. Like a pointless merry-go-round spinning with no end, the doctrine of Eternal Return essentially poses the question, "What is the universe for?" Suppose the human species survives its current state of technological adolescence. What then? Suppose we continue to explore our solar system with the Russians. What then? Suppose we "embark with cunning sails on unexplored [cosmic] seas," go where no human has gone, venturing and adventuring our way through our galaxy and beyond. And then? According to our present theories, then the universe either has enough matter in it to collapse again, crushing everything in existence back into a singularity, or it continues to expand such that all matter diffuses and all existence deteriorates into widely separated lonely atoms, never again to coalesce into a sun, a planet, a person. And then? What was the point of our venturing and adventuring?

Where the meliorist sees progress, harmony, and increasing hope in the universe, the nihilist sees only a mockery of our pretentiousness. We want to be important. We want our act of life to count. But if nothing else, the

According to Nietzsche . . . there is no world outside [Plato's] cave. . . . Where does the climb lead? Nowhere. There is only the climber and his knowledge that, however far and long he climbs, he is doomed to repeat the same ascent endlessly.

W. T. JONES, A HISTORY OF WESTERN PHILOSOPHY

The more the universe seems comprehensible, the more it also seems pointless.

STEVEN WEINBERG, THE FIRST THREE MINUTES

---

[1] It is likewise a mistake to think of Nietzsche's Superman as some ideal state of the development of human potential that humankind is moving toward. For Nietzsche, it is not necessarily better to be aware to this extreme, and it surely does not make one happier.

nihilist raises the question of honesty in proclaiming our importance. Where is the proof? Reason cannot be used to justify being rational; awareness cannot be used to justify that awareness is a good thing; being a mammal cannot be used to justify mammalian values.

For Nietzsche, life is not a matter of logical demonstration; it is an issue of choice and power. We do not use reason to make our choices; we choose our reasons. And as Camus pointed out, Nietzsche was consistent to the bitter end. After using an impeccable logic to tear apart the velvety edifice of the meliorism of his time, he offered the world his vision of life as just that, a perspective, a point of view that one either sees or does not. Nietzsche's vision is not better; the Superman is not an evolutionary improvement as Hitler thought. In the mind's eye we can see Nietzsche presenting to the world his vision and saying "Here, take it or leave it!" and then retreating into the dark depths of the soul with "a laughter which was no human laughter."

Neitzsche and Socrates

The nihilist has introduced no new issues. Philosophy begins with the most practical problem there is: "How do I know what is the right course of action for living a happy life?" Since the Greeks first posed this question, and suggested that reason, rather than authority, tradition, and popularity, was the essential tool in seeking a solution, all Western philosophical developments have been simply abstract trails that thoughtful, concerned men and women have followed in attempting to answer this question. The essential purpose of this book has been to show the connection between the philosophical contemplation of science and the more immediate issues of our time. As such this book is just another trail on the broad path of this rationalist heritage.

Consider the connection among (1) our personal concerns and the daily decisions we must make, (2) social-global perspectives and issues, and (3) the technical epistemological issues with which this book has been concerned. The same problem of uncertainty exists at all three levels. How do we know what is the right thing to do (personal level)? Should we continue to use science and technology to solve the problems science and technology, at least in part, have created (social level)? How do we separate the reasonable from the conceivable (epistemological-philosophical level)? How can we say that we possess knowledge when we do not have certainty? Being educated does not guarantee that one will be a nice person. Awareness does not guarantee survival. Gaining lots of evidence does not guarantee that a hypothesis is true; that a hypothesis has been successful does not guarantee that it will continue to be so.

Even if we agree that science has been successful, and the nihilist is not willing to agree even to this, there is no guarantee that it will continue to

Photo by Carmen Bitonio.

*That we are part of
something awesome,
we know; that we are
an important part of it,
we do not know. It
should be enough that
we are part of it all.*

*That all the labors of
the ages, all the devo-
tion, all the inspira-
tion, all the noonday
brightness of human
genius, are destined to
extinction in the vast
death of the solar
system, and that the
whole temple of Man's
achievement must
inevitably be buried
beneath the debris of a
universe in ruins—all
these things, if not
quite beyond dispute,
are yet so nearly
certain, that no phi-
losophy which rejects
them can hope to
stand. Only within the
scaffolding of these
truths, only on the firm
foundation of unyield-
ing despair, can the
soul's habitation hence-
forth be safely built.*

BERTRAND RUSSELL

Alberto Giacometti, *City Square*, 1948. Bronze, 8½" × 25⅜" × 17¼". Collection,
The Museum of Modern Art, New York, Purchase.

be so. The scientific method cannot be used to justify its own epistemological assumptions. Thus, we cannot be sure that we are on the right track in using the scientific method. For all we know, if there is a nuclear war and somehow humans survive, their very first act in starting over may be to ban science and technology. For all we know, someday we will receive a message from an intelligent extraterrestrial culture, but it will be a warning from a culture on the verge of extinction not to develop technology. For all we know, intelligence is a characteristic that evolves repeatedly throughout the universe, but also promptly extinguishes itself within a few million years.

Consider the quote from Carl Sagan at the opening of this chapter. In it Sagan summarizes what many would consider the most enlightened theme of our time, a theme with which this book has been very sympathetic. The cosmic perspective implies that we "ought" to survive. The billions of fortuitous circumstances that had to be just right will never be repeated. No variation is possible in the laws of nature and the many physical constants, such as the charge of an electron and the gravitational force; otherwise the development of intelligence and our ability to reflect on these laws and constants would have not been possible. With any variation, a much different universe would exist—one without human beings, perhaps no universe at all. In Sagan's statement is the belief, not much different from that of the ancient Greeks, that embedded in the cosmic perspective is a meaning to life and a self-explanatory and self-evident value system. Implied is the belief that reason is the connective glue that joins our factual understanding of the universe and how we ought to behave. It reflects a philosophy that is stirring, joyous, and hopeful. But it commits the naturalistic fallacy.

Ought-statements cannot be derived from is-statements. No logical connection can exist between the factual state of the universe and how we ought to behave. The cosmic perspective is just as consistent with the perspective of the nihilist, who feels despair from the thought that either the universe will end or that it will last forever. It is just as consistent with the person who decides that if this is all there is to the universe, then why be moral? Why not grab all one can before this short, insignificant existence is over? As Heinz Pagels notes at the opening to this chapter, the cosmic perspective reveals only a neutral script with which we can express any passion or attitude, and it supports them all.

We have come then full circle, back to the ancient debate between Protagoras and Socrates. After several thousand years of intellectual probing, which has pushed our factual understanding to the Planck time (10^{-43} second), a tenth of a thousandth of a millionth of a billionth of a trillionth of a trillionth of a second after the Big Bang, we still do not know whether this knowledge is a good thing or whether our existence is but a brief, tragic aberration. Perhaps it is best at this point to recall what answer Socrates gave to the relativism of Protagoras and the sophists. He was not able to "solve" the intellectual problem the sophists posed. He claimed to know only that he knew nothing. However, Socrates concluded that even though we will never be completely confident in everything we say about this matter, if we believe that it is better to seek the truth, even knowing that the truth will forever be elusive, we will be "better, braver, and less idle." He supplied no rational answer to the sophists, only an emotional one. And thus, to Nietzsche likewise we must say that if we can be equally consistent in believing that knowledge is not a good thing and that our existence is an aberration, then we can be equally consistent in *not* believing this.

As an adult, Socrates was very "irresponsible." Instead of supporting his family, he relentlessly walked the city streets of Athens probing the minds of his fellow citizens. Like a child who can play the simplest game over and over again without getting bored, he played with the universe through thought and debate. Like a child who can find an infinite fascination with almost any object, he followed trails of thought with great pleasure while many a mature adult would become impatient with his impractical questions.

Consider the different responses of a Nietzsche and a Socrates, or a child and an adult, to the following fantasy. Imagine a very long magical path of no known length. As we travel this path, at every turn a new, beautiful surprise awaits; at every moment something enchanting and alluring emerges. The adult, who perhaps enjoys himself at first, begins to worry about where the path is going and what the purpose is of traveling it. The child just plays. Suppose both finally learn that no purpose to the path exists, and as far as anyone knows, it does not end. One could walk this path as long as one desired to do so. Would not the adult be terrified of such a prospect? The next turn, and then? The next game, and then? Just one "and then" after another with no end? Is that all there is to life? Knowing this, could the adult enjoy the path? The child just plays.

From an individual point of view this fantasy path analogy is, of course, not completely accurate. In real life suffering exists along the path, and for each of us an inevitable terminal waterfall looms ahead. As individuals we cannot travel the path forever. But as a species, as a collective mind probing to uncover one fantastic secret after another from this wondrous universe, our epistemological situation is not unlike the play of a child. Who should care if there is no end in sight? Who should care if we have been condemned to play along an infinite path? Who should care if our fate is to forever join the pieces of an infinite puzzle? Is this not a cause for rejoicing?

The ancient Greeks, whom Socrates personified so well, believed that an end to the path would result in boredom. Heaven as a symbol of rest, security, and completion was considered undesirable. To live the life of a human being in this incomplete earthly existence, to seek, explore, and meander passionately through the endless possibilities of this life is what we are joyously condemned to do. It is better to travel than to arrive. As long as unrealistic expectations about the purpose of the path do not meddle with the journey, traveling the path remains exhilarating.

Considered from this point of view, the intellectual history of Western science has been like a beautiful flower slowly opening its petals, each petal

To live only for some future goal is shallow. It's the sides of the mountain which sustain life, not the top. Here's where things grow. . . . But of course, without the top you can't have any sides. It's the top that defines the sides. So on we go . . . we have a long way . . . no hurry . . . just one step after the next.

ROBERT M. PIRSIG, ZEN AND THE ART OF MOTORCYCLE MAINTENANCE

beginning to point now in an endless number of directions and the flower as a whole struggling to proclaim its significance in the midst of this vast expanse of space. Considered from this point of view, the meanderings of the collective mind discussed in this book—the serendipitous genius of Plato and Eratosthenes, the mystical perseverance of Kepler, the thoughtful arrogance of Galileo, the pious mathematical insights of Newton, the relentless fastidiousness of Darwin, the Spinozistic faith of Einstein, and the ego-shattering results of quantum physics—are like the child on our fantasy path. We do not know why we are here, why we have made these discoveries, or where they will take us. Will science help us out of our present predicament or will it provide the final touch to our extinction? We do not know. But birds sing, flowers bloom, and children naturally play. Unlike the Greeks, we cannot profess to know that the universe is a good place and that the potential of developing a cognitive intelligence is beneficial. We do not know if this potential is a gift or a curse. We know only that developing this potential seems to be as natural as a child at play.

The uncertainty of our knowledge and the incurable insecurity of our existence is also like a great romance. We cannot be sure the universe, our partner, will not betray us someday. But we will be better and less idle if we believe it never will, and we will at least have a chance at passion if we do not betray our partner with our extinction.

Science as a Religion Revisited

At a fundamental level the nihilist is correct: Science is a religion. To believe in the goodness of the universe and the goodness of our knowledge, to believe in the purposefulness of our cognitive pursuits while knowing that the scientific method cannot justify its own meaningfulness—such beliefs constitute an act of faith. In the sixteenth and seventeenth centuries science was considered a method of worshipping an indubitable God and revering His creation, the physical universe. In the twenty-first century, battered by the confidence-shattering events of the twentieth century, belief in harmony and the ability of our minds to reflect this harmony will be considered an existential leap of faith. The individual scientist, and our culture as a whole, must maintain the childlike mind while knowing of the waterfall. Of the many attitudes that we can choose, we can choose the attitude of the playful lover while knowing of the possibility of betrayal. It is a difficult, but honest, way of living.

When philosophers cannot give a purely logical justification for an important belief—a belief that seems intuitively obvious and is fundamental for some beneficial activity—they will often fall back upon citing the heuristic value of accepting the belief. Although the truth of a particular belief cannot be established, it is reasonable to accept it, because believing it creates a beneficial psychological outlook that helps guide activity to a desired end. Thus, even though we cannot use the results of our inquisitiveness and cognitive intelligence to prove a particular meaning in life or a particular value system as better, even though the naturalness of our inquisitiveness does not reveal whether our nature is a divine gift, a lucky break, or a destructive aberration, even though the scientific attitude cannot be shown absolutely to be better than a life dedicated to the simplicity of ignorance, we can make a case for the scientific attitude as a reasonable faith with which to confront the twenty-first century.

As Nietzsche said we must, we can make a case for a reasonable religion. Like the Greeks we will believe that we are part of something awesomely purposeful and beautiful, that it is good and that our understanding of it is good; we will believe that, in the long run, thinking critically produces ideas that are better and that we have an obligation to preserve this gift of life and find out more about it. We must live as if there is a reason for the universe without a guarantee.

And what are the heuristic results of this religion? Left untouched by this appraisal of science is the same element of mystery and awe experienced by primitive humankind—a sense of living at the mercy of a wondrous foreboding power, which modern science has been accused of falsely hiding behind a facade of gadgetry. Mystery in turn reveals the humbling ignorance Socrates thought was so important to acknowledge if we were ever to understand ourselves. Ignorance in turn leaves unhindered a childlike wonder at why there is something rather than nothing. Wonder produces an unpretentious outlook and a reverence for existence necessary for the expression and generalization of mammalian values.

In a certain sense, science is myth-making just as religion is. . . . My thesis is that what we call "science" is differentiated from older myths not by being something distinct from a myth, but by being accompanied by a second-order tradition— that of critically discussing the myth.

KARL POPPER, CONJECTURES AND REFUTATIONS

The Anthropic Principle

For some, however, a religion is not enough. Recently, a minority of scientists have sought an interpretation of our knowledge that would return us to a more meaningful and centralized role in the grand scheme of things. Arguing for what is called the *anthropic principle*, some scientists have attempted to make a case that because so many things had to be just right to produce consciousness, perhaps no universe could exist at all unless consciousness exists. In short, the universe exists, there is something rather than nothing, because we exist.

As one example among many, consider gravity. If the gravitational force were only slightly stronger than it is, then the universe could not have expanded to its present state, and the formation of galaxies, stars, and planets would be impossible. The universe would have long ago collapsed into a singularity. If the gravitational force were slightly weaker, then the universe would simply diffuse into lifeless, lonely molecules. In either case, life and consciousness could not have evolved.

Combined with quantum physics, the anthropic principle places human existence at an even more mystically centralized position. The universe becomes a spectrum of potential existence that becomes more real as consciousness evolves! Physical matter does not exist and then evolve through chemical and biological evolution into consciousness. Rather, physical matter evolves into a state of "definiteness" as consciousness evolves. The universe is then like a growing child. In becoming more aware of itself, it creates itself by defining its individuality.

For many scientists of a more traditional bent, such thinking is backward if not deranged. (Some have claimed that the anthropic principle is simply a new twist to the design argument and a way for an atheist to believe in God.) Cities and other structures of human habitation along the Mississippi River could not be where they are if the river had not evolved the path it did. Luxury hotels in Hawaii are often built along beautiful beaches. But few of us believe that the cities caused the path of the river or that the hotels caused the beaches to evolve. Proponents of the anthropic principle, however, argue that the results of quantum experiments show that these analogies are false and that to think that consciousness is capable of creating the present state of the universe is no more paradoxical than fine tuning a radio channel. A bare existence of some sort is out there, but it needs to be defined.

> *The most beautiful experience we can have is the mysterious. It is the fundamental emotion which stands at the cradle of true art and true science. He to whom this emotion is a stranger, who can no longer wonder and stand rapt in awe, is as good as dead.*
>
> ALBERT EINSTEIN

Voyager and the Second Death

A more important practical consideration, however, is not whether the universe is here because we are here but that all these improbable things have taken place and we know our existence could be otherwise.

The most sobering chapter in Schell's *The Fate of the Earth* is "The Second Death." In it he makes the obvious, but unheralded, point that "extinction is the death of death." Our own death in a nuclear holocaust is one tragedy, but with a nuclear holocaust we risk a second death, a far greater tragedy than that of our individual deaths. We risk a death of an infinite number of possible people, a cancellation of "the numberless multitude of unconceived people." In terms of our discussion here, we risk the death of the collective mind and its playful meandering down the path of knowledge; we risk an end to our romance with the universe, this delightful dialectic of understanding forever dissolved by an irreconcilable divorce. If the participatory interpretation of quantum physics is correct, then perhaps we risk the death of the universe itself. That we are part of something awesome, we know; that we are an important part of it, we do not know. It should be enough that we are part of it at all.

Perhaps Freeman Dyson is right. We should not underestimate the sagacity of the human endeavor, our uncanny ability to cleverly muddle through the worst situations. It has been over forty years since Robert Oppenheimer, the leader of the team that developed the first atomic bomb, recorded his state of shock at the unexpected power of the first atomic explosion with words from the Hindu bible, the *Bhagavad-Gita*, "I have become death, the destroyer of worlds." Nuclear weapons have not been used in a World War III, yet. Perhaps we will continue to muddle through.[1]

But what if the end does come? As a final *exercise* in understanding the full ramifications of the human prospect in the light of a cosmic perspective, each of us must think of what we would feel (assuming we had time to feel anything), of what we would do during the last few minutes of our lives, perhaps of our species. This exercise is important, because thinking about the momentousness of the occasion in the full light of our cosmological, biological, and cultural roots is a step toward avoiding the second death.

For the reader of this book, from myself as the author, we have had a "gesture" together. As a final gesture I would like to share with you what I hope I would be able, if I were brave enough, to feel and do during the last moments of my life knowing that the end of everything human was obviously near. I would hope that rather than whimpering, whining, and complaining of "Why me?" of indulging in a paralyzed state of self-pity, I would instead be able to adopt something like the attitude of the warrior's last stand Carlos Castaneda describes in his book *Journey to Ixtlan*. According to don Juan, the Indian sorcerer attempting to teach Castaneda, a UCLA graduate student in anthropology, the ancient ways of the Sonoran Indians, there are in life a few rare mental warriors who have truly "seen" the full power and freedom that is our potential. There is, he says, reality upon reality right in front of us, and it is only a matter of power in calling them up. The man of knowledge knows that neither the world of the sorcerer nor the world of the ordinary man is real, even though both can act on us. The man of knowledge knows how to "stop" these worlds and "sneak between the worlds" and see. The path to this seeing requires a special boldness of character, because insanity and entrapment between the worlds is a constant possibility. But for the warrior who has traveled this arduous path, a "path with a heart," the fortitude of his character demands that death allow him a special gesture.

I decline to accept the end of man. It is easy enough to say that man is immortal simply because he will endure: that when the last ding-dong of doom has clanged and faded from the last worthless rock hanging tideless in the last red and dying evening, that even then there will still be one more sound: that of his puny inexhaustible voice, still talking. I refuse to accept this. I believe that man will not merely endure: he will prevail.

WILLIAM FAULKNER, from speech of acceptance upon being awarded the Nobel Prize for Literature

---

[1] In 1987 the Soviet Union and the United States began negotiations that would lessen the number of warheads to 6,000 for each side, a reduction from 19,500 to 12,00, an easing of our backpacks from 15 tons to about 9. Is this the next step in a major transformation of the arms race? A lowering of the gun without implying a dangerous weakness? In the INF agreement we haggled, debated, postured, and celebrated all for only a 4 percent reduction. Comparatively, the issues to be resolved for the next reductions are much more difficult.

When it is time for the warrior to die, when "he feels the tap of death on his left shoulder," the warrior is allowed to return to a special place, "a place of his predilection which is soaked with unforgettable memories, where powerful events left their mark, a place where he has witnessed marvels, where secrets have been revealed to him, a place where he has stored his personal power." Here the warrior is allowed to take his last stand and perform his last dance. Even the man of knowledge cannot alter the design of his death, but because of the impeccable spirit of the warrior, death must wait and witness this last dance. If the warrior's power is great, his dance will be long and magnificent. It will be a dance of humility as the inevitable foreboding power of his impending nonexistence surrounds him; it will be a dance of joyous gratitude for being able to witness life's many marvels and of mystical awe over reality's infinite textures. But most of all it will be a dance of rejoicing over having been part of this something rather than nothing. According to don Juan, as the warrior ends his dance he will look at the Sun, for he will never see it again in waking or in dreaming, and then his death will point to the vastness.

I am no warrior and I have little desire to learn the ways of sorcery. But if the end comes, if the second death seems assured, I hope I will be brave enough to do a mental dance of appreciation for having been a small part of the human experience, of being able to rethink the thoughts of Plato, Eratosthenes, Kepler, Newton, Einstein, Bohr, and Bohm, of meandering with them down the path of scientific understanding. I hope too that following along with them on this path with a heart that my spirit will have become impeccable enough to think of the *Voyager* spacecraft billions of miles away from our exploding, agonized planet, like a seed from a dying flower, silently and relentlessly carrying the message of our pitiful, curious existence. I hope I will be able to think of this focused testament to the best of our humanity, and I hope I will be composed enough to remember that a record of our existence will remain for a billion years, that a record of our romance with the universe will be secure, and the flower of our awareness will remain alive a little longer.

Concept Summary

According to Aristotle, we are rational animals. But we are also evolution's paradox, the "glory and shame of the universe," the "greatest achievement" or perhaps its "worst mistake." We have come to understand the atom, and now some of our best-educated people calmly and steadfastly use our "lucidity" to plan for World War IV, even though the destructive force of a nuclear World War III could equal 15 tons of dynamite for every person on this Earth. One nuclear bomb is incredibly destructive, easily destroying a major city, yet both the Soviet Union and the United States have enough bombs to target the major cities of the other many times. Experts disagree whether a full nuclear exchange between the Soviet Union and the United States would lead to human extinction, but according to Schell we have at least reached a zone of unprecedented uncertainty regarding this question.

Yet nuclear preparations continue, each side believing that it must negotiate from a position of strength, and each side with sound historical, political, and strategic reasons for its policies. Like a knife, reason cuts up the possible to form a rational world view for each side, each unfortunately seeming to feed on the other's fear. The United States says "Never again" to policies that risk a new Pearl Harbor; the Soviets say "Never again" to invasion without striking first. Hence we stand on the edge of a dangerous abyss, playing a dangerous psychological game, where the United States has declared it will *use* nuclear weapons first if it thinks it needs to and the Soviets will *strike* first if they think the United States thinks it needs to.

We have perhaps one hope. The same cognitive ability that has produced a zone of extinction might be our salvation. We may be a "bit of stellar matter gone wrong," but "we are that which asks the question." According to Freeman Dyson, we may be able to trick ourselves into a safer world. We may not be able to change our nature, but advancing technology, coupled with mutual understanding of each other's world views, can make our strategic differences a contest of information and return war to warriors. Not enough, says Carl Sagan. We need much more than just different weapons. It is our perception of being different, with different needs and fears, that pushes the button of our reptilian history. This perception can be ameliorated only by adopting a larger, more stirring perspective—a perspective that is true, the cosmic perspective.

These meliorist philosophies, both presupposing that our lives can improve through human effort, could rest, however, on quicksand. It could be worse, there could be no hope at all. Out of the libraries may come the astronomer and the concerned technologist, but they are disguised schizophrenics, according to Nietzsche. There is no meaning to life for the nihilist. Our rational ability makes us aware of this, causing a psychic trembling, and then makes psychological consolations and dangerous metaphysical cases that in the long run produce insanity in groups, nations, and epochs. For Nietzsche, objectivity, progress, and God are myths. There is no truth, only interpretation; no direction, only freedom; no purpose, only a circular pointlessness.

Adopting a large perspective of our scientific culture, we see a meandering path, a pageant of ideas and responses to an awesome universe, a mélange of intellectual embraces. We don't know whether the path has a purpose; perhaps it should be enough that we are on it; perhaps we should just continue on. We don't know if our curiosity is a gift or a curse, but it is reasonable to continue our faith in the self-correctiveness of science all the while knowing that it may be an illusion and that there may be no epilogue to our search. Science presupposes assumptions it can never prove, but we can justify its method heuristically: It forces us to be intimate with the world and to critically discuss our myths. Clearly, other ways of embracing the world exist, but just as clearly, science is one of the best and is *the* best given the proper domain and certain questions. This less pretentious view of science humbles us, tempers our egocentricity, and preserves a childlike wonder of and respect for our place in the universe.

Our species is now at an awesome crossroad—a serious business that few think about. Test this thought. Think what it would mean if there were no more human thought, devotion, inspiration, or holy curiosity, no more passion of the lover or the discoverer, only the silence of nothingness.

Suggested Readings

The Arms Race and Nuclear War, by David P. Barash (Belmont, Calif.: Wadsworth, 1987).

An introductory textbook with the expressed, but understated goal of accomplishing "a minimal level of nuclear literacy." Although the author admits his "dovish" sympathies, the book is a well-balanced presentation of essential information for every voting citizen. From how a nuclear bomb works to the arms race, negotiated agreements, and possible futures, the book also includes useful summaries, pro and con policy discussions, key terms, study questions, and suggested readings at the end of each chapter. Particularly relevant to our discussion in Chapter 10 are the sections on the description and balance of U.S.–Soviet delivery systems, the different policies and views on deterrence and counterforce, and negotiated agreements (completed and in process up through 1986).

Weapons and Hope, by Freeman Dyson (New York: Harper & Row, 1984).

This book and Dyson's thoughts are covered in Chapter 10. Probably neither doves nor hawks will like his sincere, fair-minded analysis.

The Fate of the Earth, by Jonathan Schell (New York: Avon Books, 1982).

Although this book has been criticized for its lack of objective evidence and for "overstating" the nuclear threat, critics seem to have glossed over Schell's essential point: Nuclear war is something very new, we just don't know what will happen; because the evidence on the side of total destruction is compelling, this "uncertainty" deserves the immediate attention of every person on Earth. The book is useful as an existential counter to the usual cold technocratic debates and presentations on nuclear defenses.

Thinking About the Unthinkable in the 1980s, by Hermann Kahn (New York: Simon and Schuster, 1984).

Speaking of cold technocratic presentations, we can easily believe that we are doomed after reading this book. "One can imagine," says Kahn, "circumstances in which rational leaders rationally decide to initiate a thermonuclear war . . . as the least undesirable available option." Written by the intellectual leader for the view that "nuclear weapons exist, they will not go away, and one day they might even be used." Kahn says that people must continue to think of war as an experience that can be survived and recovered from if we make proper preparations, and national leaders must recognize that they will be judged by how well they help their country prevail. According to Kahn, we must be "reasonable"; we must not only acknowledge the role nuclear weapons play in international politics, military strategies, perceptions of power, and arms control negotiations but also prepare for the possibility that deterrence may fail. Also, according to Kahn "exaggerated assumptions," by Schell and others, of the apocalyptic results of nuclear war, and their romantic, morally simplistic "nonsolutions," are "impractical, illusionary, and dangerous." (For a direct criticism of Schell's book, also see Kahn's article, "Refusing to Think About the Unthinkable," *Fortune,* June 28, 1982, pp. 113–116.) For those who still believe that reason can be guided to morally acceptable and humane conclusions by the addition of more facts, perhaps they can take consolation from the fact that this treatise was written before (Kahn died unexpectedly in 1983) the hypothesis of "nuclear winter" was supported by a number of scientists.

Debating Counterforce: A Conventional Approach in the Nuclear Age, by Charles-Phillipe David (Boulder, Colo.: Westview Press, 1987).

A view of U.S. defense strategy, and the internal debate from which it is being formed, from a Canadian defense specialist. Especially revealing is the up-to-date perspective the book gives of the disconcerting emergence of what the author calls the "conventional approach" in U.S. thinking: the notion that nuclear war is just an extension of conventional war, that nuclear war can involve defense, strategy, survival, and winners. Contains an almost terrifying bibliography with many citations from the new technological elite of defense intellectuals.

The Logic of U.S. Nuclear Weapons Policy: A Philosophical Analysis, by Corbin Fowler (Lewiston, N.Y.: Edwin Mellen Press, 1987).

This is Volume 4 in the continuing series *Problems in Contemporary Philosophy.* Although the book attacks what the author calls cynical realism and advocates a practical idealism, and hence is not a dispassionate philosophical or logical analysis of weapons policy, it does attempt to weave some philosophical analysis into the nuclear debate. Also, the author's quick lists on "nukespeak" jargon, delivery systems, treaties, abbreviations, possible effects of nuclear war, and sources of fear are worth the price of admission.

A World Beyond Healing: The Prologue and the Aftermath of Nuclear War, by Nicholas Wade (New York: Norton, 1987).

By showing different ways a nuclear war might occur—due in large part to the rapid and uncontrollable advance of nuclear technology—and by discussing the plausible and horrible effects, Wade attempts to disturb the complacent who believe that we can continue the present policies.

The Caveman and the Bomb: Human Nature, Evolution, and Nuclear War, by David P. Barash and Judith Eve Lipton (New York: McGraw-Hill, 1985).

Barash (author of *The Arms Race and Nuclear War*), a zoology professor, and his wife, a psychiatrist, attempt to apply the "wisdom" of sociobiology and psychiatry to an understanding of the biological roots of human destructive tendencies. Although this book has been criticized, along with Barash's *The Hare and the Tortoise: Culture, Biology, and Human Nature* (New York: Viking, 1986), as an example of "pop sociobiology" ("sperm-bearers tend to be spear-bearers") and "a theodicy for the ultra doves," and although its use of the concept of "Neanderthal mentality," an admitted anthropologically inaccurate but "convenient caricature," has more shock value than educational value, and although the book tends to be preachy and drawn out in places, it is nevertheless full of thought-provoking descriptions and analogies of our messy nature.

The authors' main argument deserves special attention: that precisely because of the mixed baggage of our biological heritage (we are programmed to survive, reproduce, and love as much as we are to be hostile and destructive) and its interaction with the forces of culture, a global psychotherapy is possible. Such an awareness will allow us to transcend knee-jerk biological inclinations, and our reason can serve us more than just witnessing our inevitable destruction. One particularly relevant chapter discusses how "deterrence" leads to short-run cooperation, but inevitable increased anxiety, long-term increased hostility, possible retaliation, and less future cooperation.

The Tangled Wing: Biological Constraints on the Human Spirit, by Melvin Konner (New York: Holt, Rinehart & Winston, 1982).

Although this book deals with biological themes similar to those of Barash's books, it is much more humble and less ideological in its professed understanding. Nevertheless, it is a masterful, often poetic, and authoritatively documented attempted synthesis of all recent discoveries of the biological sciences (including evolutionary biology, biological anthropology, neuroanatomy, ethology, and neuro-ethology) as applied to the biological bases of human behavior. According to Konner, a professor of biological anthropology at Harvard University, the book tries to set the stage for a rational dialogue on our essential biological nature, of who we are, why we behave as we do, and how we may shape our future. Konner also reminds us that a cosmic perspective must involve a biological perspective, that space exploration is "a mere conjurer's trick" compared to understanding and mastering ourselves, that the "wings" of humankind (a symbol of our ability to transcend physical limitations) are "tangled" in the ignorance of forces, many self-destructive, that we must understand if we are to survive. An important work.

Zen and the Art of Motorcycle Maintenance: An Inquiry into Values, by Robert M. Pirsig (New York: Morrow, 1974).

How can we separate the reasonable from the merely conceivable? How can the best hypothesis be selected from an infinite number of possible hypotheses, and how is this related to the big philosophical questions and little, but important, personal predicaments of life? This book must be read, and perhaps to completely understand it, read a minimum of four times: from the standpoint of (1) personal relationships and problems, (2) social-political relationships and problems, (3) abstract epistemological problems, and (4) to see the connections among these three levels. From Plato to Pirsig, the question remains the same: How do we decide what is best? The best decisions in our personal lives? The best scientific hypotheses? The best solution to our nuclear predicament?

Irrational Man: A Study in Existential Philosophy, by William Barrett (Garden City, N.Y.: Doubleday, 1958).

Still one of the best introductions to existential philosophy; includes a section on Nietzsche. For primary sources, the following are recommended: *Basic Writings of Nietzsche,* ed. by Walter Arnold Kaufmann (New York: Modern Library, 1968), *Beyond Good and Evil: Prelude to a Philosophy of the Future* (New York: Penguin, 1973), *Thus Spake Zarathustra* (New York: Heritage Press, 1967), and *Human, All Too Human: A Book for Free Spirits* (Lincoln: University of Nebraska Press, 1984).

Index